CERAMIC COATINGS – APPLICATIONS IN ENGINEERING

Edited by **Feng Shi**

Ceramic Coatings – Applications in Engineering
Edited by Feng Shi

Published by InTech
Janeza Trdine 9, 51000 Rijeka, Croatia

Publishing Process Manager Bojan Rafaj

Technical Editor Teodora Smiljanic

Cover Designer InTech Design Team

Additional hard copies can be obtained from orders@intechopen.com

Ceramic Coatings – Applications in Engineering, Edited by Feng Shi
p. cm.
978-953-51-0083-6

For EU product safety concerns:
IN TECH d.o.o., Prolaz Marije Krucifikse Kozulić 3, 51000 Rijeka, Croatia,
info@intechopen.com or visit our website at intechopen.com.

Meet the editor

Dr Feng Shi is an associate professor in College of Physics and Electronics, Shandong Normal University, P.R.China, whose research addresses novel founctional materials with an eye toward practical applications, i.e., microwave dielectric ceramics - bulk ceramics as well as ceramic thin films. In the past 5 years he has published more than 40 papers in ceramics: in Applied Spectroscopy Reviews, Dalton transactions, Journal of Alloys and Compounds, Applied Surface Science, Journal of Materials Science-Materials in Electronics and International Journal of Materials Research. Presently, Feng Shi is the reviewer of many journals like Cryst. Growth & Des.; J. Am. Ceram. Soc.; J.Colloid Interf. Sci.; J. Nanopart. Res.; J. Raman Spectrosc.; Curr. Nanosci.; Mater. Sci. Eng. B; and Mater. Sci. Semicon. Proc.

Contents

Preface XI

Part 1 EPD Process 1

Chapter 1 **Ti-O Film Cathodically-Electrodeposited on the Surface of TiNi SMA and Its Bioactivity and Blood Compatibility 3**
Zhu Weidong

Chapter 2 **Ceramic Coatings Obtained by Electrophoretic Deposition: Fundamentals, Models, Post-Deposition Processes and Applications 43**
M. Federica De Riccardis

Part 2 Physical Deposition Process 69

Chapter 3 **Magnetron Sputtered BG Thin Films: An Alternative Biofunctionalization Approach – Peculiarities of Bioglass Sputtering and Bioactivity Behaviour 71**
George E. Stan and José M.F. Ferreira

Chapter 4 **Erosion Behavior of Plasma Sprayed Alumina and Calcia-Stabilized Zirconia Coatings on Cast Iron Substrate 99**
N. Krishnamurthy, M.S. Murali, B. Venkataraman and P.G. Mukunda

Part 3 Coating Engineering 127

Chapter 5 **Investigations of Thermal Barrier Coatings for Turbine Parts 129**
Alexandr Lepeshkin

Chapter 6 **Thermal Spraying of Oxide Ceramic and Ceramic Metallic Coatings 167**
Martin Erne and Daniel Kolar

Chapter 7 **Ceramic Coating Applications and Research Fields for Internal Combustion Engines 195**
Murat Ciniviz, Mustafa Sahir Salman, Eyüb Canlı, Hüseyin Köse and Özgür Solmaz

Part 4 Pigment 235

Chapter 8 **Ceramic Coatings for Pigments 237**
A.R. Mirhabibi

Part 5 Application in Foundry 259

Chapter 9 **Ceramic Coating for Cast House Application 261**
Zagorka Aćimović-Pavlović, Aurel Prstić, Ljubiša Andrić, Vladan Milošević and Sonja Milićević

Preface

Ceramic Coating is a protective thin film for all kinds of parts and can dramatically increase their useful life, and therefore, productivity increases by reducing maintenance downtime and by achieving longer run times before repairs are needed. Ceramic coatings are inorganic, non-metallic solid and inert thin films made by deposition, heating and subsequent cooling, which may be crystalline. That is, the definition of ceramic coatings is often restricted to inorganic crystalline thin films. About the fabrication of ceramic coatings, the main methods are physical deposition (like magnetron sputtering deposition and plasma spraying process) and chemical deposition, such as electrochemical deposition, which can affect the microstructures and properties of the ceramic coatings.

In order to know more about the ceramic coatings in the past several years, this book is organized by Intech open access publisher and wroten by different authors, who are focused on the modern ceramic coatings—applied in engineering. A feature of this text is that we keep in mind that many of today's high-tech ceramic coatings and processing routes have their origin in the microstructures and properties. Throughout the text we will make connections to these related fields. The text covers ceramic coatings from the fundamentals to industrial applications including a consideration of safety and their impact on the modern technologies, including EPD-process, physical deposition process, applications in engineering, pigment and foundry.

At last, we thank all the authors and all the editors who contribute greatly to this book and we hope that every readers are interested in this book.

Dr. Feng Shi
Associate Professor in Materials Physics & Chemistry,
College of Physics & Electronics, Shandong Normal University,
P.R. China

Part 1

EPD Process

1

Ti-O Film Cathodically-Electrodeposited on the Surface of TiNi SMA and Its Bioactivity and Blood Compatibility

Zhu Weidong
College of Materials and Metallurgy, Guizhou University, China

1. Introduction

1.1 Preface

It is easy to prepare TiO_2 film by electrochemical deposition method technologically. There are some other electrodeposition methods, such as electrophoresis method, anodic-electrodeposition method and cathodic-electrodeposition method.

Tan Xiaochun[1] et al. set up a film formation model by electrophoresis method based on experiments and made an exposition of electrophoresis mechanism. The results showed that colloidal concentration and size, DC bias and time are the major factors to form a film. Hayward[2] et al. made a investigation of electrophoresis method also.

Nanometer TiO_2 film prepared by cathodic-electrodeposition is formed in the form of microcrystalline piling up. Kavan[3] et al. took $TiCl_2$ solution as an electrolyte, to get amorphous Titanium (IV) hydrate film on the anode, and the TiO_2 film after heat treatment. Cui Xiaoli[4] et al. prepared TiO_2 film on the ITO glass substrate with anodic-electrodeposition method, and investigated the effect of the anodic current and deposition time on nano TiO_2 structure and adhesive force.

Natarajam[5] et al. used Titanium powder as a raw material, which was dissolved with H_2O_2 and NH_3 to get colloid, so that can accomplish the cathodic-electrodeposition on ITO glass in the aqueous solution. Karuppuchamy[6] et al. took $TiOSO_4$ as a raw material, to prepare TiO_2 film by cathodic-electrodeposition, the SEM testing results showed that the TiO_2 film is of porous. When a electrode was photosensitized with dyestuff, the incident light current transform rate of TiO_2 electrode could reach up to 35%.

Most of cathodic-deposition are accomplished on ITO glass. Chi-Min Lin[7] et al. took $TiCl_4$ ethylbenzene as an electrolyte, and accomplished the cathodic-electrodeposition on pure Ti. But there are few reports on the study of cathodic-electrodeposition on TiNi SMA. In this experiment, a self-prepared aqueous solution of $Ti(SO_4)_2$ was taken as the electrolyte, the cathodic-electrodeposition of TiO_2 film on TiNi SMA was achieved. This method is with more convenient in operation, lower cost and easier to accomplish technologically.

1.2 Experimental method

1.2.1 Experimental materials and media

TiNi shape memory ally (SMA) (Ti: 49.3 at%, Ni: 50.7%) was used as the experimental material. A sample's size was 20mm×10mm×1mm. The TiNi SMA samples were ground by degrees with abrasive papers (No. W40-W10). After washed ultrasonically with absolute ethyl alcohol and acetone solutions, the samples were put in absolute ethyl alcohol to await for using.

The experimental media were the self-prepared $Ti(SO_4)_2$ solution (PH0.71) of 0.606mg/ml, Hank's solution (making up diluted hydrochloric acid and NaOH solutions into PH7.45) to simulate human body balanced salt solution, and simulated saliva[8] (making up lactic acid and ammonia water into PH6.13).

The method of self-preparation of $Ti(SO_4)_2$:

Weighing out 1.0108g of Titanium dioxide (99.9%) pre-burned at 800°C for 2h, putting it in a platinum crucible with a platinum cover, adding to 3g of potassium pyrosulfate to melt, then putting it in a beaker of 400ml after taken out and cooled; adding sulphuric acid solution of 200ml, leaching the liquid melt at an electric heater, then taking the platinum crucible out and clearing with water, continuing to heat until to the solution clarify entirely, cooling, moving it into a volumetric flask of 1000ml, diluting to the graduation with deionized water, and finally, shaking the flask to be even, in order to get ready to use. And this Ti concentration was 0.06 mg/ml.

Hank's solution composition:

$NaCl$ 8g/l + Na_2HCO_3 0.35g/l + Na_2HPO_4 0.0475g/l + KCl 0.4g/l + KH_2PO_4 0.06g/l + $MgCl_2 6H_2O$ 0.10g/l + $MgSO_4 7H_2O$ 0.10g/l + $CaCl_2$ 0.18g/l + Glucose 1g/l.

Fusayama solution composition:

$KCNS$ 0.52g/l + $NaHCO_3$ 1.25g/l + KCl 1.47g/l + NaH_2PO_4 0.19g/l + lactic acid 0.75ml.

1.2.2 Experimental methods

1.2.2.1 Cathodic-electrodeposition of Ti-O film

A standard trielectrode system was used in the experiment. The device sketch is as shown in figure 1-1. The power supply was DJS-292 type of potentiostat; the reference electrode was saturated calomel electrode (SCE); a TiNi SMA sample was connected with copper wire, its working area was 10mm×10mm, and the non-working area was coated with silica gel; the electrolyte was the self-prepared $Ti(SO_4)_2$ solution.

Before cathodic-depositing the PH value of the $Ti(SO_4)_2$ solution was modified to PH1.2 with 300g/l of KOH solution, adding to 20g/l KNO_3. Under this circumstance of that let sample be cathode, at a constant current density of 5mA/cm^2 for 4 min of electrodepositing, a layer of Ti-O film was formed on the surface of TiNi SMA. After finished deposition, the samples were exposed to air for drying naturally, then put them in drying oven for 1h at 120 °C, and then awaited for use.

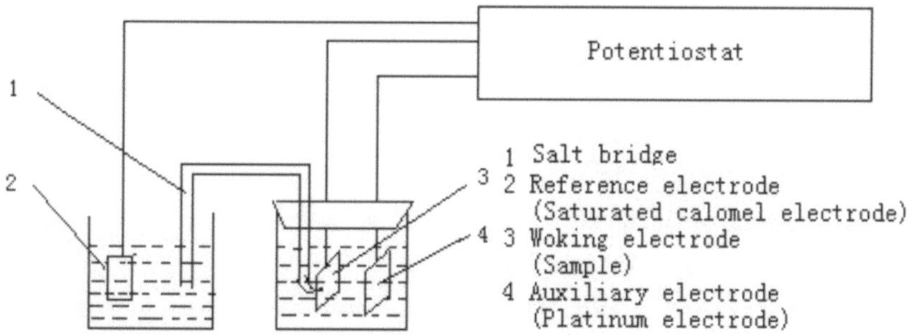

Fig. 1-1. Schematic plan of the trielectrode system

1.2.2.2 Composition analysis of Ti-O film

The surface of the sample cathodically-electrodeposited for 4 min at 5mA/cm^2 was observed with XL30 s-FEG type of scanning electronic microscope (SEM). Ti and Ni contents of Ti-O film were analyzed with Finder 1000 type of energy dispersive spectroscopy (EDS). The elemental composition and valent state of Ti-O film were analyzed with MII type of XPS, in order to analyze the film's depositional effect.

The deposited sample was put in a muffle furnace to experience crystallization treatment (annealling) for 1h at 300°C and 450°C respectively, furnace heating and cooling. The phase analysis of Ti-O film was with D/max 2000 type of X-ray diffractormeter (XRD).

1.2.2.3 Corrosion potential test of the material in simulated physiological environment

Corrosion potential was measured with DJS-292 potentiostat. The connecting diagram is as shown in figure 1-2.The reference cathode was saturated calomel electrode (SCE). The sample was connected with copper wire, its working area was 10mm×10mm, and non-working area coated with silica gel. The samples both before and after cathodic-electrodeposition were dipped into Hank's solution (PH7.45) and Fusayama solution (PH6.13) respectively, then recording the electrode potential and time, potential value recorded once every one minute, until the potential no changing basically. Finally a curve of corrosion potential changed with time could be charted. The experimental parameters were set as in table 1-1.

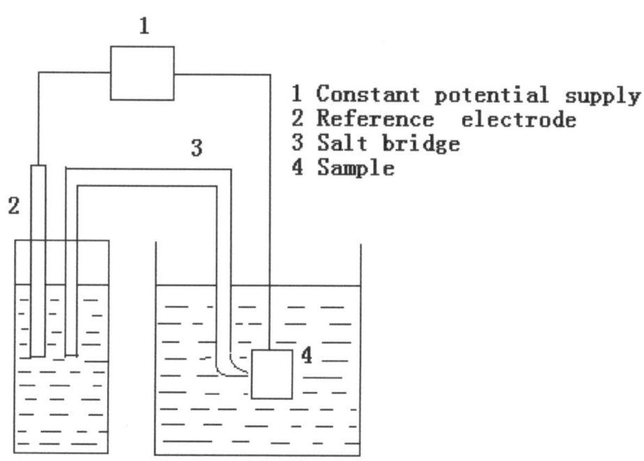

Fig. 1-2. Schematic plan corrosion potential test

Start potential (V/SCE)	Stop potential (V)	Sample interval (mV)	Scanning rate (mV/min)
-0.2	2.0	2	20

Table 1-1. Parameters of dynamic polarization test

1.2.2.4 Potentiodynamic polarization test of the material in simulated physiological environment

The experiment device used a standard trielectrode system. The signal generator used MEC-12 type of multi-function electrochemical analyzer. The reference electrode was SCE. The auxiliary electrode was a Pt one. The sample was connected with a wire, working area 10mm×10mm and non-working area coated with silica gel. The electrolytes were Hank's solution (PH7.45) and Fusayama one (PH6.13). The experiment used the staircase voltage scanning method, to compare and analyze the corrosion resistances of the TiNi SMA before and after cathodic-electrodeposition respectively in simulated physiological environment.

1.2.2.5 Effect of current density on the surface morphology and corrosion resistance of Ti-O film

With an SZX-STAD2 type of OLYMPUS metallurgical microscope, the Ti-O films obtained by cathodic-electrodeposition at different current densities for 4min respectively of TiNi SMA were observed, and the effect of the current densities on surface morphology of Ti-O film were investigated.

The corrosion potential and anodic polarization curve of SMA deposited at 20mA/cm^2 and 5mA/cm^2 in hank's solution and Fuayama solution were respectively tested, in order to investigate the effect of current density on corrosion resistance of Ti-O film.

1.2.2.6 Effect of crystallization treatment on the surface morphology and corrosion resistance of Ti-O film

The samples of TiNi SMA deposited at 20mA/cm^2 for 7min and 5mA/cm^2 for 4min respectively were crystallized at 450°C, with an SZX-STAD2 type of OLYMPUS metallurgical microscope, the Ti-O films before and after depositing respectively were observed, and the effect of crystallization treatment on Ti-O film morphology was investigated.

The TiNi samples deposited at 5mA/cm^2 for 4min before and after crystallizing respectively were taken into Hank's solution (PH7.45) and Fusayama solution (PH6.13) to test their corrosion potential and polarization curve, in order to investigate the effect of crystallization treatment on corrosion resistance of Ti-O film.

1.2.2.7 Effect of deposition time on the surface morphology and corrosion resistance of Ti-O film

Depositing TiNi SMA respectively at a constant current density of 20mA/cm^2 for different times and at another constant current density of 5mA/cm^2 for different times, with an SZX-STAD2 type of OLYMPUS metallurgical microscope, the Ti-O films were observed, to investigate the effect of deposition times at the current densities of 20mA/cm^2 and 5mA/cm^2 respectively on Ti-O film morphology.

1.2.2.8 Effect of electrolytes' PH values and NO_3 concentration on cathodic-electrodeposition

With the SZX-STAD2 type of OLYMPUS metallurgical microscope, observation of the Ti-O films obtained by cathodic-electrodeposition for 4min in the electrolytes of PH0.71 and PH1.2 respectively, and the electrolyte without NO^-_3 added in and electrolyte with 0.02M and 0.2M of NO^-_3 adding in respectively, were carried out, in order to investigate the effect of PH value and NO^-_3 concentration on cathodic-electrodeposition.

1.2.2.9 Effect of air flow curtain on surface morphology of Ti-O film

Depositing TiNi SMA setting in different bearings, making direction of air flow produced during the TiNi SMA deposition different directions, with the SZX-STAD2 type of OLYMPUS metallurgical microscope, the effect of air flow on Ti-O film morphology was investigated.

1.3 Experimental results and discussion

1.3.1 Cathodic-electrodeposition of Ti-O film

Under trielectrode system, in mixing well the self-prepared $Ti(SO_4)_2$ solution electrolyte, catodic-electrodeposition was carried out at a constant current density of $5mA/cm^2$ for 4min. Because NO_3^- and H_2O on the surface of cathode sample obtained electrons to be reduced, making sample surface PH value get high, moreover Ti (IV) could be easily hydrolyzed, so that a layer of amorphousstate Ti-O film was deposited on the surface of TiNi SMA. Its producing process is as following:

$$2H_2O + 2e \rightarrow H_2 + 2OH^- \quad (1.1)$$

$$NO_3^- + 6H_2O + 8e \rightarrow NH_3 + 9OH^- \quad (1.2)$$

$$Ti^{4+} + 4OH^- \rightarrow Ti(OH)_4 \quad (1.3)$$

The $Ti(SO_4)_2$ gel reactively produced was deposited on the surface of the sample.

1.3.2 The composition analysis of Ti-O film

Figure 1-3 shows the SEM topography (at different amplifications) of a Ti-O film cathodically-electrodeposited. It can be seen from the figure that under this experimental parameter, the Ti-O film obtained was well-distributedly deposited on TiNi SMA surface. This Ti-O film is closely composed of a lot of tiny particles, the particle's size is in about dozens of nanometers. This close bonding of the particles in nanometer class is conducive to strengthen the TiNi SMA surface properties, and to prevent Ni release from substrate.

Figure 1-4 shows an EDS analysis of the Ti-O film after cathodic-electrodeposition. It can be seen from the figure that this film mainly contains Ti element, as well as a small of amount of Ni element, and the ratio of Ti and Ni reachs 3:1. It illustrates that there is a lot of Ti element in the film. Because EDS might possibly puncture the Ti-O film, thus reflecting the substrate's element, so it could be preliminarily verified that by cathodic-elcetrodeposition, a layer of film mainly including Ti element indeed has been deposited on the surface of TiNi SMA.

Figure 1-5 shows an XPS (X-ray photoelectron spectroscope) analysis result of Ti-O film surface after cathodic-electrodeposition of TiNi SMA. It can be seen from the figure that Ti-O film mainly contains the elements of O and Ti, besides element C (an extraneous pollutant), no Ni element exists.

Figure 1-6 shows a high resolution XPS analysis result of Ti2p in Ti-O film. It can be seen from the figure that there are two peaks, at 458.8eV and 464.6eV respectively, both are Ti quadrivalentcations, corresponding with Ti-O bonding energy at the spin state of $2p_{3/2}$ and $2p_{1/2}$.

Figure 1-7 shows a high resolution XPS analysis result of Ni2p in Ti-O film. It can be seen from the figure that there are no clear peaks emerging, illustrating that no Ni element existing in the Ti-O film.

Figure 1-8 shows a high resolution XPS analysis result of O1s in Ti-O film. It can be seen from the figure that there are two peaks existing, one 530.2eV, this is O^{2-}; another one 530.7eV, and this peak is an integrated OH^-. So, the Ti-O film obtained by cathodic-electrodeposition should exist in the form of TiO_2 or hydrate of $Ti(OH)_4$.

Fig. 1-3. SEM result of Ti-O film

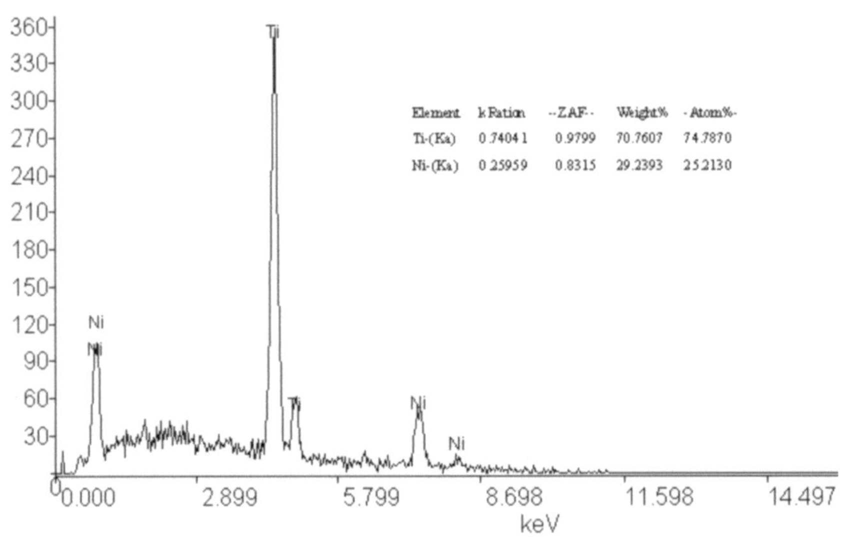

Fig. 1-4. EDS result of Ti and Ni in Ti-O film

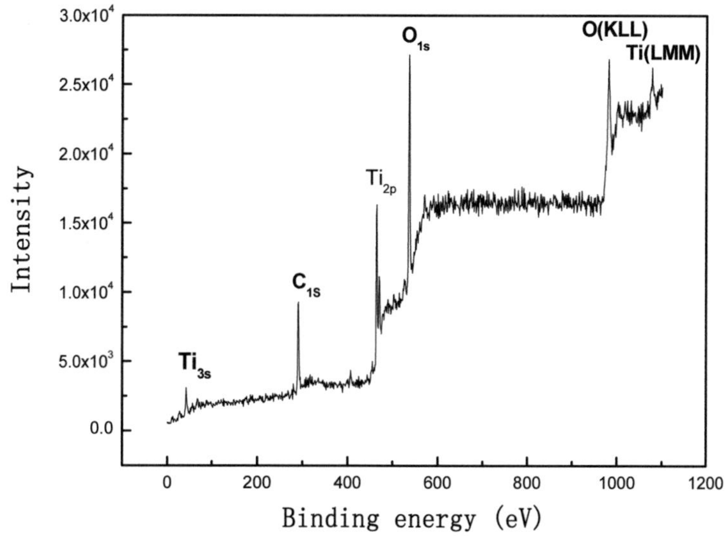

Fig. 1-5. XPS survey spectra of sample cathodically-electrodeposited

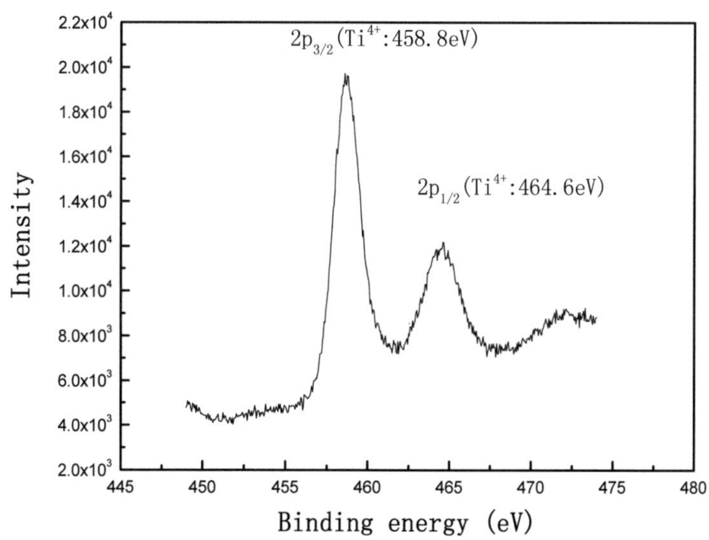

Fig. 1-6. Ti2p XPS energy spectrum of sample cathodically-electrodeposited

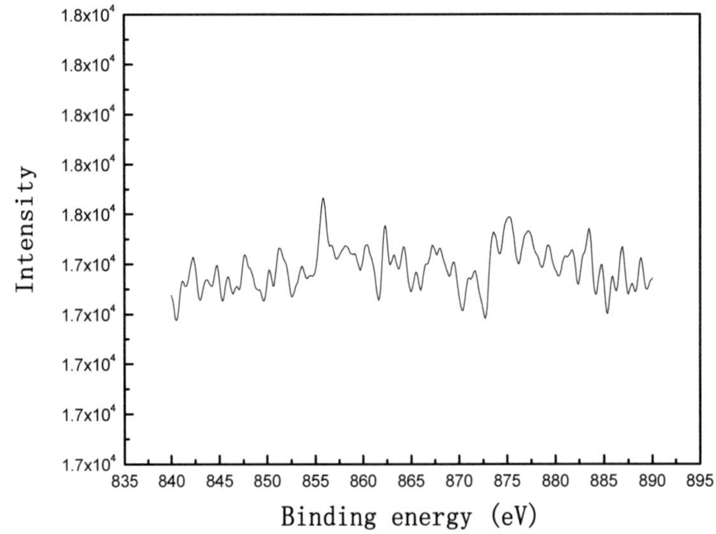

Fig. 1-7. Ni2p XPS energy spectrum of sample surface cathodically-electrodeposited

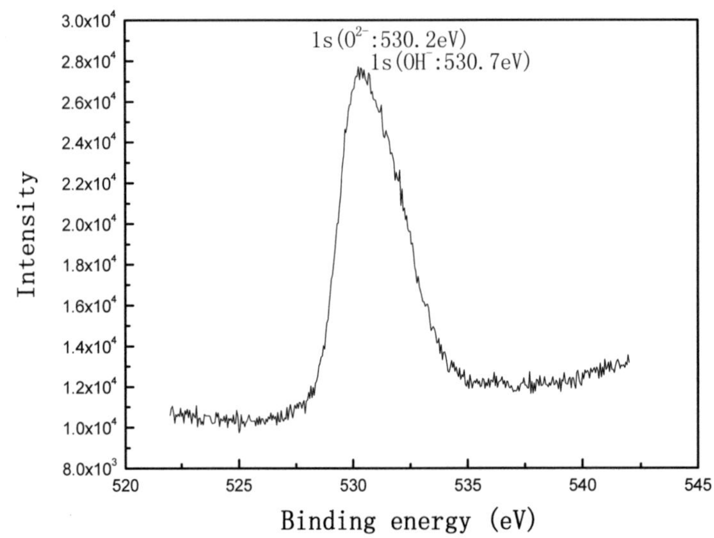

Fig. 1-8. O1s XPS energy spectrum of sample cathodically-electrodeposited

Figure 1-9 shows an XRD analysis result of TiNi SMA deposited after heat treatments at different temperatures. It can be seen from the figure that in the Ti-O film crystallized at 300°C, no TiO_2 existing was found besides the main peak reflecting the substrate. But after crystallized at 450°C, there is the peak emerging of anatase TiO_2 in the figure. Because the substrate also has Ti element existing, to investigate if the anatase TiO_2 formed by substrate's thermo oxidation, the same parameters' XRD for the TiNi SMA blank sample before depositing was done. But in the result, no clear TiO_2 peak merged after crystallized at 450°C, this illustrates that the film obtained by cathodic-electrodeposition is truly an amorphous-state Ti-O film, its composition is mainly in the form of hydrate $Ti(OH)_4$. The formation of TiO_2 is as the reaction formula:

$$Ti(OH)_4 \rightarrow TiO_2 + 2H_2O \tag{1.4}$$

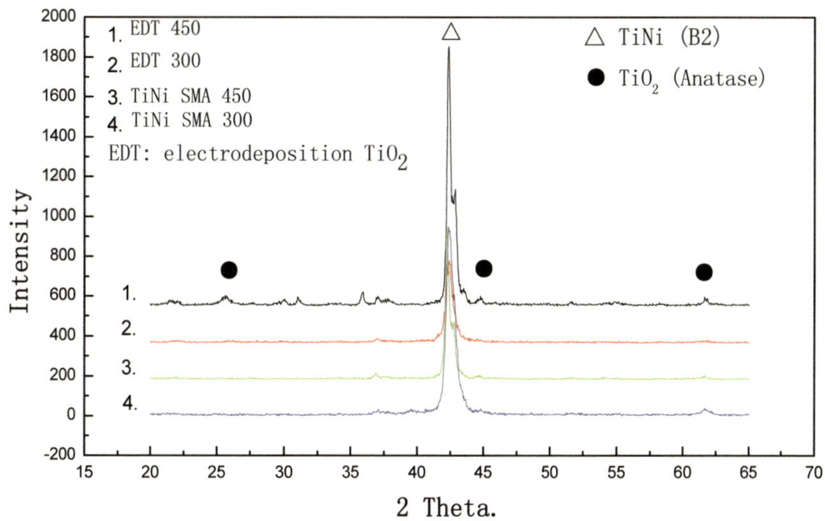

Fig. 1-9. XRD result of heat treated Ti-O film

1.3.3 Corrosion potential test of the materials in simulated physical environment

Figure 1-10 shows a relation of corrosionpotential-time of the samples before and after depositing in Hank's solution (PH7.45). It can be seen from the figure that the corrosion potential of the deposited sample dipping in solution went down quickly, about 5 minutes after, the corrosion potential did basically not change any more, stabilizing at -0.4V. Furthermore the stable corrosion potential was more positive than the non-depositing sample's, at about -0.2V. This illustrates that the deposited sample has a higher thermodynamic stability than the non-depositing sample's one at PH7.45 in Hank's solution.

Figure 1-11 shows a graphofarelation of corrosionpotential-time of the samples before and after depositing in Fusayama solution (PH6.13). It can be seen from the figure that the corrosion potential of the deposited sample dipping in solution went down quickly also, 7 minutes after, basically stabilizing, although somewhat undulating, the potential did not change on the whole, keeping at -0.12V. But the non-depositing sample's potential basically stabilized at -0.15V, which was more negative than the deposited one's. This illustrates that the deposited sample has a higher thermodynamic stability than the non-depositing sample's one at PH6.13 in Fusayama solution.

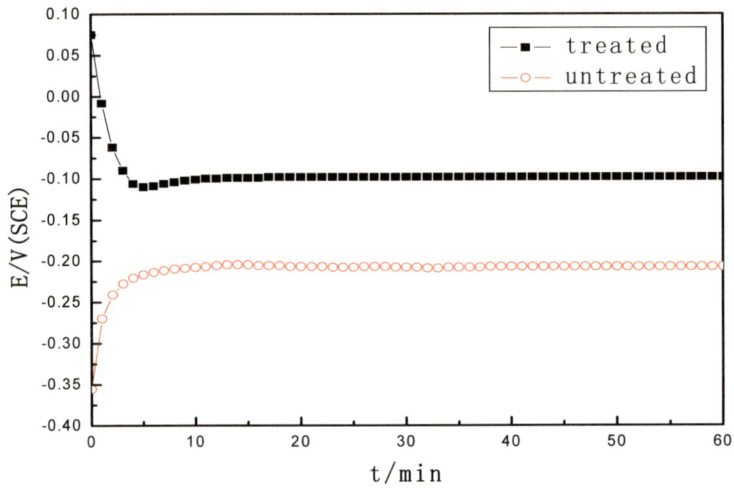

Fig. 1-10. The corrosion potential change plotted as a function of time of samples in Hank's solution (PH7.45)

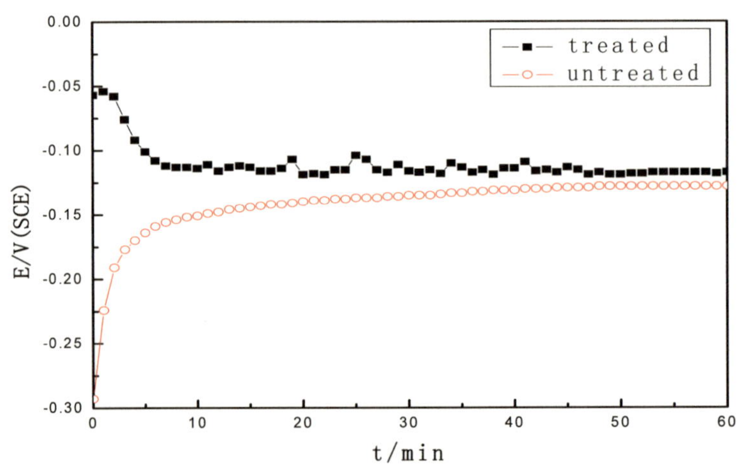

Fig. 1-11. The Corrosion potential change plotted as a function of time of samples in Fusayama solution (PH6.13)

1.3.4 Test of potentiodynamic anodic-polarization of the material in simulated physiological environment

Figure 1-12 shows the polarization curves of TiNI SMA before and after cathodic-electrodeposition in Hank's solution (PH7.45). It can be seen from the anodic-polarization curves that the current density of the deposited sample is always lower than that of the one before depositing, even if at passive region, the current density of the deposited sample is similarly lower than that of non-depositing one, and the broken potential of the deposited sample was about 1.2V, and that of non-depositing one about 1.1V, thus the broken potential of the deposited sample was somewhat higher than that of non-depositing one. It can be seen from cathodic-polarization curves that the difference between the two samples' densities is not very great. This illustrates that the cathodically-electrodeposited sample in Hank's solution of PH7.45 has better properties of anodic-polarization and anti-corrosion, and the two properties are approximately close in cathodic-polarization. This is because of the Ti-O film existed on the surface of the deposited sample, with a protective role to substrate, that to a certain extent prevented Ni ions release from substrate, thus having a better corrosion resistance.

Figure 1-13 shows the polarization curves of TiNi SMA before and after cathodic-electrodeposition in Fusayama solution (PH6.13). It can be seen from the polarization curves in the figure that the current density of the deposited sample began to lower than that of non-depositing one from 0.3V, and even at passive region, the current density of the deposited sample was still lower than that of the one before depositing. It can be seen from cathodic-polarization curves that the difference between the two samples' current densities was not great. This also illustrates that because of the Ti-O film existed on the surface of the deposited sample with a protective role to substrate, that to a certain extent prevented Ni ions release from substrate, thus having a better corrosion resistance.

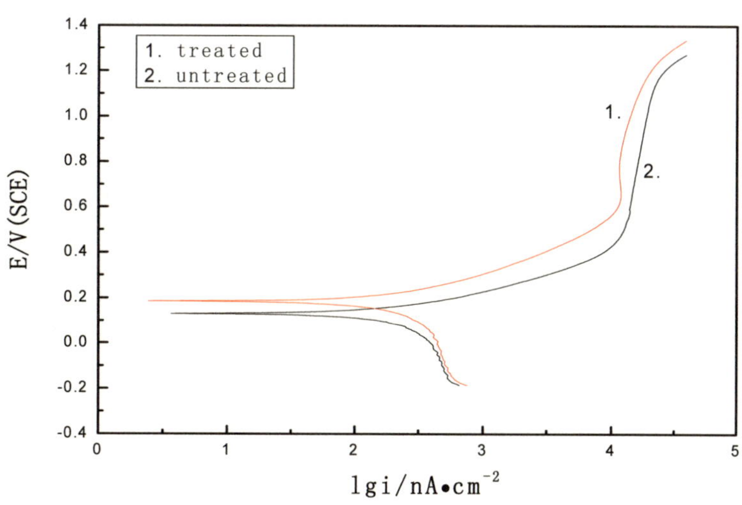

Fig. 1-12. The anodic-polarization curves of samples in Hank's solution (pH7.45)

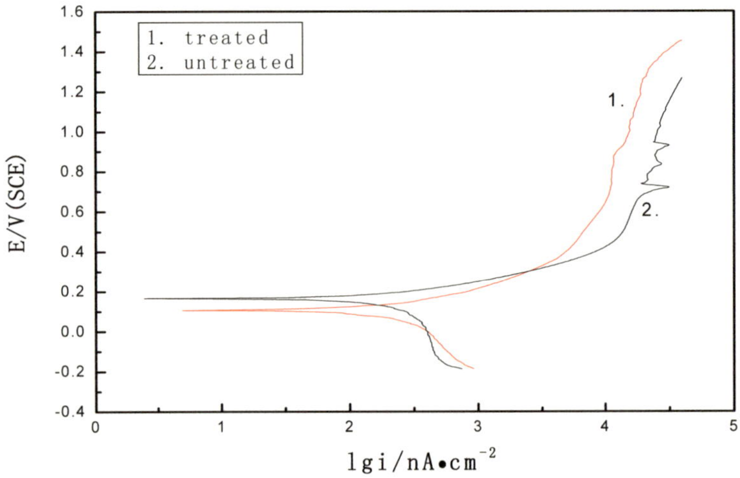

Fig. 1-13. The anodic-polarization curves of samples in Fusayama solution (PH6.13)

1.3.5 Effect of current density on surface morphology and corrosion resistance of the film

Figure 1-14 shows the morphologies of under optical microscope at 1000× of amplification of the Ti-O film cathodically-electrodeposited at different current densities. It can be seen from figure a and b that the Ti-O films obtained at $40mA/cm^2$ and $30mA/cm^2$ were uneven, lumpedly distributed on the surface of TiNi SMA, and with some big cracks; it can be seen from figure c that Ti-O film was fairly even at $20mA/\ cm^2$, distributed as floc-like on the surface of Ti-O film, but with some considerable cracks; it can be seen from figure d that the Ti-O film obtained at $5mA/cm^2$ was fine and close, and with small cracks; from figure e Ti-O film's existence could hardly be seen, only the scratch on the substrate. This illustrates that the Ti-O film obtained at $5mA/cm^2$ was the best. Because current density got major, the driving force for Ti(IV) to hydrolyze got major too, Ti-O film quickly formed and grew on the surface of TiNi SMA within a short time. Owing to massedly-produced hydrogen at depositing time, Ti-O film continuously peeled off, thus causing surface very uneven. Besides the film produced at this time was very thick, a stress difference produced due to shrinking in the process of drying, so that many big cracks were brought about. With current density going down, the driving force for Ti(IV) to hydrolyze also getting down, the particles formed getting small, and this made the Ti-O film surface more even, cracks less. But when the current density went down to a certain extent, the driving force for Ti(IV) to hydrolyze would not be enough to form a film on the surface of TiNi SMA.

Figure 1-15 shows the corrosion potential-time curves of the samples athodically-electrodeposited at different current densities in Hank's solution (PH7.45). It can be seen from the figure that the sample deposited at the current density of $20mA/cm^2$ had a more positive corrosion potential about -0.2V; but the stable corrosion potential of non-depositing sample was -0.1V. This illustrates that the deposited sample has a better thermodynamic stability at current density of $20mA/cm^2$ in the solution. This is because of the Ti-O film deposited at the current density of $20mA/cm^2$ was thicker, bonding more closely, that made it have a better thermodynamic stability.

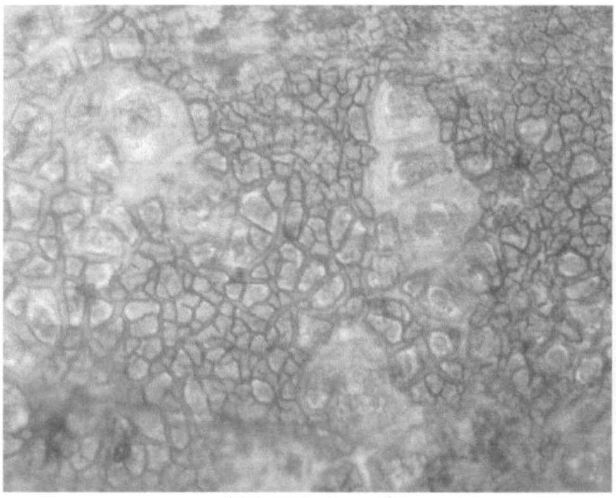

a (40mA/cm2)

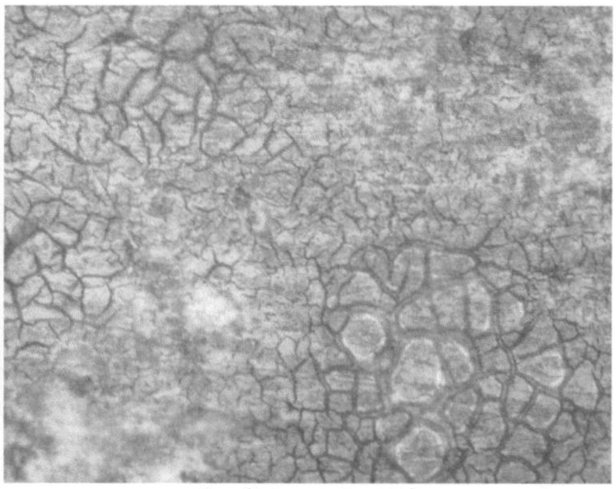

b (30mA/cm2)

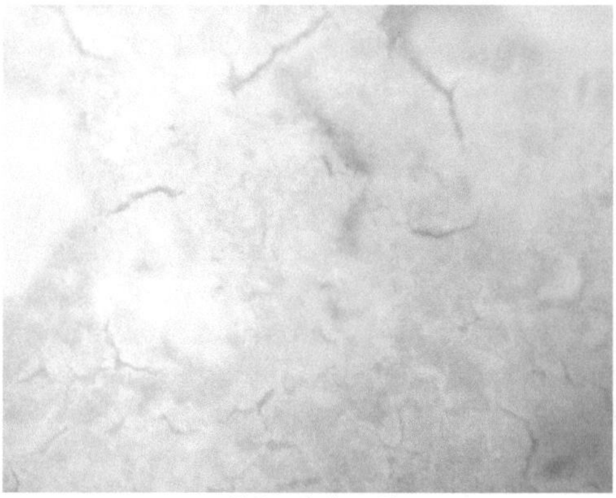

c (20mA/cm2)

d (5mA/cm2)

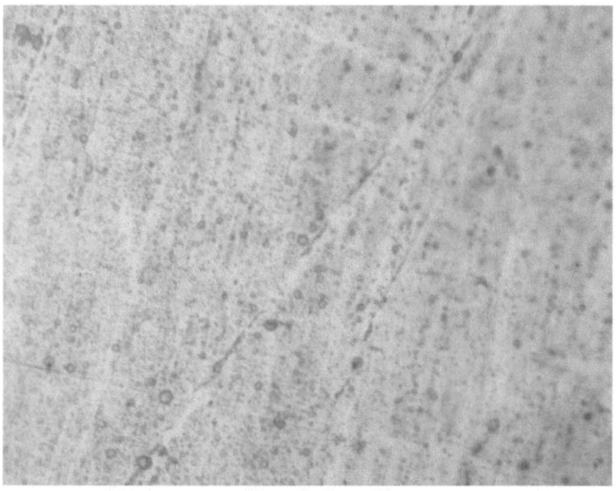

e (2mA/cm2)

Fig. 1-14. Optical microscope of Ti-O film by different current densities (1000×)

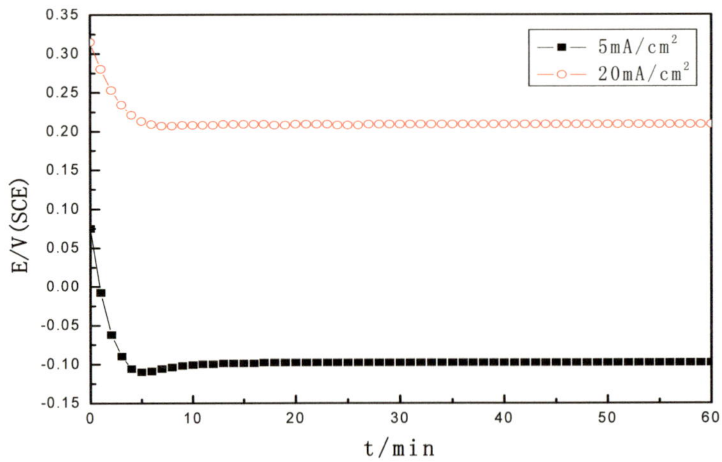

Fig. 1-15. The Corrosion potential change plotted as a function of time of samples with different current densities in Hank's solution (PH7.45)

Figure 1-16 shows the anodic-polarization curves of the samples athodically-electrodeposited at different current densities in Hank's solution (PH7.45). It can be seen from the anodic-polarization curves in the figure that the deposited sample's at current density of 20mA/cm^2 was much less than the one's at current density of 5mA/cm^2, but the passivation region of the deposited sample at current density of 5mA/cm^2, was longer than that of the one at high current density. This is because of the Ti-O film of deposited sample at high current density was thicker, which could effectively prevent the ions from transfer, thus making the current density lower. Because the surface of the deposited sample at high current density was very rough, that would bring about the passive film not enough stable after the surface passivated, it is easy to be punctured, so that made the passive region short. But from the figure of view, the slope of the deposited sample at current density of 20mA/cm^2 is large, it illustrates that the deposited sample has a better cathodic-polarization property.

Figure 1-17 shows the corrosionpotential-time curves of the deposited samples at different current densities in Hank's solution (PH7.45) after crystallized at 450°C. The corrosion potential of the deposited sample at the current density of 20mA/cm^2 after crystallized was not very stable, undulating much, finally the current potential was about -0.205V, which was more negative than that of the one at the current density of 5mA/cm^2 after crystallized, -0.195V. This illustrates that the deposited sample at the current density of 5mA/cm^2 after crystallized is with a better thermodynamic stability.

Figure 1-18 shows the polarization curves of the deposited samples at different current densities after crystallized at 450°C. It can be seen from the polarization curves in the figure that the voltage of the deposited sample at current density of 5mA/cm^2 after crystallized was far lower than that of the one at 20mA/cm^2. The breakdown potential of the deposited sample at current density of 5mA/cm^2 after crystallized was 1.2V, but the one at 5mA/cm^2 had no the passivation phenomenon emerging. This illustrates that the deposited sample at current density of 5mA/cm^2 after crystallized displayed a very good anti-corrosion. Because the granularity of the deposited sample at5mA/cm^2 was small, the surface even and very thin, with less crystal defects, after crystallized and heavy thermalstress, it was easy to corrode, thus presenting a poor corrosion resistance. From the cathodic-polarization curve of view, the deposited sample under current density of 5mA/cm2 after crystallized presented similarly a better cathodic-polarization property.

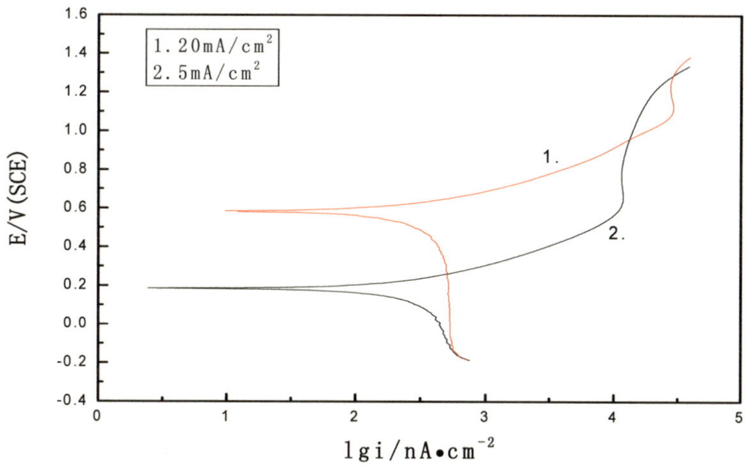

Fig. 1-16. The anodic-polarization curves of samples with different current densities in Hank's solution (PH7.45)

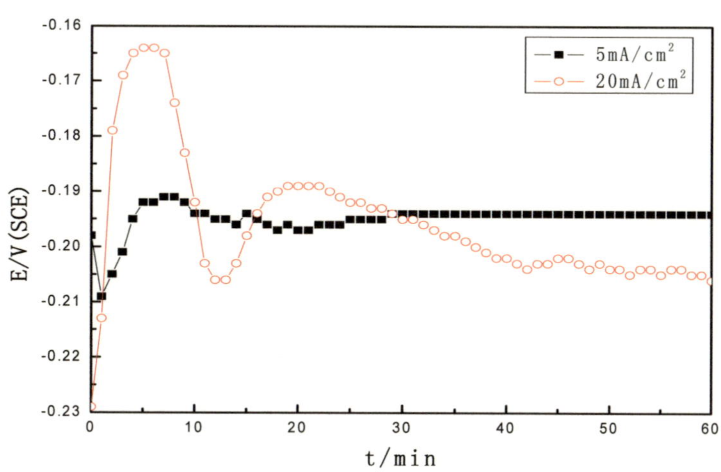

Fig. 1-17. The Corrosion potential change plotted as a function of time of heat treated sample by different current densities in Hank's solution (PH7.45)

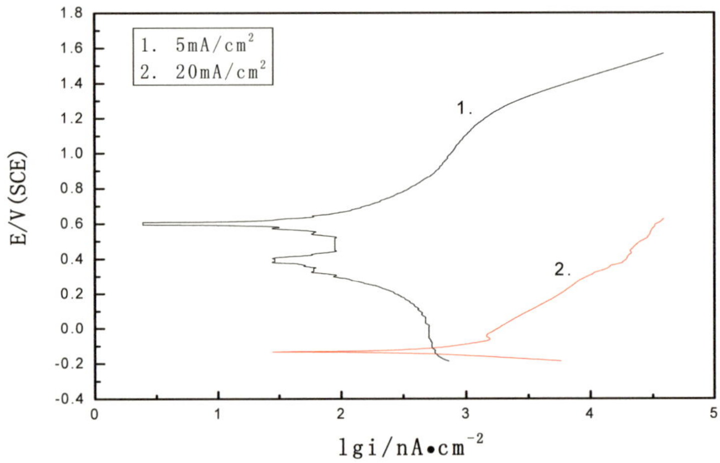

Fig. 1-18. The anodic-polarization curves of heat treated sample by different current densities in Hank's (PH7.45)

Figure 1-19 shows the corrosionpotential-time curves of the deposited TiNi SMA samples at different current densities in Fusayama solution of PH6.13. It can be seen that the potentials of both samples moved to more positive direction. The current density of the deposited sample at 20mA/cm2 had a positive potential, 0.15V, and the stable corrosion potential of the deposited sample at 5mA/cm2 was -0.12V. This illustrates that the deposited sample at the current density of 20mA/cm2 in Fusayama solution of PH6.13 has similarly a better thermodynamic stability.

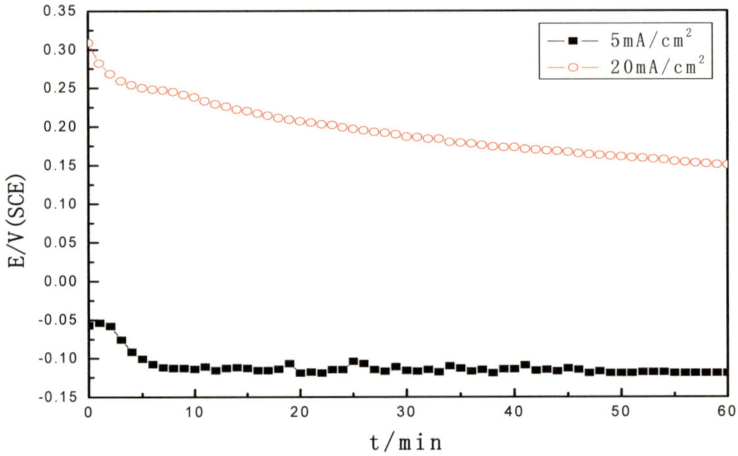

Fig. 1-19. The Corrosion potential change plotted as a function of time of samples with different current densities in Fusyama solution (PH6.13)

Figure 1-20 shows the polarization curves of the deposited TiNi SMA samples at different current densities in Fusayama solution of PH6.13. It can be seen from the polarization curves that in the process of polarization of the deposited sample at the current density of 5mA/cm^2, the current density was far higher than that of the one at 20mA/cm^2. Although the passivation region of the deposited sample at the current density of 5mA/cm^2 was longer, its passivation current density was also higher than that of the deposited sample in large range. From the anodic-polarization curves of view as a whole, the deposited sample at the current density of 20mA/cm^2 had a better corrosion resistance. The reason caused was similar to that the Ti-O film obtained at current density of 5mA/cm^2 was very thick. From cathodic-polarization curves of view, the deposited sample at the current density of 20mA/cm^2 similarly presented a better cathodic-polarization property.

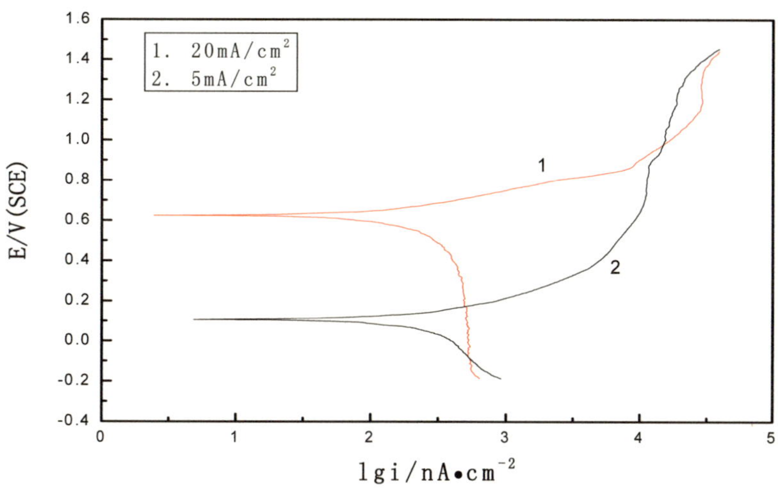

Fig. 1-20. The anodic-polarization curves of samples with different current densities in Fusayama solution (PH6.13)

Figure 1-21 shows the corrosionpotential-time curves of the deposited TiNi SMA samples at different current densities after crystallized at 450°C in Fusayama of PH6.13. It can be seen from the figure that the corrosion potential of deposited sample at the current density of 5mA/cm^2 after crystallized went up with time, and the potential did not change basically at about -0.19V, but the one at higher current density was in a more negative potential, about -0.21V. This illustrates that the sample depositing at the current density of 5mA/cm^2 after crystallized in Fusayama solution was a continuous formation of a passivation film, its corrosion potential was more positive, and the film presented a better thermodynamic stability.

Figure 1-22 shows the polarization curves of the deposited TiNi SMA samples at different current densities after crystallized at 450°C in the fusayama solution of PH6.13. It can be seen from the polarization curves that the breakdown potential of deposited sample at the current density of 5mA/cm^2 after crystallized was 0.7V, and the current density in the process of polarization was always lower than that of the one at 20mA/cm^2, but there was no the passivation phenomenon emerging in depositing the sample. This illustrates that the deposited sample at current density of 5mA/cm^2 after crystallized had a better property of anti-corrosion. Similarly, this is because of that the deposited sample at current density of 5mA/cm2 was with small granularity, even surface and less crystal defects. From the cathodic-polarization curves of view, the deposited sample at current density of 5mA/cm2 after crystallized displayed a better cathodic-polarization property.

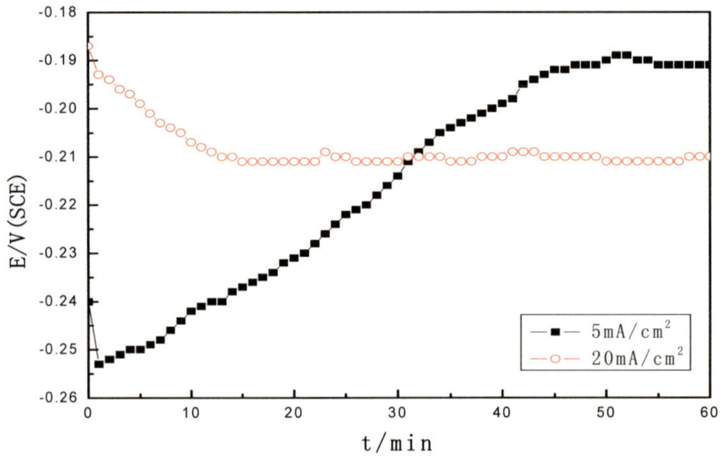

Fig. 1-21. The Corrosion potential change plotted as a function of time of heat treated samples by different current densities in Fusayama solution (PH6.13)

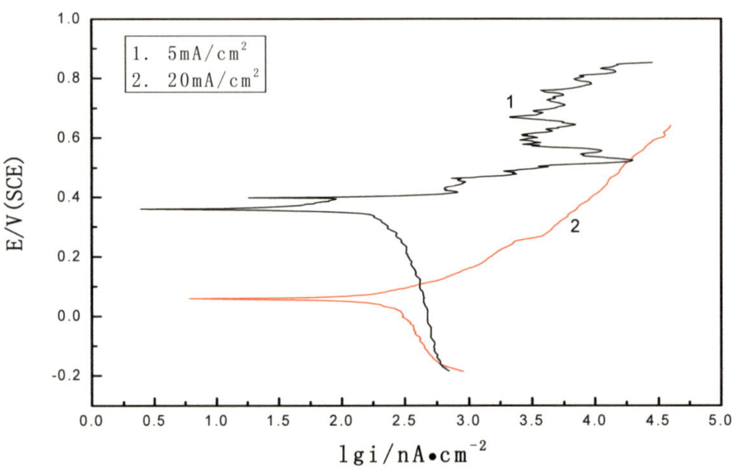

Fig. 1-22. The anodic-polarization curves of heat treated sample by different current densities in Fusayama solution (PH6.13)

1.3.6 Effect of crystallization treatment on surface morphology and corrosion resistance

Figure 1-23 shows the optical microscope morphologies of the deposited samples at the current density of $20mA/cm^2$ for 7min before and after crystallization. It can be seen from figure 1-21a and figure 1-21b that before crystallization the surface Ti-O film, presenting bigger lump in shape, covered on the surface of TiNi SMA; it can be seen from figure 1-21c and figure 1-21d that the crystallized Ti-O film in smaller lump in shape distributed on the substrate surface. This is because of that the crystallization treatment made the hydrate (mainly $Ti(OH)_4$) of amorphous Ti dehydrate water molecules and the crystal lattice recombinate to form the anatase type of TiO_2. Due to that the current density in cathodic-electrodeposition was higher, the Ti-O film obtained was thicker too, so, in the process of crystallization, there might be uneven dehydration and stress distribution existing, thus bringing about serious crystal defects to affect the properties of Ti-O film to a large extent.

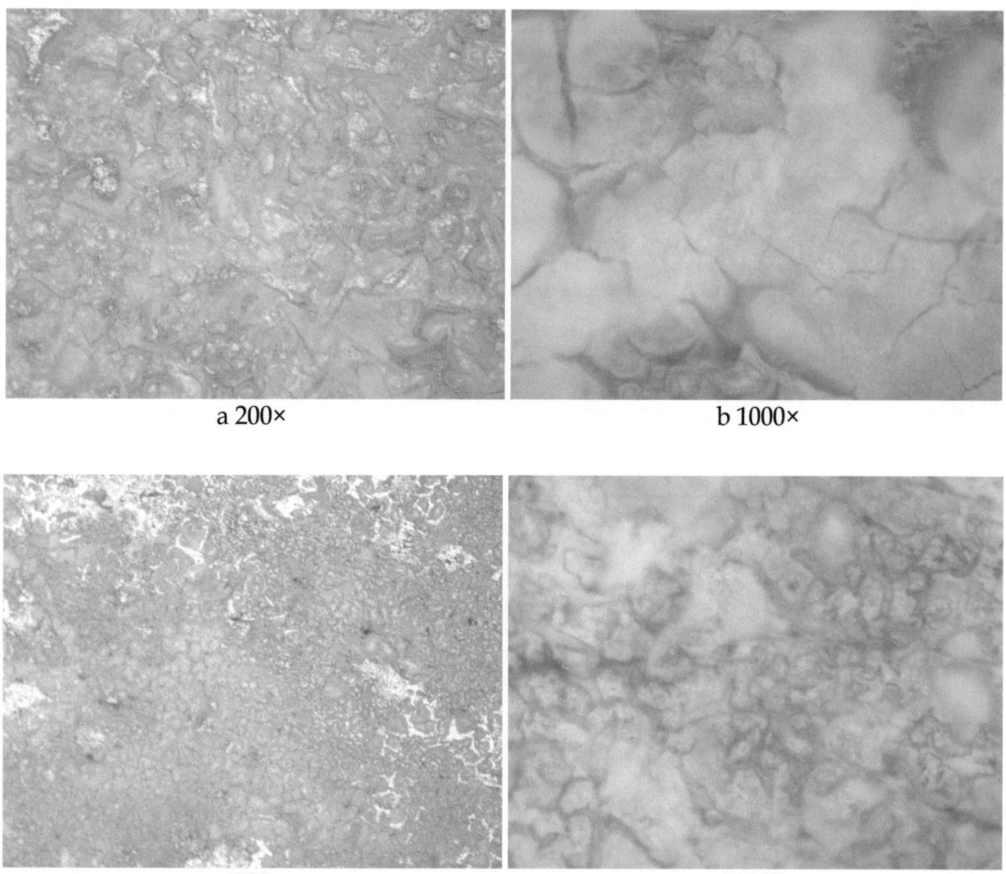

a 200× b 1000×

c 200× d 1000×

Fig. 1-23. Optical microscope result by current $20mA/cm^2$ for 7 min (a, b untreated; c, d heat treated)

Figure 1-24 shows the optical microscope morphologies of the deposited samples at current density of 5mA/cm^2 for 4min before and after crystallization. It can be seen from figure 1-22a and 1-22b that Ti-O film was even and smooth, and from 1-22c and 1-22d that the film distributed in type of tiny particles. Similarly, this is because of that the crystallization treatment made the hydrate (mainly $Ti(OH)_4$) of amorphous Ti dehydrate water molecules and the crystal lattice recombinate to form the anatase type of TiO_2. Because the film was thinner, the crystal defects caused in the process of crystallization was less, so that the effect on the film's properties would not be serious.

Figure 1-25 shows the corrosionpotential-time curves of the deposited samples at current density of 5mA/cm^2 before and after crystallization in Hank's solution of PH7.45. it can be seen from the figure that the corrosion-potential of the sample before crystallization basically reached a stability at 5min, about -0.1V; and that of the crystallized one was about -0.19V, presenting a more negative potential. This illustrates that in the Hank's solution of PH7.45, the deposited TiNi SMA sample at current density of 5mA/cm^2 before crystallization displayed a better thermodynamic stability.

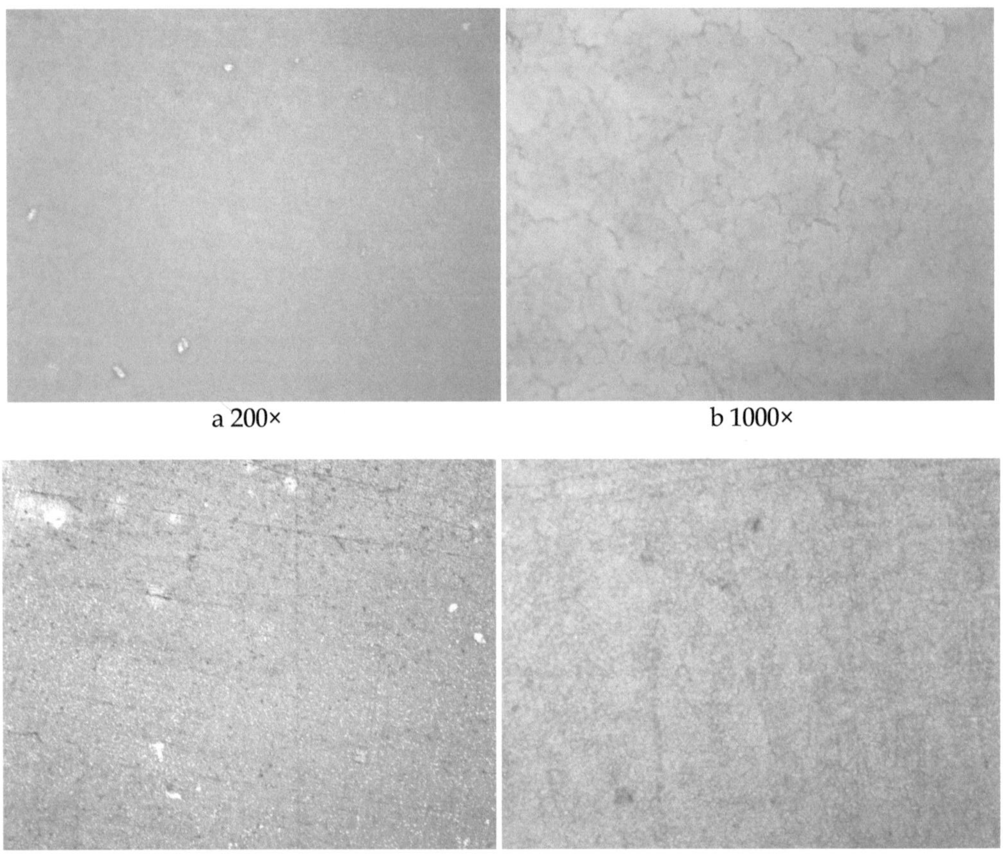

a 200× b 1000×

c 200× d 1000×

Fig. 1-24. Optical microscope result by current 5mA/cm^2 for 4 min (a, b unannealed; c, d annealed)

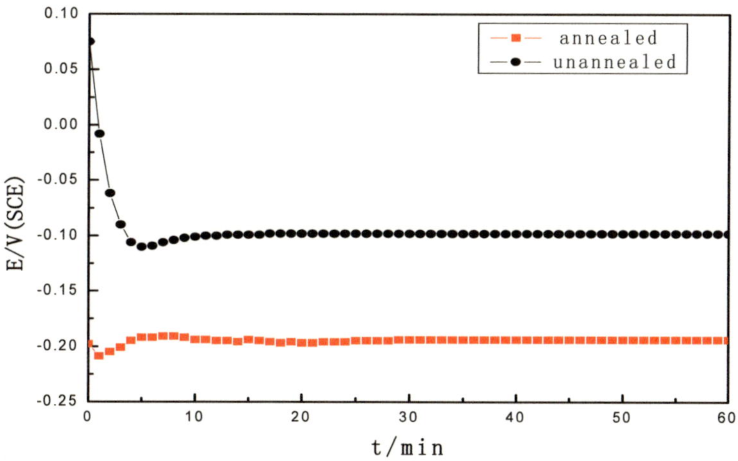

Fig. 1-25. The Corrosion potential change plotted as a function of time of annealed and unannealed samples in Hank's solution (PH7.45)

Figure 1-26 shows the polarization curves of the deposited samples at current density of 5mA/cm^2 before and after crystallization at 450°C. It can be seen from the polarization curves that the breakdown potential of the uncrystallized sample was 1.3V, the current density in the process of polarization always was lower than that of the crystallized sample, and a longer passivation region emerging illustrates that the deposited sample at current density of 5mA/cm^2 before crystallization presented a better corrosion resistance in Hank's solution of PH7.45. Due to the single structure of unannealed Ti-O film, it is easy for the crystallized sample to arise more crystal defects, and bring about the crystal structure not unitary, thus presenting a poor corrosion resistance. From the cathodic-polarization curves of view, the slope of the sample 's cathodic-polarization curves before crystallization was larger, but the current density was lower than that of the crystallized sample, thus presenting a better cathodic-polarization property.

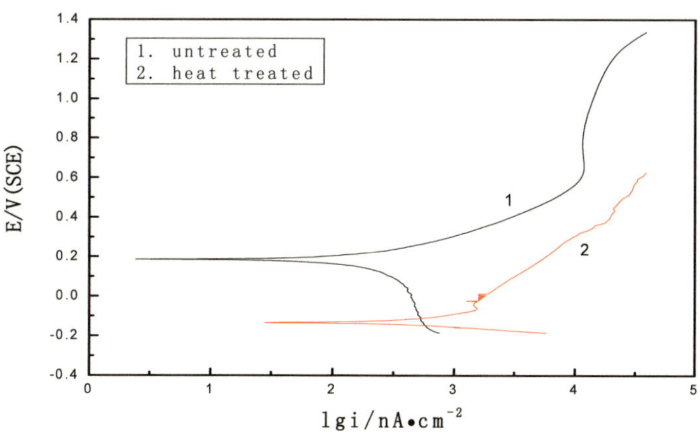

Fig. 1-26. The anodic-polarization curves of annealed and unannealed samples in Hank's solution (PH7.45)

Figure 1-27 shows the corrosionpotential-time curves of the cathodically-electrodeposited TiNi SMA samples before and after crystallization at 450°C in Fusayama solution of PH6.13. It can be seen from the anodic-polarization curves that the stable corrosion potential of uncrystallized sample was more positive relatively, about -0.12V, but that of the crystallized one was more negative. That illustrates that the uncrystallized sample in Fusayama solution of PH6.13 possessed similarly a better thermodynamic stability.

Figure 1-28 shows the polarization curves of cathodically-electrodeposited TiNi SMA samples before and after crystallization at 450°C in Fusayama solution of PH6.13. It can be seen from the anodic-polarization curves that at lower potential, the crystallized sample's in the process of anodic-polarization was lower than the uncrystallized one's, but the uncrystallized sample's passivation region was shorter, from 0.5V to 0.7V, with a 200mV span. On the other hand, the passivation region of uncrystallized one was very long, from 0.4V to 1.3V, with a 900mV span. It can also be seen that the passivated film of crystallized sample began to breakdown from 0.8V, its current density increased quickly and far away exceeded that of the uncrystallized one. The sample before crystallization was still in a passive state, until 1.3V, the current density began to increase apparently, and its breakdown potential was 500mV higher than that of crystallized sample, presenting a better property of anti-corrosion. Similarly, this is because of that the crystal defects caused by crystallization brought about the drop of the property of anti-corrosion. From the cathodic-polarization curves of view, the uncrystallized sample similarly presented a better cathodic-polarization property.

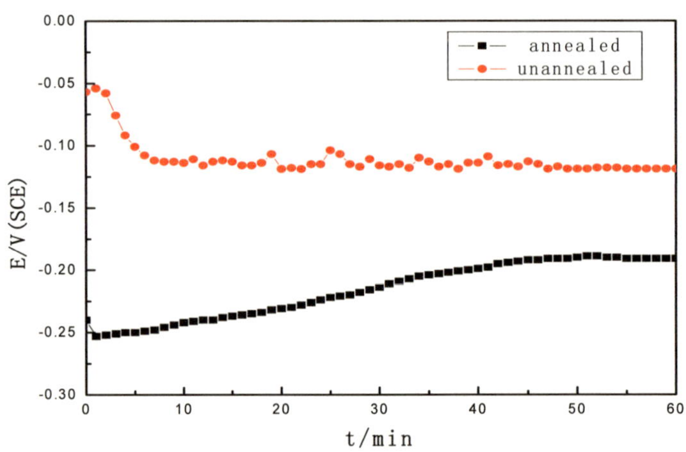

Fig. 1-27. The Corrosion potential change plotted as a function of time of annealed and unannealed samples in Fusyama solution (PH6.13)

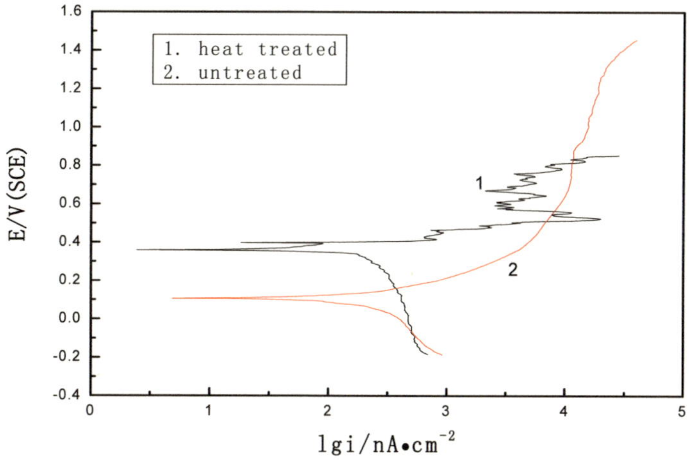

Fig. 1-28. The anodic-polarization curves of annealed and unannealed samples in Fusayama solution (PH6.13)

1.3.7 Effect of cathodic-electrodeposition time on surface morphology

Figure 1-29 shows the optical microscope morphologies of the cathodically-electrodeposited samples at current density of 20mA/cm^2 for different times. It can be seen from the figure that after deposited for 3min, obviously the Ti-O film began to emerge on the surface of TiNi SMA, fine, close and even, but there were some holes, which might be caused by the gas produced on the sample surface when cathodically-electrodeposited at the high current density. After deposited for 5min, the film thickness of the sample surface got increased apparently, but there were some big cracks emerging, adhered lumpily on the sample surface. After deposited for 7min, the Ti-O film fluffily adhered to the sample surface, bonding closely, with less cracks relatively, but uneven. The difference between the morphology of the sample deposited for 9min and that of the one for 7min was not great, fluffily adhered to the sample surface the same. Apparently, the surface of the sample deposited for 2h was different from that of the ones for short times, and its surface had no thicker fluffy Ti-O film, but with some holes. Similarly, the sample deposited for 12h had no fluffy Ti-O film emerging on the surface, but with some small cracks and porosities in different sizes and deepnesses. This is because of that at beginning of depositing, the Ti-O film nucleated and grew quickly on the sample surface, and its thickness increased quickly too. At the same time, with the increase of the thickness in the process of nucleation and growth, the Ti-O film on TiNi SMA surface continuously dissolved and peeled off, besides a lot of gas produced on the sample surface, that brought about uneven stress emerging internal Ti-O film, thus causing distributing lumpily. Because current density was higher, Ti-O film continuously formed and peeled off with time, thus gradually forming the fluffy distribution, and uneven very much. When to a certain time, because of reduction of Ti^{4+} in electrolyte, that made the level of Ti-O film peeling off far higher than that of forming, until the fluffy film disappearing completely, so some porosities in different sizes and deepnesses emerging. The cracks emerged in figure 1-27f might be caused by long time depositing, when drying, under the uneven stress, thus producing cracks. In short, at current density of 20 mA/cm^2, Ti-O film was a growth process of "thickening-cracking-lumping-peeling off-cracking again".

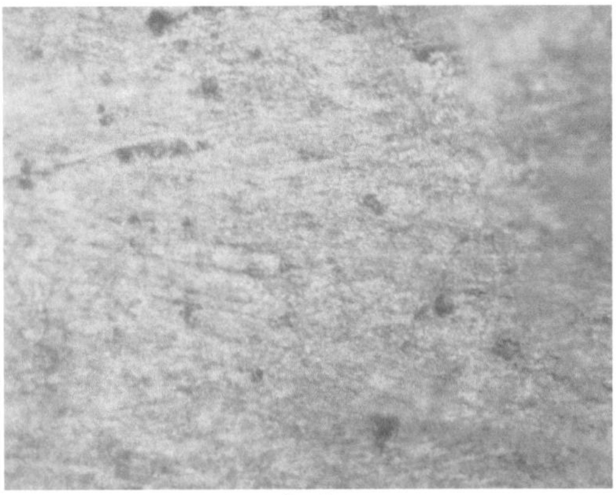

a 3min

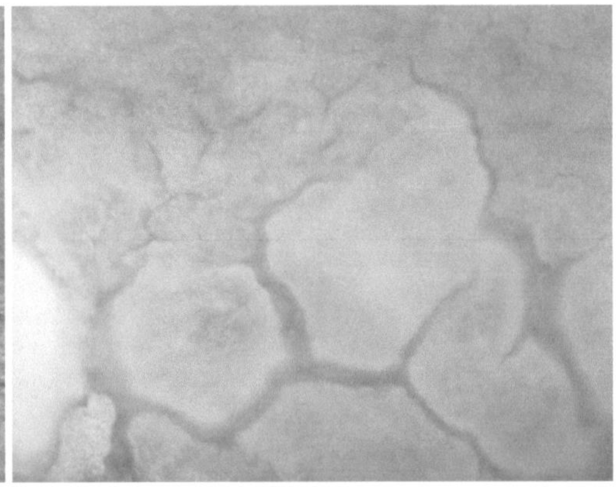

b 5min

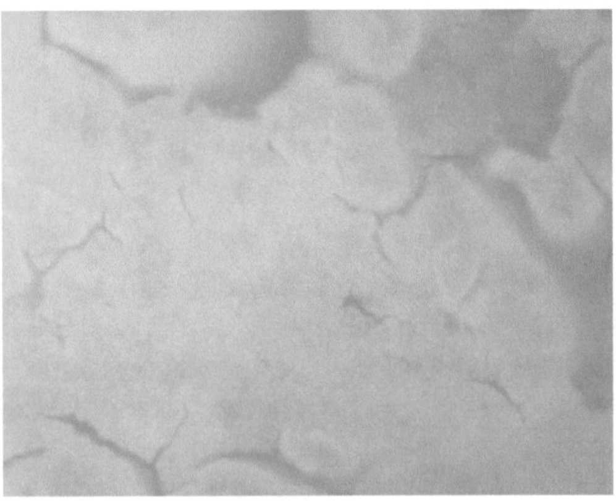

c 7min

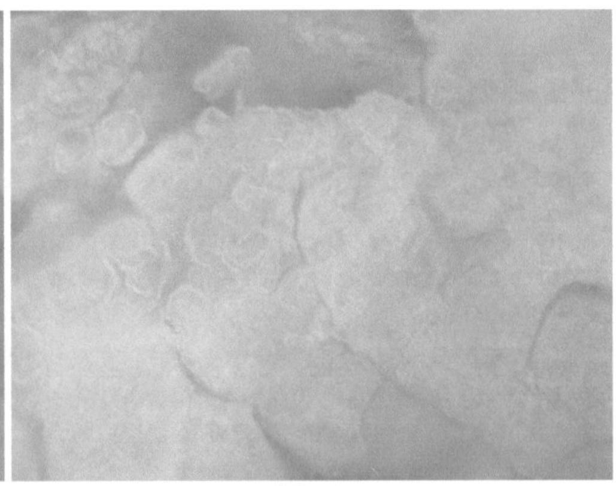

d 9min

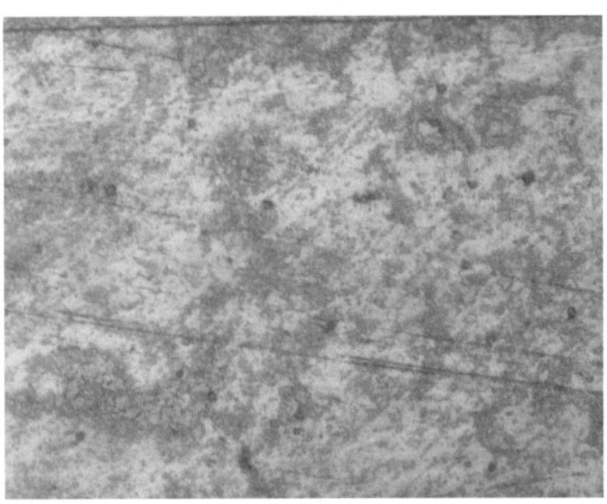

e 2h

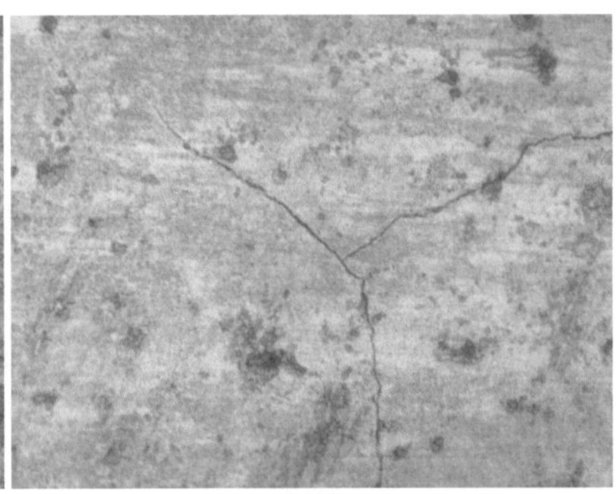

f 12h

Fig. 1-29. Optical microscope result of sample by current 20mA/cm^2 for different times (1000×)

Figure 1-30 shows the optical microscope morphologies of cathodic-electrodeposition at low current density (5mA/cm^2) for different times. It can be seen from the figure that at low current density, the Ti-O film deposited on the sample surface for 1min had formed in large area, but not even, there were some places where undeposited or deposited very thin. The surface Ti-O film of the sample deposited for 2min had been very even, and adhered to the substrate's surface in large area. The surface Ti-O film of the sample deposited for 3min got thickening, distributed well laminally, there were a lot of fine holes and cracks in disorder. On the surface of the sample deposited for 4min, the lamina Ti-O film(s) combined and grew with each other, that made the holes number reduce, and fine cracks in disorder get long. Deposited for 5min, surface Ti-O film(s) continuously combined with each other, and thickened, with long surface cracks, Ti-O film adhered in large laminas to substrate surface. The surface Ti-O film of the sample deposited for 6min continued to thicken, and emerge big cracks and distributed not well lumpily on the substrate's surface. The big cracks in the Ti-O film of the sample deposited for 7min began to get less, fine and shallow, and there were some fine cracks emerging within the Ti-O film in big lumps. The surface cracks of the sample deposited for 8min got more and fine, and Ti-O film in big lumps got smaller. The surface Ti-O film of the sample deposited for 30min distributed well on the substrate surface in fine particles was big relatively. From the change of the surface morphology of Ti-O film in the figure of view, that was a process of "forming-evening-cracking-aggregating-cracking again". It can be seen that the particles of sample cathodically-electrodeposited at low current density were fine and even. This is because of that the drying force for nucleation of Ti-O film at low current density was not so great, the film would slowly nucleate on the substrate surface, besides at low current density, the gas produced on the sample surface was not strongly, so that made the Ti-O film adhere evenly to the substrate surface. Among the samples, the Ti-O film of the sample deposited at $5mA/cm^2$ for 4min had the best levels of evenness and thickness.

a 1min

b 2min

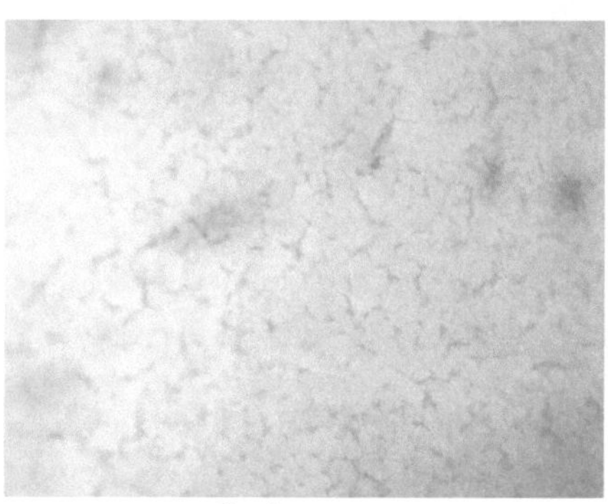

c 3min

d 4min

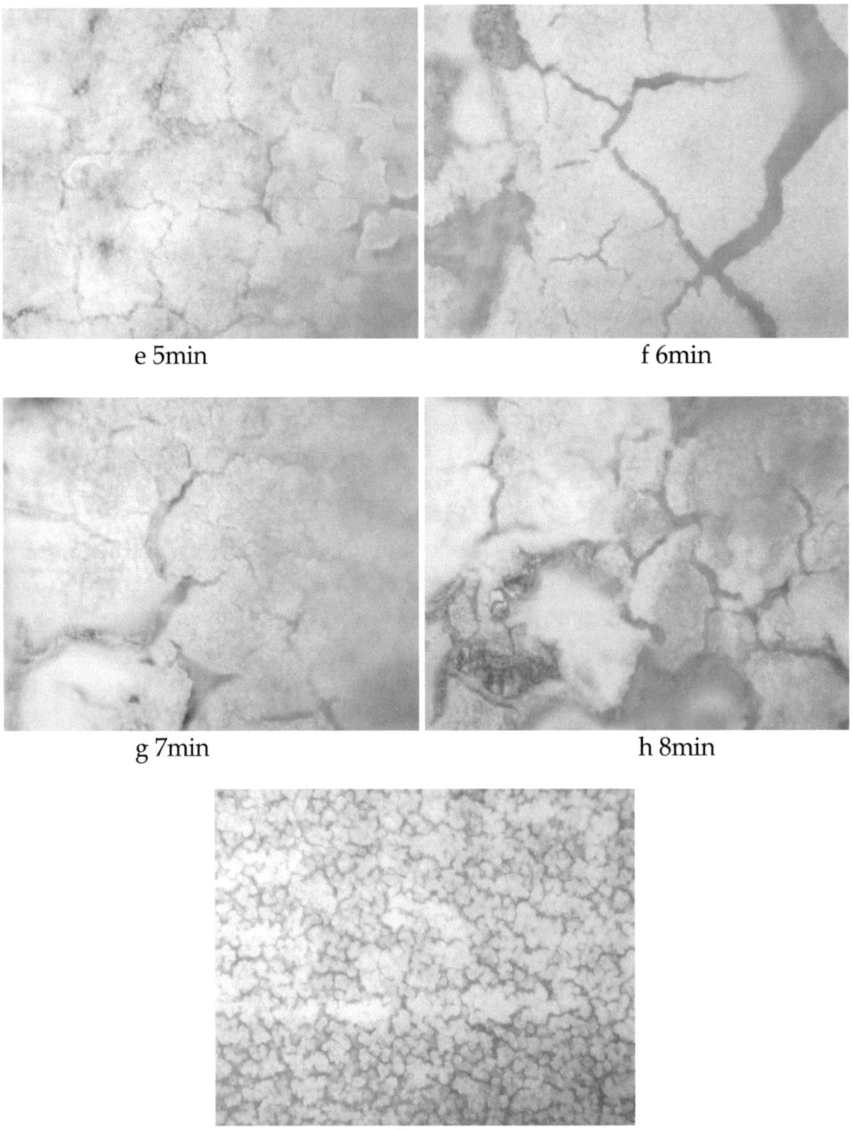

e 5min f 6min

g 7min h 8min

i 30min

Fig. 1-30. Optical microscope result of samples by current $5mA/cm^2$ for different times (1000×)

1.3.8 Effect of electrolyte PH value and NO_3^- concentration on cathodic-electrodeposition

Figure 1-31 and table 1-2 are the optical microscope morphologies, technical parameters and results affected by cathodic-electrodeposition in different PH values of electrolyte and NO_3^- concentrations. It can be seen from the figure 1-29a that there was no obvious Ti-O film existing on the sample surface. This is because of that acidity of the solution was too high, although OH^- produced on the cathodic sample surface, then quickly dissolved, it did not form a Ti gel to adhere to the sample surface within a short time. It can be seen from figure 1-29b that there was an obvious Ti-O film produced. This because of that when PH value increased, Ti gel could be formed within a short time. It can be seen from figure 1-29c that there was less Ti-O film formed. This is because of that under the circumstance of higher PH value, although there was no NO_3^- reduced to OH^- yet, at driving of high current density, H_2O was reduced to OH^- which could form less Ti gel within a short time. But at low current density, even if small amount of NO_3^- was added in, it was not enough to form a Ti gel within a shorter time, this can be seen from figure 1-29d and e. It can be seen from figure 1-29e that under higher PH value, adding in enough NO_3^-, it could form fine and close Ti-O film at low current density within a shorter time. So, electrolyte PH value and NO_3^- concentration have played a vital role in the process of TiNi SMA cathodic-electrodeposition.

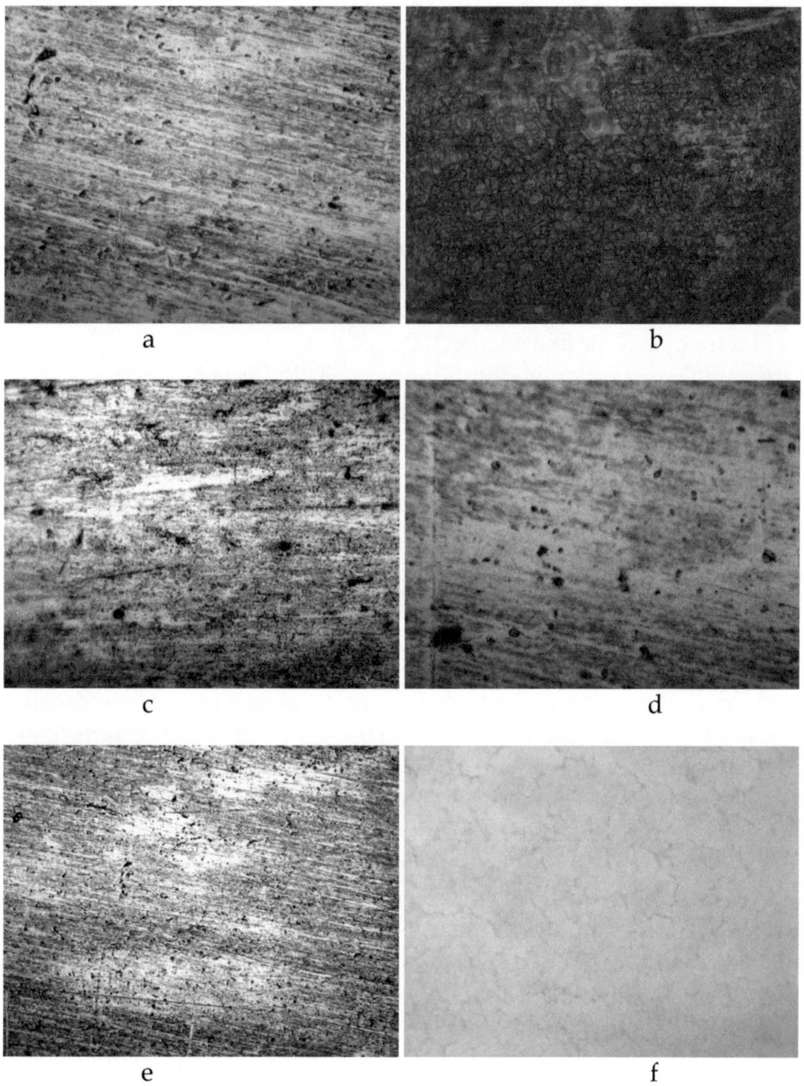

Fig. 1-31. Optical microscope of sample in electrolyte of different PH value and NO_3^- concentration (1000×)

Picture	PH	NO_3^- (M)	i(mA/cm²)	t(min)	Result
a	0.71	0.02	20	4	No film
b	1.2	0.02	20	4	Thick film
c	1.2	0	20	4	Thin film
d	1.2	0	5	4	No film
e	1.2	0.02	5	4	No film
f	1.2	0.2	5	4	Thick film

Table 1-2. Processing parameter and electrodeposition result in electrolyte of different PH values and NO_3^- concentrations

1.4 The summary of this chapter

1. In self-prepared $Ti(SO_4)_2$ solution of electrolyte (PH1.2), at a constant current density of 5mA/cm², cathodically-electrodeposited for 4min, a layer of amorphous Ti-O film could be obtained on TiNi SMA. This Ti-O film is composed of a lot of fine and close particles gathered, the particles' size in dozens of nanometers. And the atomic ratio of Ti and Ni reached to 3:1 through a primary detection by EDS. This illustrates that there were a lot of Ti element included in the film, and this closely bonding of fine particles has contributed to strengthening the surface properties of TiNi SMA.
2. Through XPS analysis, Ti-O film elements of TiNi SMA cathodically-electrodeposeted existed in form of TiO_2 or hydrate $Ti(OH)_4$.
3. After the sample cathodically-electrodeposited Ti-O film was crystallized at 300°C, no TiO_2 existing was found, but crystallized at 450°C, there was anatase TiO_2 peak emerging. After the non-deposited blank TiNi SMA sample was crystallized at 450°C, there was no TiO_2 peak emerging. This illustrates that the film obtained by cathodic-electrodeposition was certainly an amorphous Ti-O film, and its composition was existed in form of hydrate $Ti(OH)_4$.
4. The tests of corrosion potential and electrochemical corrosion illustrated that the TiNi SMA cathodically-electrodeposited presented the better thermodynamic stability and corrosion resistance in Hank's solution (PH7.45) and Fusayama solution (PH6.13).
5. At different current densities, there was much different in roughness, even level, thickness and crack of Ti-O film surface morphology of cathodically-electrodeposited TiNi SMA, among them, the film obtained at the current density of 5mA/cm² was the best one.
6. The thermodynamic stability and electrochemical anti-corrosion of the cathodically-electrodeposited samples at different current densities in Hank's solution (PH7.45) and Fusayama solution (PH6.13) were different. Those of the sample deposited at current density of 20mA/cm² were better. But for the one crystallized at 450°C, and cathodically-electrodeposited at the current density of 5mA/cm², its thermodynamic stability and electrochemical anti-corrosion were better.
7. The surface morphologies of Ti-O film before and after crystallization at 450°C were different. The crystallized Ti-O film was distributed in fine and close particles. Moreover the sample before crystallization had better thermodynamic stability and corrosion resistance in Hank's solution (PH7.45) and Fusayama solution (PH6.13).
8. The morphologies of the Ti-O film(s) obtained which were cathodically-electrodeposited for different times were different. At the current density of 20mA/cm², The Ti-O film change with time was a growth process of "thickening-cracking-

agglomerating-peel off-cracking again". At that of 5mA/cm^2, the change was a process of "forming-evening-cracking-agglomerating-cracking again".

9. The electrolyte's PH value and NO_3^- concentration played a vital role in the process of TiNi SMA cathodic-electrodeposition. Only under PH1-3 and about 0.2M of NO_3^- concentration, a better effectiveness of cathodic-electrodeposition could be obtained.

2. Chapter II

2.1 Preface

The key, for a plant material to bonding surface, is that the material must be with excellent bioactivity, that is, the biomaterial should have capability of closely bonding with surrounding living tissue under physiological fluid environment[9]. The bioactive bonding called, is that after a bone plant is planted into human body, there will be a layer of bioactive hydroxyapatite (HA) forming on the material's surface. HA is a main inorganic substance, occupying about 69%, and about 41.8% in volumetric fraction of human bone. Through HA thin layer, chemical bonding at a molecular level is formed. After planting into human body for 3-6 months, the interface strength of the bioactive bonding is equal to or higher than that of surrounding bone tissue[9]. Under the circumstance of guaranteeing other properties, promoting coating's bioactivity will be helpful to shorten osseointegration period, make patient's injured site(s) recover the functions early, and alleviate suffering[10].

Blood compatibility includes quite wide content. There are the reactions at cell level, such as thrombosis (platelet adhesiveness, agglomerating, and deformation) when connecting with material, haemolysis and leucopenia; there are the reactions at plasma protein level, such as coagulation system and fibrinolytic system activation; and there are the reactions at molecular level, such as immune ingredient change, platelet receptor and ADP (adenosine diphosphate)[11].

When a material connects to blood, a competitive protein adsorption will come about first to form a complex protein adsorption layer, then blood cell and platelet adhesiveness, and then thrombosis, if which can not be repaired or cured, it is easy to bring about coagulating. So, avoiding thrombosis and preventing coagulation are the most important factors of a biomaterial's blood compatibility. Approach a mechanism of thrombosis to promote material's blood compatibility is a research hot point in the academic circle[12,13], and it has been made a great progress recent years.

In this chapter, evaluations for the biomaterials' blood compatibility in vitro have been experimentally carried out in following two aspects[14]:

1. Haemolysis rate test For a plant apparatus directly connected to blood, it is necessary to carry out the experiment on haemolysis rate in vitro. Under normal circumstance, the average life of erythrocyte is 120d. Under the same reason, the life of erythrocyte shortens, it is called that a haemolysis process comes about. Through haemolysis rate experiment the rank of toxicity produced by erythrocyte in blood connecting to material can be evaluated, the erythrocyte in blood will be damaged in varying degrees because of the toxic substance(s) of the material, to release hemoglobin, and to bring about haemolysis. By testing the amount of hemoglobin released from the material, the haemolysis rate for the material can be obtained. Generally speaking, the concentration

of a toxic substance which can produce haemolysis reaction is higher than that of the one which can only produce toxic effect. For the biomedical materials, it is extremely important to alleviate erythrocyte damage.

2. Dynamic coagulation time test Blood would be activated to coagulate and to affect platelet's formation and function as a material connects with blood. And blood coagulation is a result from a series of reaction in blood. There are two processes which would initiate coagulation, that is, intrinsic coagulation (activated by coagulation factor XII) and extrinsic coagulation reactions. Firstly, prothrombin factor XII was changed into active factor XII a; the active XIIa made the coagulation factor XI change into an activation factor XIa, and furthermore made factor IX activate to be IXa; then the activated IXa combined with the factor VIII, phosphatide and C^{2+} ions to form a complex compound, which would make factor X change into activation factor Xa; furthermore, the active factor Xa combined with factor Va, phosphatide and C^{2+} to form the complex compound which would make prothrombin change into thrombin; finally, the thrombin made fibrinogen molecules change into fibrinogen monomer, then into the fibrin colloid under the action of activated factor XIII a, thus forming the blood clots. On the other hand, the extrinsic clotting system, activated by tissue factor and VII factor, directly made the factor X activate to be Xa, the next process emerged was as the same as the intrinsic one's[15].

The process of blood coagulation shows that globin and fibrin in blood on material's surface would bring about activation of the coagulation factor, thus causing fibrin to form and to coagulate. At the same connecting time, by the experiment of dynamic coagulation time in vitro, the activation degrees of different materials on intrinsic coagulation factor can be compared. The activation degrees on coagulation are different, so are the coagulation degrees. With the time of material connecting to blood, coagulation degrees increases correspondingly. The dynamic coagulation time curve is a relation curve to absorbance and time. The more smooth the curve is, the longer the coagulation time lasts, it shows, and the lower the degree activated by coagulation is. And the longer the coagulation time lasts, it shows, the lower the degree for a material to bring about thrombus is, and the better the material's anti-coagulating property is.

2.2 Experimental methods

2.2.1 Experimental materials and media

The experimental materials were the examples of TiNi SMA, TiNi SMA after anodic-oxidation and TiNi SMA after cathodic-electrodeposition respectively. The size of all the samples was in 10mm×10mm×1mm.

SBF (Simulated body fluid) solution[16] (making up to PH7.40 with diluted hydrochloric acid and NaOH solution)[17] and fresh blood of New Zealand rabbit.

SBF solution ingredient: Na_2SO_4 0.07g + Na_2HPO_4 0.36g + $NaHCO_3$ 0.35g + NaCl 7.88g + KCl 0.37g + $MgCl_2$ 0.31 + $CaCl_2$ 0.28g + H_2O 1L.

2.2.2 Experimental methods

2.2.2.1 Evaluation of biocompatibility of TiNi SMA surfacely-modified

The samples before and after anodic-oxidation respectively were dipped into 25ml SBF solution of PH7.40. The working area was 10mm×10mm, and the non-working one was coated with silica gel. Dipping for 7d in a water bath at 37°C. The deposition of the Ca/P layer on the surface was detected with Finder 1000 type of EDS.

The samples before and after cathodic-electrodeposition respectively were dipped into 25ml SBF solution of PH7.40. The working area was 10mm×10mm, and the non-working one was coated with silica gel. Dipping for 7d in a water bath at 37°C±1°C, the solution was changed every two days. The deposition of the Ca/P layer on the surface was detected with Finder 1000 type of EDS. And the hydroxyapatite (HA) on the surface was analyzed with D/max 2000type of X-ray diffractormeter (XRD).

2.2.2.2 Test of hemolysis rate of TiNi SMA after surfacely- modified

8ml fresh rabbit blood selected was anti-coagulated with 1ml heparin solution, and diluted with 10ml physiological saline. The samples were put in 10ml physiological saline in the water bath at 37℃, then putting in 0.2ml dilution blood, mixing well gently, and continuing to keep the temperature for 60min. Then the fluid was poured into a test tube, separated at a speed of 1000r/min with LXJ-64-01 type of separator, and the up layer solution was taken out to test absorbance value at wave length of 545nm with Lambda 35 type of spectrophotometer. Anode control group used 10ml distilled water + 0.2ml dilution blood, and cathode one 10ml physiological saline + 0.2ml dilution blood. By the formula:

$$\alpha(\%) = \frac{D_t - D_{nc}}{D_{pc} - D_{nc}} \times 100\% \tag{2.1}$$

D_t : Sample absorbance; D_{nc} : Cathode control group absorbance; D_{pc} : Anode control group absorbance hemolysis rate was taken out.

2.2.2.3 Test of dynamic coagulation time of TiNi SMA after surface modification

1. To compound an ACD blood and 0.47g citric acid, 0.3g glucose and 1..22g sodium citrate were dissolved in 100ml distilled water to compound the blood preservation solution (ACD). Taking that fresh rabbit blood : ACD as 1 : 4 in proportion to compound ACD blood.
2. Taking 0.2ml ACD blood and dropping on the surface of the test material after clearing. Adding to 20μl from the 0.2ml $CaCl_2$ solution, then mixing well and then recording time.
3. At 5, 10, 20, 30, 40, 50 and 60min of given times, respectively having 100ml distilled water flowed through the material surface slowly, the fluid was gathered up to a beaker, the absorbance values (O. D.) of the solution at different times were got at 540nm wavelength with Lambda 35 type of spectrophotometer, and plotting the O. D.-t curves to compare. Taking the connecting time at 0.100 absorbance as the dynamic coagulation time for different materials.

2.3 Experimental results and discussion

2.3.1 Evaluation of biocompatibility of TiNi SMA surfacely-modified

Figure 2-1 shows the EDS results of Ca/P deposition on TiNi SMA before and after anodic-oxidation respectively in SBF solution for 7d. It is seen from Figure 2-1a that there was small amount of the elements of Ca, P and O on the sample before oxidation, this coincides with the Hanawa's investigation[18]. But there was a lot of the elements of Ca, P and O on the sample after oxidation, as shown in figure 2-1b.

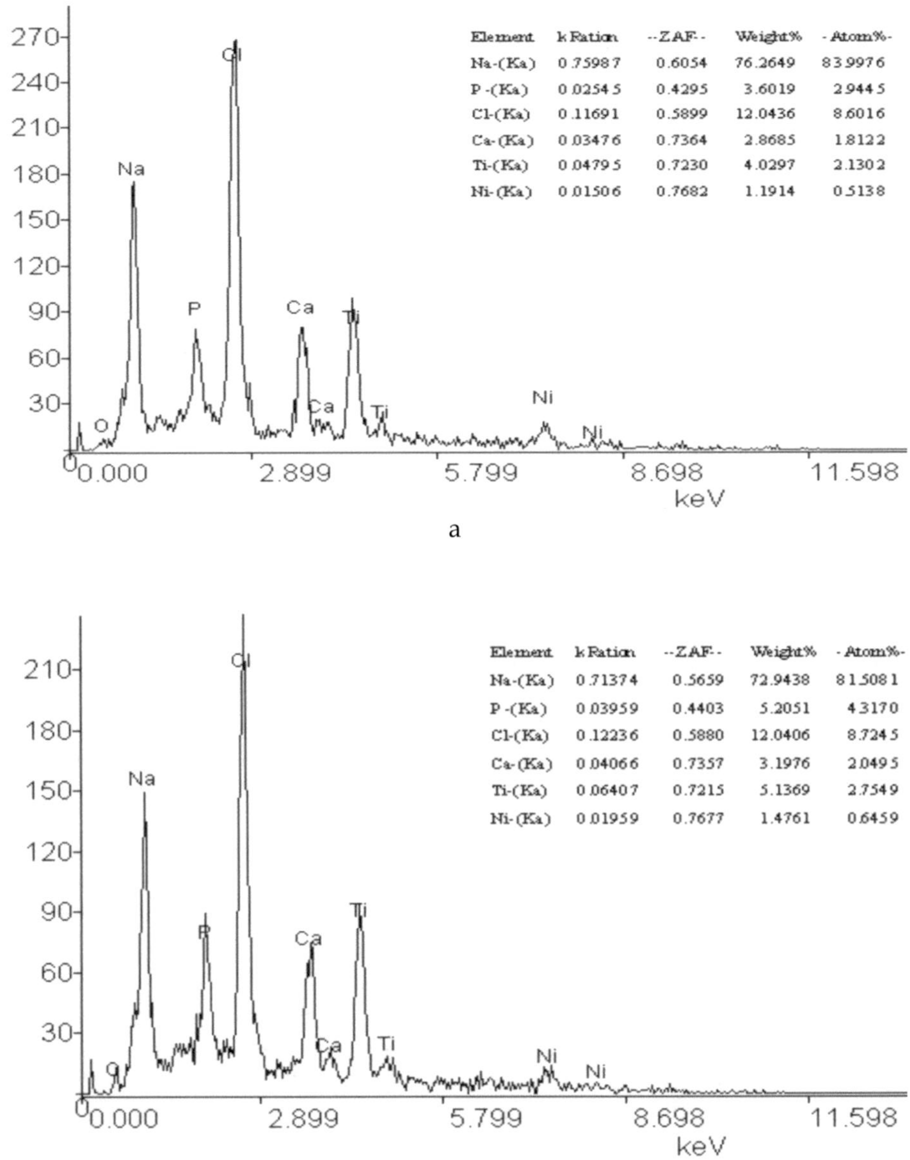

Fig. 2-1. EDS result of Ca/P coating in SBF solution (PH7.40) (a TiNi SMA; b anodic-oxidation)

This is because of that Titanium dioxide's isoelectric point has a great influence on the behavior of Titanium in organism, the isoelectric point of TiO_2 is about 6.2[19], which is somewhat lower than PH value (7.4) in physiological environment. This illustrates that under physiological environment, the Titanium surface may be with some weak negative charges. And this kind of surface with negative charges has a close relation to osseous integration between plant and its peripheral live bone[20-23]. In the body fluid, Ca^{2+} will combine with the negative charge on the TiO_2 surface under the action of Coulombforce. And the OH^- on the surface will adsorb PO_4^{3-} through hydrogen bond to gather forward to surface, thus it can introduce calcium phosphorus layer to deposit on the surface. The concrete adsorption process is as following:

$$Ti\,(OH)^{3+}_{(ox)} + H_2PO_4^- -- Ti^{4+}_{(ox)}HPO_4{}^{2-}_{(ads)} + H_2O \quad (2.2)$$

$$Ti^{4+}_{(ox)}HPO_4^{2-}{}_{(ads)} + OH^- -- Ti^{4+}_{(ox)}PO_4{}^{3-}_{(ads)} + H_2O \text{ or} \quad (2.3)$$

$$Ti\,(OH)^{3+}_{(ox)} + HPO_4^{2-}{}_{(aq)} -- Ti^{4+}_{(ox)}PO_4^{3-}{}_{(ads)} + H_2O \quad (2.4)$$

Among them, (ads), (ox) and (aq) respectively stand for ions adsorbed on the alloy surface, substances in oxidation layer and ions in solution.

The self-produced oxidation film of TiNi SMA is too thin, not stable, but the TiO_2 film produced by anodic-oxidation is thicker, so the capability to adsorb Ca/P layer is stronger. This illustrates that anodic-oxidized TiNi SMA may have some bioactivity, but it needs further verification.

Figure 2-2 shows an EDS result of Ca/P deposition in SBF solution of TiNi SMA before and after cathodic-electrodeposition. It can be seen from the figure that after dipping for 7d, there were higher contents of Ca and P elements deposited on the sample surface. Ca and P mole ratio reached to 1.5 : 1 in figure 2-2a; 2.4 : 1 in figure 2-2b; and 1.8 : 1 in figure 2-2c. The material's after cathodic-electrodeposition is the highest, the one's crystallized came second and the un-treated sample's was the lowest. This is because of, that the Ti-O film obtained after cathodic-electrodeposition was mainly $Ti(OH)_4$, because of OH^- existence, it was easy for apatite on the surface to nucleate. Li et al[24] found in research for bioactivity of titanium metals that the surface titanium oxide layer including Ti-OH group as well as being with electronegativity was one of the major factors to introduce HA nucleation. Because HPO_4^{2-} and PO_4^{3-} generally exist more easily under alkaline environment, crystallization treatment cause OH^- loss, so nucleation is comparatively more difficult. There are researchs showing that the materials with high mole ratio are more stable than ones with low mole ratio under physiological environment. High Ca and P mole ratio is advantageous to prevent from host organization reaction of fast dissolution of small molecules after material planted and it can also prevent from the acute inflammation reaction caused due to fast degradation[25].

At constant temperatures, solubility product of $Ca_{10}(PO_4)(OH)_2$ (hydroxyapatite, HA) is far less than that of other phosphates, so HA oversaturation in calcification solution at the same concentration is higher than that of other phosphates. It can be inferred that surface crystal may mainly be HA.

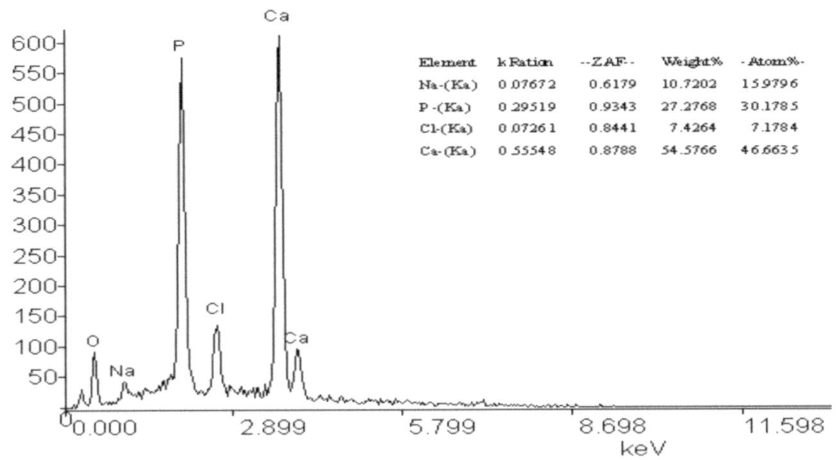
Ca
P
Cl
O
Na
Ca
600
550
500
450
400
350
300
250
200
150
100
50
0
0.000
2.899
5.799
8.698
11.598
keV
Element k Ration --ZAF-- Weight% -Atom%-
Na-(Ka) 0.07672 0.6179 10.7202 15.9796
P -(Ka) 0.29519 0.9343 27.2768 30.1785
Cl-(Ka) 0.07261 0.8441 7.4264 7.1784
Ca-(Ka) 0.55548 0.8788 54.5766 46.6635

a

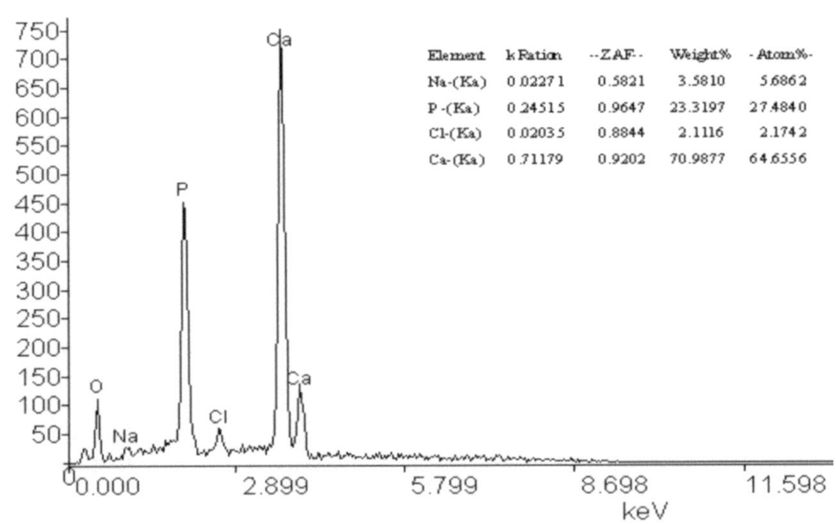
Ca
P
Cl
O
Na
Ca
750
700
650
600
550
500
450
400
350
300
250
200
150
100
50
0
0.000
2.899
5.799
8.698
11.598
keV
Element k Ration --ZAF-- Weight% -Atom%-
Na-(Ka) 0.02271 0.5821 3.5810 5.6862
P -(Ka) 0.24515 0.9647 23.3197 27.4840
Cl-(Ka) 0.02035 0.8844 2.1116 2.1742
Ca-(Ka) 0.71179 0.9202 70.9877 64.6556

b

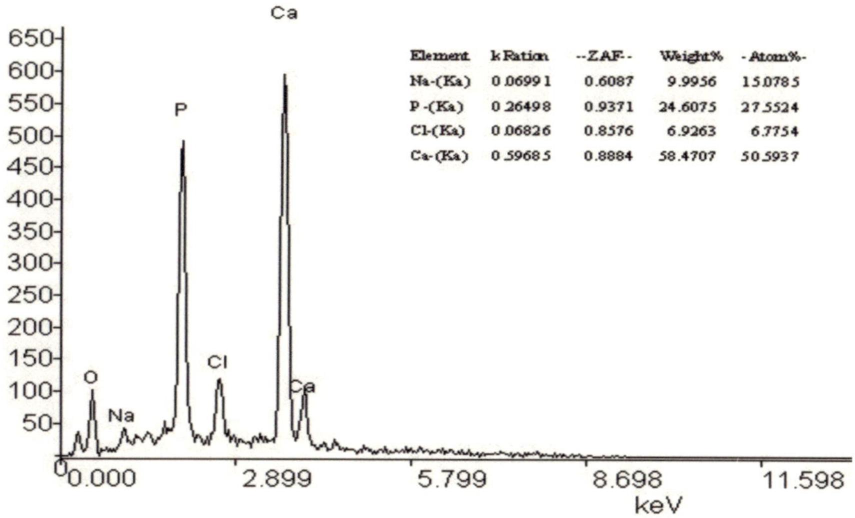

c

Fig. 2-2. EDS result of Ca/P coating in SBF solution (PH7.40) (a TiNi SMA; b electrodeposition; c annealed)

Figure 2-3 shows the surface XRD analysis results of cathodically-electrodeposited samples in SBF solution before and after crystallization respectively. It can be seen from the figure that HA existed on both samples. This illustrates when an amorphous Ti-O film riching in Ti-OH group was dipped in SBF solution for 7d, HA could form on the surface; this also illustrates that the obtained TiO_2 after crystallization treatment still preserved some hydroxyl groups with bioactivity. The formation process of hydroxyapatite on a material surface is actually a formation and growth one of a new phase. The process can be summarized with two stages[26].

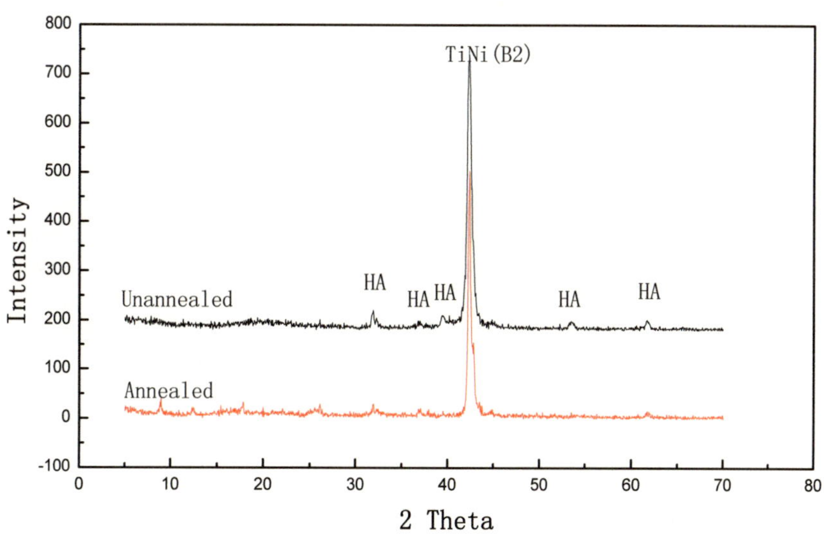

Fig. 2-3. XRD result of HA on the surface of sample cathodically-electrodeposited

2.3.1.1 Nucleation formation

When a material connects to SBF solution, because of electrostatic attraction, Ca^{2+}, HPO_4^{2-} and PO_4^{3-} directly acting on solid surface are adsorbed on the surface, to form calcium-phosphorus compound, which interacts with HPO_4^{2-} and CO_3^{2-} in SBF solution, to form a new phase nucleus. According to two-dimension nucleation theory, as soon as a nucleus forms, the reactant ions will continuously nucleate on the deposited surface, and the crystal will continuously grow.

The factors to affect nucleation are those as following:① Calcium-phosphorus concentration at localization on surface plays a very key role to nucleation. For a sample cathodically-electrodeposited, the particle distribution on the surface film makes surface roughness of the sample increase, and surface area increase too. Thus the concentration at localization on the surface would be comparatively higher than that on a smooth surface. So, under the same conditions, it is easier for Ca^{2+} and HPO_4^{2-} concentration on the roughness surface to reach the nucleation critical value. ② The interfacial energy of material surface. Apatite crystals form and directly grow on the material surface. It may be thought that is increasing ionic concentration in the saturated solution to form the low interfacial energy surface. According to Ostward's nucleation theory, the free energy to nucleate depends on solution's oversaturability (S), pure interfacial energy (σ), temperature (T) and particle surface area (A):

$$\Delta G = -T \ln S + \sigma A \tag{2.5}$$

This nucleation theory illustrates that: the increase of solution's oversaturability and decrease of pure interfacial energy are advantageous to interface heterogeneous nucleation. S increasing., will make nucleation free energy decrease. As long as S is high enough, even if a surface with low energy, which has not been subjected to any treatment, it can also introduce heterogeneous nucleation. Because ionic concentration is high, it can overcome the barrier of material's surface nucleation, to nucleate on the surface. ③ The geometric morphology of material surface. It is reported that a crystal nucleus firstly occurred at the roughness places on a surface. Because higher localization concentration can be kept in these regions, the critical value to nucleate would quickly be reached, at the same time these locations supply the nucleation spots.

2.3.1.2 Growth of a crystal nucleus

In the process of hydroxyapatite formation, crystal nucleus formation is a homogeneous nucleation, that is, to form a nucleus on the certain substrate. Through the nucleation process, changes in crystal structure and composition accomplish simultaneously. A main role for grain boundaries to nucleate is to decrease boundaries' area and interface energy, thus to low the nucleation work. The decrease of interface energy drives crystal to grow. The deposition process of hydroxyapatite on sample surface in SBF solution is as following[27]:

$$10Ca^{2+} + 6PO_4^{3-} + 2OH^- \longleftrightarrow Ca_{10}(PO4)_6(OH)_2 \tag{2.6}$$

As a consequence, HA begins to deposit on sample, as long as apatite crystal nuclei form, the nuclei will consume Ca and P in solution, hydroxyapatite continuously deposit on sample surface.

2.3.2 Haemolysis rate test of TiNi SMA after surface modification

When blood connects with a foreign material surface some red cells will be destroyed to release hemoglobin, that is, haemolysis occurring. A material with good blood compatibility should have a lower haemolysis rate. Figure 2-4 shows a haemolysis rate tested result of TiNi SMA samples, which were untreated, anodically-oxidized and cathodically-electrodeposited respectively. It can be seen that the TiNi SMA and its surface-modified materials have met the medical requirement (<5%) for a biomaterial. Haemolysis rate of the modified material has decreased somewhat. And haemolysis rate of anodically-oxidized sample was lower than that of cathodically-electrodeposited one. This is because of that, TiO_2 crystals have a good blood compatibility. Besides, the material surface smoothness and film quality have effects on haemolysis rate. Haemolysis rate of cathodically-electrodeposited sample is higher than that of anodically-oxidized one. This is relevant to the smoothness differences on sample surface.

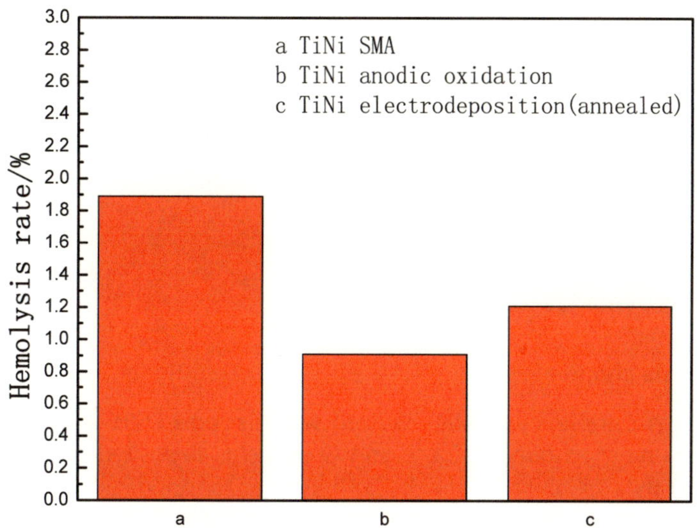

Fig. 2-4. Haemolysis rate of modified TiNi SMA

2.3.3 Dynamic coagulation time test of TiNi SMA after surface modification

The experiment on dynamic coagulation time is to test the degree of activating endogenous coagulation factor, the more smooth the dynamic coagulation curve is, the higher the absorbance at ordinate will be. This shows that, the longer the coagulation time, the lower the degree of activating coagulation factor. Generally, the time at O.D.=0.100 is often settled as the coagulation time, in order to comparing. Figure 2-5 shows the change curves of absorbance and blood connecting time of TiNi SMA, which were surface modified by anodically-oxidized and cathodically-electrodeposited respectively. It can be seen from the figure that the absorbance of the anodically-oxidized sample is highest, and the coagulation time is the longest. This illustrates that it has an excellent anti-coagulation property. The secondary is the cathodically-electrodeposited sample, and the last one is TiNi SMA sample (un-treated). Besides, material surface smoothness also has an effect on dynamic coagulation time. Generally the higher the surface smoothness, the longer the dynamic coagulation time.

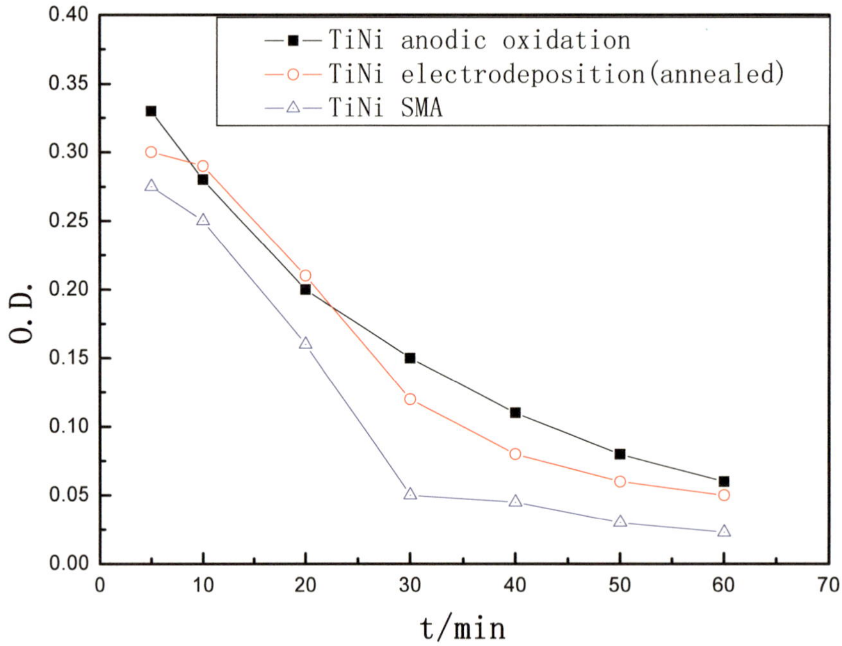

Fig. 2-5. Absorbance of modified TiNi SMA versus time

2.4 Summary of this chapter

1. After TiNi SMA sample anodically-oxidized was dipped in SBF solution (PH7.40) for 7d, there were more amount of Ca, P and O elements deposited on the surface. This illustrates that the capability to adsorb Ca/P layer of the anodically-oxidized sample is stronger, so TiNi SMA after anodic oxidization may have some bioactivity.
2. After TiNi SMA sample cathodically-electrodeposited was dipped in SBF solution (PH7.40) for 7d, there were higher contents of Ca and P elements deposited. The mole ratio of Ca and P reached to 2.4 : 1. And Ca and P mole ratio after crystallization at 450°C was 1.8 : 1. XRD analytical result showed that both crystallized and un-crystallized samples cathodically-electrodeposited have HA in existence. It illustrates that after the amorphous Ti-O film riching in Ti-OH group was dipped in SBF solution for 7d, HA could be formed on the surface; it also illustrates that TiO_2 obtained after crystallization has some bioactivity.
3. In the experiment of haemolysis rate, the haemolysis rate of surfacely-modified TiNi SMA got decreased somewhat. But the haemolysis rate of cathodically-electrodeposited sample was higher than that of anodically-oxidized one. And this was relevant to surface smoothness differences of the samples.
4. In the experiment of dynamic coagulation time, the surface modified TiNi SMA samples had a better anti-coagulation property. But the absorbance of the anodically-oxidized one was the highest, coagulation time longest, showing the best anti-coagulation property.

3. Conclusions of this paper

1. After anodic oxidization, TiNi SMA displayed a better thermodynamic stability and anti-corrosivity in simulant physiological environment, and improved blood compatibility. This laid a foundation for the biomaterial to apply in oral cavity.
2. In the self-prepared and blended well $Ti(SO_4)_2$ electrolyte solution (PH1.2), at a constant current density of 5mA/cm^2, by cathodic-electrodeposition for 4min, a layer of Ti-O film in size of dozens of nanometers had been obtained, after annealing at 450°C, an anatase TiO_2 with bioactivity was obtained.
3. After cathodic electrodeposition, TiNi SMA displayed a better thermodynamic stability and anti-corrosivity in simulant physiological environment, and improved blood compatibility.
4. Through the discussion for Ti-O film's morphologies and properties of TiNi SMA at different technics, the optimal technical parameters obtained were: blending $Ti(SO_4)_2$ solution to PH1-3, adding about 0.2M of NO_3^-, current density at 5mA/cm^2 and time for 4min.

4. References

[1] Tan Xiaochun, Huang Songyu. Studies on preparation of ultrafine TiO_2 membrane through electrophoretic method. Chinese journal of chemical physics[J], 1998, 11(5): 416-421
[2] Hayward R C, Saville A, Aksay A. Electrophoretic assembly of colloidal crystals with optically tunable micropatterns[J]. Nature, 2000, 404(6733): 56-59
[3] Kavan L, O'Regan B, Kay A, et al. Preparation of Titania (anatase) films on electrodes by anodic oxidative hydrolysis of titanium trichloride[J]. Electrical chemistry, 1993, 346(1-2):291-307
[4] Cui Xiaoli, Jiang Zhiyu. Studies on the preparation and characteristics of titanium dioxide nano thin film. Electroplating and finishing, 2002, 21(5): 17-21
[5] Natarajam C, Nogarni G. Cathodic electrodeposition of nanocrystalline titanium dioxide thin films[J]. J Electrochem Soc, 1996, 143(5): 1547-1550
[6] Karuppchamy S, Amalnerkar D P, Yamaguchi K, et al. Cathodic eledtrodeposition of TiO_2 thin films for dye-sensitized photoelectrochemical applications[J]. Materials science and engineering C, 2006, (26): 54-64
[7] Z. D. Cui, H. C. Man, X. J. Yang. The corrosion and nickel release behavior of laser surface-melted NiTi shape memory alloy in Hanks solution[J]. Surface & coatings technology, 2005, (192): 347-353
[8] Leung V W H, Darvell B W. Artificial salivas for in vitro studies of dental materials[J]. Journal of dentistry, 1997, 25(6): 475-484
[9] Heng L L. Biomaterials: a forecast for the future[J]. Biomaterials, 1998, 19: 1419-1423
[10] Xiong Xinbo, Li Hejun, Huang Jianfeng. Research progress in surface bioactivity modification of Ti-based metals[J]. Rare metals letters, 2004, 23(3): 4-9
[11] Zhang Jingchuan. Research progress in evaluation of biomaterials' compatibility. International journal of biological engineering, 1988, 11(5): 330-334
[12] Guo Haixia, Liang Chenghao. Research progress in materials' blood compatibility[J]. Shanghai biological engineering, 2001, 22(3): 44-48

[13] Wang Chuanhua, Leng Xigang. Research progress in mechanism of thrombosis induced by biomaterial[J]. International journal of biological engineering, 1998, 21(1): 1-7
[14] Liu Jingxiao. Research on preparations and structures of 316stainless and NiTi alloy surface films and their blood biocompatibilities. Doctoral thesis, Dalian university of technology, 2001
[15] Buddy D. Ratner. Allan S. Hoffman, Frederick J. et al. Biomaterials Science: an introduction to materials in medicine, 1996, 195-197
[16] Zhang Xiaokai, Liu Wei, Chen Xiaofeng. Mophology character of the sol-gel derived bioactive glass in SBF solution, Chinese journal of chemical physics[J], 2004, 17(4): 495-498
[17] Liang Chenghao, Cheng Bin, Cheng Bangyi. Corrosion behavior of surface passivatied Cu-Zn-Al shape memory alloy in salne solution[J]. Corrosion science and protection technology, 2005, 17(5): 304-306
[18] Hanawa T. In vivo metallic biomaterials and surface modification[J]. Materials Science and Engineering, 1999, A (267): 260-266
[19] Park G A. The isoelectric point of solid oxides, solid hydroxides, and aqueous hydroxy complex system[J]. Chem Rev, 1965, 65:177-198
[20] Hench L L. Bioderamic: from concept to clinic[J]. J Am Ceram Soc, 1991, 74: 1487-1510
[21] Kokubo T. Bioactive glass-ceramics: properties and applications[J]. biomaterials, 1991,12: 155-163
[22] Li P, Kangasniemi I, De Groot K, et al. Bone like hydroxyapatite induction by a gel-derived titania on a titanium substrate[J]. J Am Ceram Soc, 1994, 77: 1307-1312
[23] Fendorf M, Gronk R. Surface precipitation reaction on oxide surface[J]. J Cooloid Interface sci, 1992, 148: 295-298
[24] Li P J, Ducheyne P. Quasi-biological apatite film induced by titanium in a simulated body fluid[J]. J Biomed Mater Res, 1998, 41: 341-148
[25] Wang Xuejiang, Li Yubao. Study on bionic Composite of nano-HA needle-like crystals and polyamide[J]. High technology letters, 2001, 5: 1-5
[26] Duan Yourong, Wu Yao, Wang Chaoyuan, Chen Jiyong, Zhang Xingcong. A Study of bone-like apatite formation on calcium phosphate ceramics in different kinds of animals in vivo. Journal of biomedical engineering, 2003, 20(1): 22-25
[27] Reut godley, David starosvetsky, Irena goman. Bonelike apatite formation on niobium metal treated in aqueous NaOH[J]. Journal of materials science: Materials in medicine, 2004, 15: 1073-1077

2

Ceramic Coatings Obtained by Electrophoretic Deposition: Fundamentals, Models, Post-Deposition Processes and Applications

M. Federica De Riccardis
ENEA-Italian National Agency for New Technologies, Energy and Sustainable Economic Development, Technical Unit of Materials Technologies of Brindisi, Brindisi, Italy

1. Introduction

Electrophoretic Deposition (EPD) is one of the most outstanding coating techniques to be based on electrodeposition. Nowadays, increasing interest has been gained both from academic and industrial payers, due to its wide potential in ceramic coating processing technology.

The main advantages of this technique are:

its high versatility when used with different materials and their combinations;
its cost effectiveness, because it requires simple and cheap equipment.

Moreover, it can be used both on a large scale, also to coat objects with a complex shape, and on small scale, to fabricate composite micro- and nanostructures, as well as near net-shape objects having accurate dimensions (micro- and nano-manufacturing).

The basic phenomena involved in EPD are well-known and have been the subject of extensive theoretical and experimental research. Nevertheless, further efforts have to be devoted in order to understand the fundamental mechanisms of EPD and to optimise the working parameters, especially when multicomponent suspensions are used.

EPD is a two-step process. In the first step, charged particles suspended in a liquid medium move towards the oppositely charged electrode under the effect of an externally applied electric field (electrophoresis). In the second step, the particles deposit on the electrode forming a more or less thick film, depending on the process conditions (concentration of particles in solution, applied electric field, time). The substrate acts as an electrode and the deposit of particles is the coating.

The aim of this paper is to provide an updated review of the wide potential of EPD as a technique to produce ceramic coatings, without omitting possible problems and drawbacks.

Applications of EPD, especially on nanoparticles, are in continuous expansion both in industry and academia, stimulating a great interest in developing of predictive analytical and numeric modelling of the EPD process. The proposed mechanisms have explained experimental results, but the current experts opinion is that a full understanding is still lacking, mainly due to the phenomena that are at the base of the interaction between the charged particles approaching both each other and to the electrode to form the solid deposit.

For some applications, a requirement is that the ceramic EPD deposit is dense, so a post-deposition treatment should be performed in order to densify it. Usually, this consists of a conventional heating treatment in a furnace, but some problems could occur such as delamination, cracks or residual stress due to the differential coefficient of thermal expansion. Moreover, the high densification temperature of ceramics can be detrimental for the substrate which could be damaged.

Sometimes, the sintering temperature can be decreased by adding some low melting additives. Alternative sintering methods could be considered, such as microwave, laser or electron beam.

Finally, the wide range of applications of EPD deposits will be mentioned. EPD process is very versatile, therefore porous, layered, and graded deposits can be obtained besides dense coatings. Recently, it has been clearly demonstrated the possibility of obtaining nanocomposite materials, especially those containing carbon nanotubes (CNTs), by using EPD. As a consequence, the applications are in a spread number of sectors: biomaterials, fuel cells, barrier coatings, electronics, catalysis, optical devices.

2. Fundamentals and models

Electrophoretic Deposition is a traditional processing method in the ceramic industry that is gaining increasing interest for production of new materials coatings.

EPD is achieved through the movement of charged particles dispersed in a suitable liquid towards an electrode under an applied electric field. This movement results in the accumulation of the particles and in the formation of a homogeneous deposit at the appropriate electrode (Figure 1).

The main requirement to obtain an efficient EPD process is to use suitable suspensions where ceramic particles are well suspended and dispersed. When a ceramic particle is in a liquid medium, it can be charged through four mechanisms (Van der Biest & Vandeperre, 1999):

a. selective adsorption of ions onto the solid particle from liquid,
b. dissociation of ions from solid phase into the liquid,
c. adsorption or orientation of dipolar molecules at the particle surface, and
d. electron transfer between the solid and the liquid phase due to the difference in work function.

For the analysis and discussion on charging mechanisms and particles interactions, that are at the base of this ceramics processing method, one can refer to the fundamentals of colloid science widely discussed in literature (Lewis, 2000).

Since EPD is assumed to be a two-step process, electrophoresis and deposition, each step require accurate attention. Firstly, the kinetics of the process will be described giving particular attention to the relation between process parameters and time evolution of the deposit yield. Then, the mechanisms of deposition proposed in literature will be discussed.

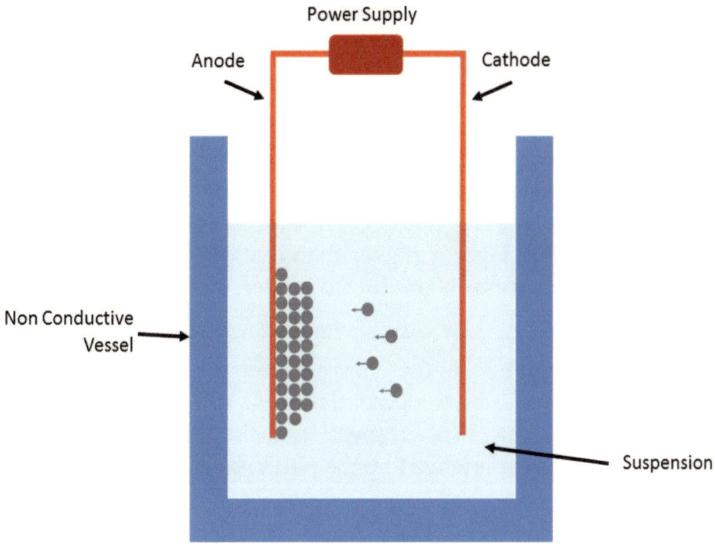

Fig. 1. Scheme of EPD process

2.1 Kinetic models

The first model used to describe EPD process is attributed to Hamaker (Hamaker, 1940) who proposed a general expression for the deposited mass per area unit (*m*, g cm^{-2}) in a cell with planar geometry:

$$m = C_s \mu A E t \tag{1}$$

where C_S, solids concentration in the suspension (g cm^{-3}); t, deposition time (s); μ, electrophoretic mobility (cm^2 V^{-1} s^{-1}); E, electric field strength (V cm^{-1}); A, surface area of the electrode (cm^2).

Several years later, Sarkar and Nicholson (Sarkar & Nicholson, 1996) considered again the first model of Hamaker and analysed the dependence of kinetics on some experimental conditions.

In eq. (1), μ and A can be evaluated numerically and one reasonably supposes that they are constant during the process. It is not so true for C_S and E, that vary as the process while the process is going on.

Sarkar and Nicholson considered the variation of the particles concentration in the suspension for long deposition time, starting with the condition that the only change in the concentration is due to the mass of powder deposited by EPD. It is equal to zero when the process starts and varies with time according to the expression:

$$m(t) = m_0(1 - e^{-t/\tau}) \tag{2}$$

where m_0, initial mass of powder in suspension (g) and $\tau = V/\mu A E$ defined as characteristic time (V is the volume of the suspension considered constant). Actually, τ is the reciprocal of k, the "kinetic parameter" that represents a key parameter in the modelling of EPD process.

In order to consider that some particles arriving to the electrode do not take part to the formation of the deposit, Sarkar and Nicholson introduced a "sticking factor", f ≤ 1, a multiplicative efficiency factor.

For short time, eq. (2) is reduced to the Hamaker model, since in the early stages of the process the variation of bulk concentration is negligible due to the small amount of powder deposited. When the process advances, the deposited mass is relevant and its effect on the evolution of the deposition process is not negligible. In this condition, under the hypothesis that the resistivity of deposited layer is higher than that of the suspension, the electric field suffers a drop, also if the process is conducted under the condition of constant applied voltage.

In eq. (1), the electric field strength is the effective strength that affects the particles in the suspension and so it is decreased by the potential drops due to the resistance of electrodes, suspension and deposited layer. With respect to the applied voltage externally, v, the expression of the electric field experienced by the particles is:

$$E(t) = (v - [\Delta V_{e1} + iR_d s(t) + iR_s(d - s(t)) + \Delta V_{e2}])/d \quad (3)$$

ΔV_{e1} and ΔV_{e2} are the potential drop (V) at the electrode 1 and 2, respectively; i, is the current (A); d is the distance between the electrodes (m); $s(t)$ is the thickness of the deposited layer (m); R_d and R_s are the resistivity (Ω/m) of the deposit and the suspension (Fig. 2).

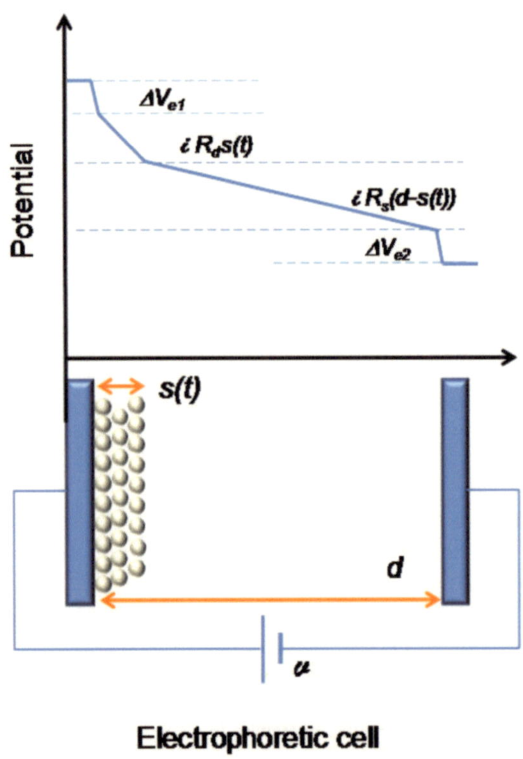

Fig. 2. Variation of the potential in an electrophoretic cell during the deposition

The hypotheses and the conditions that are at the basis of the eq. (3) were widely discussed by Van der Biest (Van der Biest & Vandeperre, 1999), who cited also much experimental evidence on the dependence of the electric field in the suspension from the parameters above mentioned.

In order to avoid the effect of deposit resistivity increasing, Sarkar and Nicholson proposed to work in galvanostatic condition rather than in potentiostatic condition. In such a way the number of particles arriving on the electrode is constant and not influenced by the electrical condition of the suspension-deposit system. Ma (Ma & Cheng, 2002) determined the relationship between the kinetic parameter of Sarkar and Nicholson and the applied current density, i, making the predictions of the deposition yield easier, when the deposition is at constant current density:

$$k = k_0(e^{i/i_0} - 1) \tag{4}$$

where k_0, a reference kinetic constant, and i_0, a reference applied current density.

Other factors that influence the drop of the electric field are a change in the polarization of the electrode and a change in the conductivity of the suspension, occurring in the progress of EPD process.

Changes in the electrode polarization can be due to a change in the concentration of the reactants at the electrode-electrolyte interface. When the process starts and a current flows towards the electrode, the concentration of reacting species at the electrode drops off. They are replaced by the species of bulk suspension through diffusion, convection or migration, but the presence of deposit layer can retard the transport of reactants to the electrode. As a consequence, the polarization at the working electrode changes.

As regards the suspension conductivity, a model introduced by Vandeperre and Van der Biest (Vandeperre & Van der Biest, 1998) affirms that both the particles and the ions surrounding the particles contribute to the conductivity. During the EPD process, the electric conductivity decreases due to the effect of the depletion of particles in the suspension and for the reduced presence of ions moving together with powder particles. This effect is more evident if the amount of deposited powder is high with respect to the powder in suspension and if the suspension ionic conductivity is relatively small.

Several studies were devoted to the investigation of the effect of suspension conductivity on the EPD process, through the study of the influence of binders, salts, stabilizers, and a liquid medium (aqueous and non aqueous)(Ferrari & Moreno, 1996; Westby at al., 1999; Moreno & Ferrari, 2000; De Riccardis et al., 2007). As a common result, in order to have an effective deposition process the electrostatic or electrosteric stabilisation of the suspension has to be such that the ionic concentration in suspension is low and the suspended particles are the main current carriers.

In order to schematize the EPD yield and according to Sarkar and Nicholson, it is possible to recognize four different behaviours, depending on the process conditions: constant voltage, constant current, constant concentration, and variable concentration.

In Curve A of Fig. 3 (constant current and constant concentration) the deposition rate is linear with respect to time. In curves B, C and D, the deposition rate decreases asymptotically with deposition time. Obviously, the decreasing concentration produces a reduction of the final yield and therefore of the rate deposition either at constant current (Curve B) or at constant voltage (curve C and D).

Comparing curve A and C (both with constant concentration), the final yield is considerably higher in curve A than in curve C. This is due to the electric resistance of the deposit layer, considered higher than that of an equal thickness taken up by the suspension. Therefore, when the process goes on, the voltage per unit length (or electric field) decreases and consequently decreases the particle velocity as function of deposition time.

In curve D, the deposition rate is lower than other curves for both the contribution of the concentration decrease and of the resistance of deposit layer. As a deduction, the nature of the deposit layer plays an important role on deposition rate.

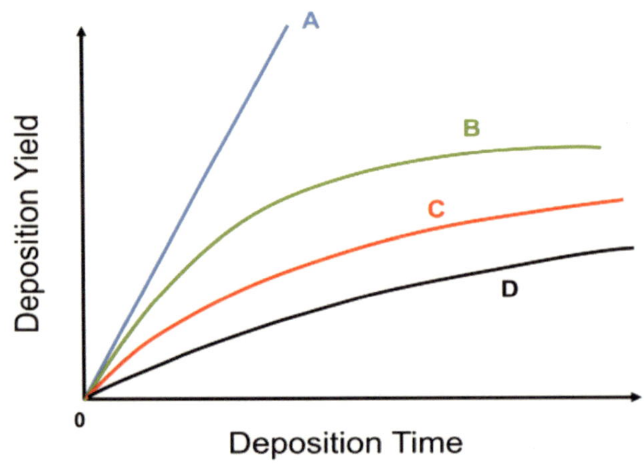

Fig. 3. Schematic representation of the kinetics of EPD process: curve A, constant current-constant concentration; curve B, constant current-variable concentration; curve C, constant voltage-constant concentration, curve D, constant voltage-variable concentration.

Adopting the standard scheme of an electrochemical cell, a combined resistive-capacitive model can be used to represent the electrical behaviour of the EPD process (Fig. 4).

In addition to the resistors, discussed above, some capacitors are added to represent the current transient that occurs when the voltage starts to be applied, generally interpreted as due to the establishment of a concentration profile under a diffusion limited regime (Van der Biest & Vandeperre, 1999).

Ferrari (Ferrari et al., 2006) proposed a resistive model for the deposition kinetics, considering a linear relationship between the suspension resistivity and the solid loading in the suspension. This model agrees with the experimental data referred to a long time deposition of yttria stabilized tetragonal zirconia polycrystalline (Y-TZP), that shows a S-shaped variation of the mass per area unit with time.

Recently Baldisserri (Baldisserri et al., 2010) proposed some assumptions to define the electrical behaviour of a potentiostatic EPD cell: a) the effect of capacitive transients on cell current is negligible, b) the faradic resistance of both the electrode/deposit interfaces is constant, c) the electrical resistance of the EPD deposit is proportional to its thickness, d) the electrical resistance of the suspension is constant.

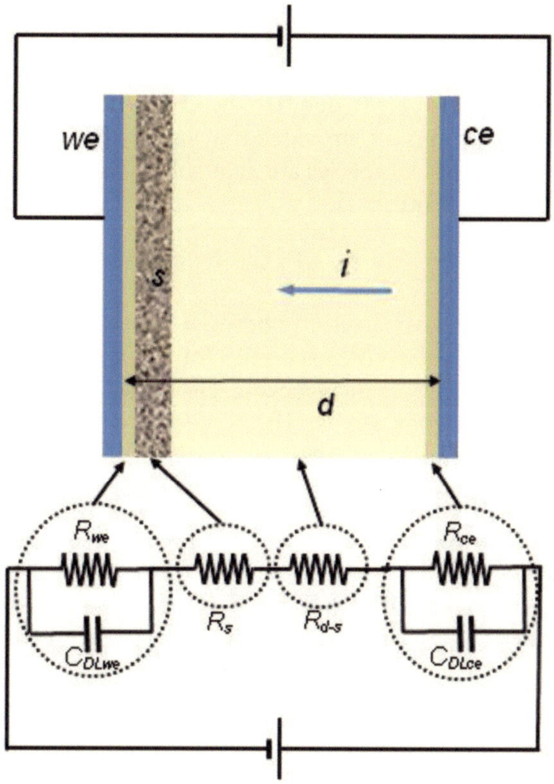

Fig. 4. Electric model of EPD process with plane electrode (we= working electrode, ce= counter electrode, DL= double layer)

The assumption a) is based on the estimate of time required for double-layer charging. The assumption b) is supported by the consideration that after the formation of the first monolayer, the interface at the working electrode/suspension should be stabilised as the charging of double-layer should be completed. Similarly at the interface between counter electrode/suspension, but it is less relevant as no deposition occurs at that electrode.

The assumption c) relies on the compliance of the physical properties of deposited layer comparable with a porous media, where the electrical resistance is proportional to its thickness. The assumption d) is based on the consideration that the deposited layer is less thick than the electrodes distance, therefore the decrease in the volume filled by the suspension is negligible.

Baldisserri applied this model to EPD of TiO_2 particles and verified experimentally that the previously mentioned assumptions were satisfied. As a result, they derived a linear correlation between deposited mass and passed charge and a non-linear kinetic equation:

$$m(t) = A\frac{VD_d}{i_0\rho_d}\left[(1+\alpha t)^{1/2}-1\right],\ \alpha = \frac{2Ki_0^2}{V}\left(\frac{\rho_d}{D_d}\right) \tag{5}$$

where i_0, current density at time $t=0$, K, the deposited mass-passed charge ratio, V, the external applied voltage, ρ_d and D_d, resistivity and density of the deposited layer, respectively. This equation at short deposition time is approximated by:

$$m(t) = Ai_0Kt \tag{6}$$

Therefore, this resistive model is able to describe the EPD process both at regime and during the transient of the deposition current, provided that suspension contains such dispersant or binder as the electroactive chemical species are available for all the deposition time, making deposition a diffusive control process.

2.2 Deposition models

Unlike the kinetic models regarding electrophoresis, the second step of EPD, that is the deposition, is still matter of discussion, especially for the arrangement of particles on the electrode surface. After arriving at the working electrode, particles are packed onto its surface reproducing the electrode shape. The aggregation and the arrangement of the particles on the electrode surface depend on surface chemistry of particles and on interactions between particles in suspension and between particles and substrate.

The fundamental condition for a performing EPD process is to use a stable suspension where particles are keep well dispersed in the liquid medium and can move towards the electrode without influencing or being influenced by other particles. Here, particles can be rearranged on the electrode surface during packing under the action of electric field.

Of course, from the electrical point of view, the suspension composition is critical because the presence of deflocculants, binders or dispersants influences the surface charge of particles and their electrical response.

The interactions between particles inside a liquid medium is largely described by the Derjaguin-Landau-Verwey-Overbeek (DLVO) theory. When an external force exists, as the driving force of an electric field, the minimum of the energy curve is shifted towards an higher value (Fig. 5). If the applied field exerts a force so great as to overcome the mutual repulsion force, two particles coagulate. Similar phenomenon occurs when a particle moves toward an electrode in an EPD cell.

Some deposition mechanisms are proposed to explain the phenomenon of particles accumulation:

- flocculation by particles accumulation,
- particles charge neutralization at the electrode,
- electrochemical coagulation of particles, and
- distortion and thinning of electrical double layer (EDL).

The last, proposed by Sarkar and Nicholson (Sarkar & Nicholson, 1996), is the most diffused and accepted. This theory affirms that when a positive charge moves toward an electrode together oppositely charged ions, its shell is distorted, thinning ahead and thickening behind, due to fluid dynamics. As a consequence, the particle feels a weak attraction toward another positively charged particle, so together they can move under the electric field. The EDL of the two particles is less wide than that of a single particle, so when another particle is approaching, it can be close enough to interact through van der Waals attractive forces and so coagulates. Similar mechanism occurs at the electrode surface where the high particles concentration allows the formation of a particles deposit.

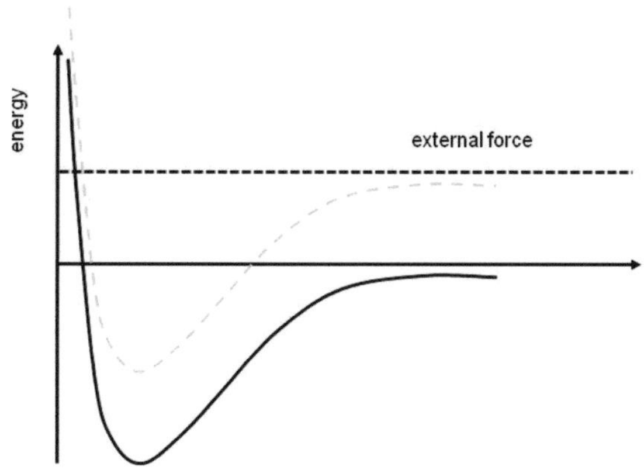

Fig. 5. Potential energy vs. separation distance for particles in absence (——) and in presence (- - -) of an electric field.

This model was integrated by Fukada (Fukada et al., 2004) with a further consideration based on the experimental observation of De (De & Nicholson, 1999). According to De, Fukada verified that H^+ are depleted at the cathode because of particle discharge, through the reaction:

$$H^+_{x=\infty} \xrightarrow{transport\ process} H^+_{x=0} + e^- \xrightarrow{charge\ trasfer} \frac{1}{2}H_2 \tag{7}$$

This implies an increase of local pH toward the isoelectric point and then the coagulation is facilitated. Fukada found an analytical expression for the variation of the H^+ concentration with time, validated by experimental results for alumina suspension in ethanol. This expression, generally suitable for suspensions containing H^+ or H_3O^+, shows that the steady state with respect to diffusion and charge transfer of H^+ ions corresponds to a reduction of zeta-potential at the cathode and consequently to coagulation.

Other deposition mechanisms, mentioned above, although explain experimental results in some conditions, are not valid in general. In fact, flocculation by particles accumulation suggests a deposit formation by electrophoresis similar to gravitation, so that the pressure exerted by the arriving particles at the electrode makes the particles close to form a deposit, overcoming the repulsion forces between particles. This hypothesis does not explain why the deposition occurs on a membrane which does not act as an electrode.

Similarly, particles charge neutralization at the electrode suggests the charged particles are neutralized by the contact with the electrode surface, but it does not explain the deposition mechanism when the deposit is thicker than a monolayer.

With respect to electrochemical coagulation of particles, the hypothesis is that an increase of electrolyte concentration around particles produces a reduction of the repulsion between particles near the electrode, where particles can coagulate. However, time required to have an increased electrolyte concentration is not negligible, experimentally estimated as inversely proportional to the square of applied voltage. Moreover, this mechanism is invalid when there is no increase of electrolyte concentration near the electrode.

Recently, new models considering particles flow from an electro-dynamic point of view are developed (Guelcher et al., 2000; Ristenpart et al., 2007). Experimental results of Guelcher confirmed a numerical prediction of clustering of colloidal particles deposited in a DC electric field by considering an electro-osmotic particles flow. By analysing the long-range attraction force intra-particles, Ristenpart demonstrated the flow direction of a particle depends on the sign of its dipole coefficient. Under particular conditions, the electro-osmotic component and the electrohydrodynamic component of flow can have the same direction and so can produce aggregation.

This overview demonstrates that the discussion on models and mechanisms of electrophoresis and deposition of ceramic particles in presence of an electric field is still open, and that many efforts have been made for some decades, from Hamaker to today, to explain and understand the large amount of experimental results.

3. Post-deposition processes

Depending on coating functions, a post-EPD process may be required in order to densify the coating and to improve its mechanical properties and adhesion to substrate. Usually, this post-EPD process is a heat treatment devoted to cure or sinter the coating, but some precautions should be followed to avoid defects inside the coating or at interface with the substrate.

As mentioned before, EPD is especially regarded as a suitable method to obtain coatings on bodies of a complex shape. To respect this feature, one of the main requirements is to have a low concentration of powder agglomerates in the ceramic suspension. The size of these agglomerates define the order of magnitude of the size of stacking defects during consolidation. Therefore, all methods for quality assurance, such as control of powder dimension and phase, impurity content, process parameters, have to be pursued in order to minimize the defects in the final product and therefore improve its functional properties.

With respect to other colloidal consolidation processes, in EPD the deposition rate is almost unrelated to particle size and thickness of the deposited layer, so high deposition rates can be achieved. In order to avoid defects during deposit coagulation, especially at high deposition rates, very fine particles should be used. Moreover, the use of suspensions in an organic solvent could minimise the incorporation of gas bubbles in the deposit due to electrolysis. Zithomirsky (Zhitomirsky & Gal-Or, 1997) deposited hydroxyapatite (HA) coatings on Ti6Al4V. They found that suspension presedimentation had a significant effect on the deposit quality, because it allowed the removal of undesirable agglomerates from the suspension and, as a result, the deposit consisted of finer particles. This reduced the porosity in the deposited layer and a denser packing was obtained.

After deposition, the coating has to be dried. While the green deposit is still immersed in the suspension, it is saturated by liquid. After removing from the suspension and drying, the green density can reach 60%. Also during drying some shrinkage occurs so that cracks can appear due to stress induced by the flow and evaporation of the liquid through the pores (Scherer, 1990). Cracks occurring during drying may be reduced if the deposit thickness is lower than a critical value which depends on the powders used for deposition. Van der Biest (Van der Biest et al., 2004) obtained coatings on stainless steel with WC-5Co, Al_2O_3, TiC, and TiB_2 powders, having an average particle size and a surface area equal to 1, 0.3, 2, and 1.5-2 μm, and 2.47, 10, 1-2 and 0.5-1.5 m^2/g, respectively. The thickness below which no cracking was observed was 125, 316 and 56 μm, for the first three powders, whereas surprisingly a layer 5 mm thick was deposited without observing cracks in the case of TiB_2. This result supports the influence of the powder characteristics on the quality of the deposit.

Generally, an EPD coating is deposited on a metal substrate that can be non resistant to high temperature which is necessary to sinter ceramics. Two approaches exist to limit damages of substrate: to use some method to lower the sintering temperature, such as powders with a fine grain or a low melting additive, or to use a different sintering treatment, such as microwave or irradiation. In the following, some of the methods cited before are reported.

In order to lower the sintering temperature, a first stratagem that could be used is the choice of precursors suitable to form a ceramic material. Some examples were those of Boccaccini (Boccaccini et al., 1996, 1997) and Kooner (Kooner et al., 2000) who prepared EPD sols based on boehmite (γ-AlOOH), fumed amorphous silica, and fumed δ-alumina with appropriate concentrations, as precursors of mullite. Mullite has a number of attractive properties for high-temperature structural applications, but its sintering temperature is higher than 1600°C. The use of nano-particles and fine mullite seeds lowered the sintering temperature by up to 1300-1400°C, making possible the formation of mullite matrix by EPD in fabrics based on silicon carbide and Nextel 720 fibres.

Reaction bonding (RB) is a forming technique developed to produce near net-shape ceramics and to overcome problems caused by shrinkage during sintering. It consists of introducing some elements or compounds that, by reacting with an oxidant or reducing atmosphere at a temperature higher than room temperature, can produce a ceramic matrix.

Aluminium particles were added to PSZ suspensions by Wang (Wang et al., 2000b; Wang et al., 2002). During heat treatment in air at low temperature (600°C), metal powder was converted to nanometer sized oxide crystals, that subsequently were sintered and bonded to PSZ at 1200°C. The volume expansion associated with the $Al \rightarrow Al_2O_3$ reaction partially compensated the sintering shrinkage. The combined use of EPD with the reaction bonding process allowed to fabricate crack-free and relatively dense ceramic coatings, maintaining the sintering temperature lower than that usual one (1350-1500°C). However, as the oxidation of the Al powder in the green form is affected by the thermal processing profile, the oxidation and sintering temperature has to be appropriately chosen to optimise the density and the quality of coatings.

Another candidate for reaction bonding is ZrN due to its low reaction temperature. Baufel (Baufel et al., 2008) utilised a suspension with zirconia and zirconium nitride to obtain an EPD coating on Ni alloys. Two different contents of ZrN were mixed together with YSZ in ethanol and milled in order to reduce the grain size of powders. After deposition and drying in ambient conditions, a heat treatment was performed in air at 1000 °C for 6 h. In XRD spectra, the treated samples showed only pronounced peaks of zirconia without evidence of ZrN peaks, so they concluded that all ZrN transformed into zirconia, within the detection limit of XRD. As a result, combining EPD and reaction bonding, Baufel obtained zirconia coatings, about 100 μm thick, sintered at 1000°C, with a microstructure comparable with the one prepared without reaction bonding and sintered at 1200°C.

EPD was successfully applied by Lessing (Lessing et al., 2000) to obtain reaction bonded joints using several compositions of silicon carbide and silicon nitride mixed to graphite and carbon black particles. The use of EPD allowed to form joints filling a large gap and coatings rounded corners or undercut sections, with structures originated by molten silicon at 1450°C.

A very interesting application of reaction bonding process is that optimised by Hang (Hang et al., 2010) who combined EPD and RB to achieve a graded coating based on hydroxyapatite (HA). HA is a material extensively studied and used as a biomaterial, thanks to its biocompatibility and osteoconductivity. EPD is potentially an attractive method to obtain HA coatings on Ti substrate, but often the bonding strength between EPD coating and substrate was not high enough for the requirement of clinic application. Moreover, a great difficulty is represented by the sintering. In fact, as high temperature produces degradation of Ti substrate and thermal decomposition of HA, sintering temperature should be ideally below 1000 °C under which HA is difficult to be fully densified. On the other hand, the thermal expansion coefficient of Ti substrate is much lower than that of HA, so a large thermal contraction mismatch could arise and tend to induce the formation of cracks when cooled from an elevated temperature. Again, a significant firing shrinkage during sintering will lead to the formation of cracks also in coatings.

As a solution, HA/Al_2O_3 composite coatings were produced by a combination of EPD and reaction bonding process at a relatively low sintering temperature of 850 °C. Reaction bonded Al_2O_3 with relatively lower coefficient of thermal expansion (CTE) was introduced into HA coating to reduce the difference of CTE between Ti substrate and HA coating, and to overcome problems caused by the firing shrinkage during sintering. Both advantages were proven to be beneficial in avoiding the formation of cracks and improving the bonding strength of ceramic coatings. On the coating containing HA and reaction bonded Al_2O_3, a further coating with a gradient of composition was deposited, up to the top coating composed only by HA. The development of this functionally gradient coating (FGC) based on HA, contributed to reduce the discontinuity in thermal expansion coefficients and, as a result, minimised the residual stress in the coatings.

Therefore, electrophoretic deposition and reaction bonding process were successfully combined to produce HA functionally gradient coatings on Ti substrate at a relatively low sintering temperature of 850 °C. HA FGC was uniform and crack-free, having a chemical composition and microstructure with a gradient variation along its cross section. The content of HA increased gradually from the inner part of HA FGC (diffusion layer) to the outer part (top layer), and proportionally the density increased from the inner part to the outer part. The HA FGC took efficiently the advantages of both the mechanical properties of Ti and the biological performances of HA ceramic.

Microwave heating is a method fundamentally different from the conventional techniques used to densify materials. The direct coupling of energy to a material with dielectric loss results in extremely rapid heating. Typical heating and cooling rates are of the order of hundreds of degrees Celsius per minute, much higher than those of conventional heating sintering. This can be very advantageous, especially when conventional temperature and time of sintering can result detrimental to the substrate material.

A combination of EPD process and microwave sintering was used by Streckert (Streckert et al., 1997) to prepare SiC composites formed by silicon carbide-based perform and SiC matrix. After achieving infiltration with SiC powder by EPD, microwave sintering at 2.45 GHz was performed under fluxing a mixture of nitrogen and hydrogen. A high density in composite was obtained by applying a load during microwave heating. Therefore, the combined use of EPD process and microwave sintering has the potential to produce good quality composite rapidly and economically, due to short process time and simple equipments required.

Dense, uniform and crack-free Al_2O_3/YSZ composite coatings on Ni based superalloy were prepared by a novel sol-gel process, optimised by Ren (Ren et al., 2010). The composite coatings were firstly prepared by the electrophoretic deposition of a suspension containing aluminium oxide sol, nano-Al_2O_3, and micro YSZ particles and then treated by Pressure Filtration Microwave Sintering (PFMS). Suspensions with several mass ratios of ceramic powders versus aluminium oxide sol were used to produce thick deposits by EPD. After drying at room temperature for 24 h, the green deposits were treated in a 2.45 GHz microwave oven for 10 minutes. Ren obtained coatings with micro-cracks, pores and a granular structure when the mass ratio of ceramic powders/aluminium oxide was the lowest. On the contrary, at maximum of the mass ratio, the coating was dense, without cracks and with a microstructure composed by YSZ particles embedded in nano-Al_2O_3 particles. Moreover, no defects and spallations were found across the interface, as an indication that the thermal match between coating and substrate was good. In the Ren's opinion, the pressure and microwave heating were doubly beneficial for the sintering and densification of YSZ/Al_2O_3 coatings at a relatively lower temperature. In fact, the compound effect of pressure and filtration in PFMS process constrained the shrinkage of coating in three-dimensional directions and therefore micro-cracks were avoided and the adhesion to the substrate was improved. Furthermore, nano-sized α-Al_2O_3 decreased the phase transition and the crystallization temperature of aluminium oxide sol and at the same time the microwave energy lowered the phase transformation temperature of α-Al_2O_3.

In general, the use of sol-gel suspensions for the EPD process has two advantages: it allows to obtain thicker deposits than those prepared by a conventional sol-gel method, and the heating treatment is performed at a temperature typical of a sol-gel process, certainly lower than that used to sinter ceramics. The sol-gel method should be suitable to obtain protective coatings on metal against corrosion, due to the low temperature of densification which are not detrimental for the metal substrate, but the typical maximum thickness of sol-gel coating is approximately 2 μm. Moreover, the presence of pores and microcracks due to drying shrinkage, limits the applicability of sol-gel coating as protection. Castro (Castro et al. 2004) incorporated silica nano-particles to an acid-catalysed SiO_2 sol in order to increase the coating thickness without increasing the sintering temperature. Then this hybrid organic-inorganic suspension was used to produce EPD coatings 5 μm thick. After sintering at 500°C for 30 min, these coatings demonstrated to have a good corrosion resistance.

Laubershiemer (Laubershiemer et al., 1998) produced a deposit by EPD using a synthesised liquid precursor of lead-zirconate-titanate (PZT) according to a modified sol-gel route. Besides the advantage mentioned before, it is worthy to note that the sol-gel process has the advantages of the conventional mixed-oxide route, such as much greater purity and compositional control with liquid-mix homogeneity and a lower process temperature. In order to obtain a PZT micro-component, Laubershiemer used a polymeric gel by introducing polymerisable chelating agents in the synthesis of the precursors. After performing EPD into a microstructured mould simultaneously to the gelation of the sol, a gel-body was obtained, that then was hardened by polymerisation induced by UV-light. After drying and removing of the mould, a heating treatment was performed in a furnace at 550°C to obtain a ceramic component. The addition of polymerisable agents allowed the creation of an organic network which was interpenetrated into the inorganic chains. This reduced the risk of cracks formation and considerably increased the mechanical stability of the gel-body. It was demonstrated that this method is suitable for production of micro components based on ceramic materials.

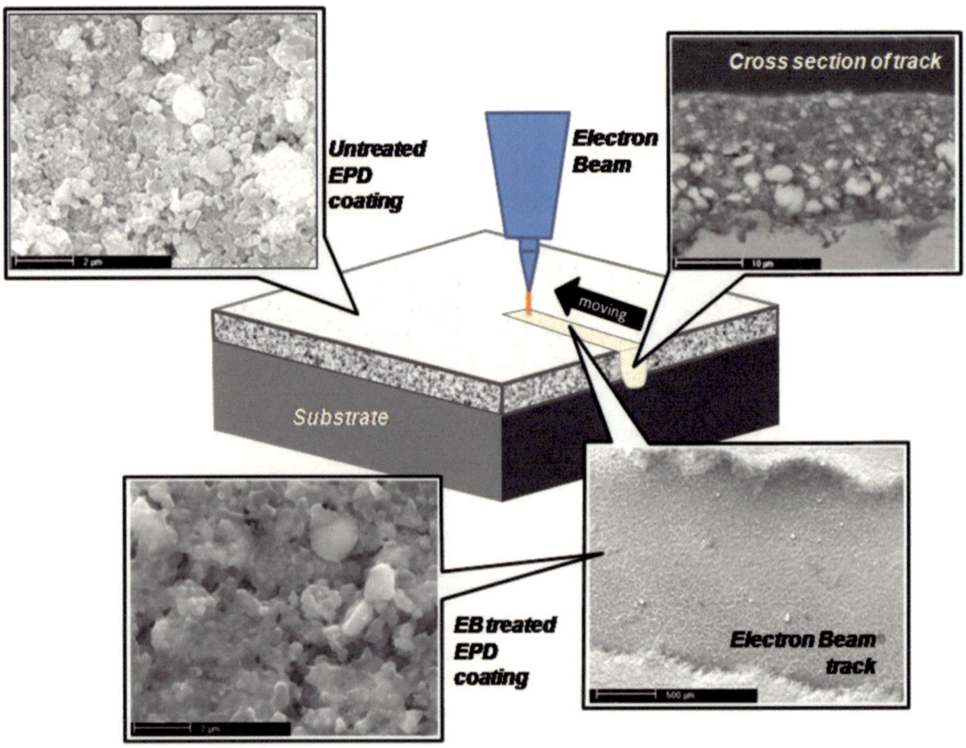

Fig. 6. Scheme of the EB treatment on the surface of Alumina-Zirconia EPD coating and SEM images referred to the zones inside and outside the EB track. The EB track presents a different contrast with respect to the untreated coating. Inside the EB track, the grains are more compact than those ones outside the track.

An alternative method was successfully used to densify the electrophoretic deposit and to make it more adherent to substrate. It was based on the irradiation of the green ceramic coating surface by a high power density energy source, such as an electron beam (EB).

Compared to laser energy sources, in EB treatment the reflectivity by the irradiated material is lower and therefore the beam efficiency is higher. Moreover, the thickness of affected material can be varied by changing the beam power. So 10 kW power is usually employed to cut or weld materials, whereas a lower power allows to treat few micrometers of the material surface.

De Riccardis (De Riccardis et al., 2008) used EB treatment up to 13.75 J/mm2 of fluence, to sinter the alumina-zirconia EPD coatings, by confining the high temperature to the ceramic material without affecting deeply the metallic substrate (Fig. 6). The EB irradiation on alumina-zirconia coatings produced a nanostructured ceramic composite material, formed by micro-particles of α-Al_2O_3 and tetragonal ZrO_2, embedded in an amorphous matrix containing also nano-sized crystals. Nevertheless some residual porosity, the electron beam treatment had two positive effects: it increased the ceramic coating density and improved the adhesion between coating and substrate. In fact, the adhesion stress values evaluated for coatings applied on sandblasted substrates were comparable to those typical of plasma-sprayed coatings. It is worthy to note this method can be applied to sinter ceramic coating deposited on metallic support having low temperature resistance. Moreover, since EB irradiation did not damage the material outside the EB track, several adjacent tracks could be performed in order to obtain densified coating on a large area.

4. Applications: From traditional to advanced materials

After its first use in 1933 to deposit thoria particles on a platinum cathode for electron tube application, EPD was mainly utilized for traditional ceramic processing. Its industrial application was the deposition of clay or vitreous enamel coatings on metals, which after firing showed an evident improvement in the finishing properties of coatings with respect to conventional dipping or spraying processes.

In the last 20 years the interest shown by the academic and the industrial world regarding EPD has increased thanks to its wide range of applications, especially due to the insertion and diffusion of nanomaterials which allow to obtain structures with characteristics never conceived before.

A numbers of reviews reported extensively on the several applications of EPD as coatings and free standing objects, based on ceramics and metals (Sarkar & Nicholson, 1996; Van der Biest & Vandeperre, 1999; Boccaccini & Zhitomirsky, 2002; Boccaccini et al., 2003; Besra & Liu, 2007; Corni et al., 2008). Particularly successful is the use of EPD to produce porous, laminated and graded ceramic coatings (Hatton & Nicholson, 2001; Put et al, 2004; You et al, 2004) as well as fibres reinforced composites (Boccaccini et al., 1996,1997; Wang et al., 2000a; Freidrich et al., 2002; Kaya et al., 1999, 2000, 2001). Moreover, EPD has proven to be an effective method to texture superconductors structures, such as BSCCO and YBCO (Hang et al.,1995; Yau & Sorrell 1997; Grenci et al., 2006) and electrodes for solid oxide fuel cells (Zhitomirsky & Petric, 2000; Hosomi et al., 2007).

The aim of this review is to present more extensively those that are the current and most appealing applications for the most recent material science: biomaterials and nanomaterials.

4.1 Biomaterials

Numerous studies have been devoted to the development of coatings based on the conventional bioactive inorganic materials, such as hydroxyapatite and bioglass, in order to improve the adhesion of tissue to the implant surface and so increase the performance of metal implants.

As calcium phosphates are the mineral component of bone, dentine and dental enamel, hydroxyapatite [$Ca_5(PO_4)_3(OH)$] is the best candidate to form a chemical bonding with hard tissues. EPD is particularly advantageous in the deposition of HA coatings because this technique allows the control of coatings composition, thickness and microstructure, that are essential to obtain the maximum benefit. As mentioned before, the first requirement to obtain a densely packed HA layer is to use very fine particles and to prepare well stable suspensions. Xiao (Xiao & Liu, 2006) optimised a HA suspension using nano-HA particles prepared by hydrothermal method, in primary aliphatic alcohols (C1-C4) with triethanolamine as a dispersant. Maximum stability was achieved when n-butanol was used as liquid medium. EPD process (30 V for 1 min) produced a deposit with a high packing density, small drying shrinkage, and therefore optimal mechanical properties.

Meng (Meng et al., 2005) studied the effect of applied voltages on the microstructure of HA coatings, both with constant voltage, equal to 20 and 200 V, and with dynamic voltage, varying between 0 and 200 V. The dynamic voltage process consisted of three increments with three different rates, slowly in the first step and faster during the third. After sintering at 800°C for 2 h, the coating prepared at low constant voltage was dense and consisted of fine particle whereas the coating prepared at high constant voltage was porous and with large HA agglomerates. Differently, the coating prepared at dynamic voltage consisted of continuously gradient particles, with smaller particles building up in the inner layer and bigger particles forming the outer layer, both with a high packing of particles.

Bioactive glasses, such as Bioglass 45S5 which is composed by SiO_2 (45 wt%), NaO_2 (24.5 wt%), CaO (24.5 wt%) and P_2O_5 (6wt%), readily react with physiological fluids forming a hydroxyapatite layer on their surface and creating tenacious bonds to hard and soft tissue. Coatings of Bioglass 45S5 were obtained on stainless steel and Nitinol (Ti 50wt% and Ni 50wt%) by EPD (Krause et al., 2006). The conditions of the coatings deposition and sintering were optimised for the two types of substrates. After sintering at 800°C, the Bioglass coating covered completely the stainless steel substrate surface even if not uniformly. When the substrate was Nitinol, a diffusion of nickel and titanium was observed in the Bioglass coating, as a consequence of sintering process performed at 950°C.

Recently, the advantages of polymers composites containing HA or bioactive glass have been highlighted. In fact, the combination of polymeric and inorganic bioactive phases is common to several natural materials, like bone which is a composite containing collagen and HA crystals. Moreover, the interfacial bonding strength of HA or Bioglass coatings on metal substrate can be improved by their combination with polymers.

For these reasons, different types of composite coatings, based on HA or Bioglass or both, were deposited, where the organic component was chitosan or alginate (Zhitomirsky et al., 2009). Both the organic component enabled the electrosteric stabilisation of HA and Bioglass particles in suspension and promoted a co-deposition of the organic and inorganic components in EPD process. The presence of chitosan and alginate allowed to obtain coatings with improved strength with respect to the pure HA or Bioglass coatings. Moreover they did not obstruct the bioactive function of HA and Bioglass.

Bioglass was successfully co-deposited with polyetheretherchetone (PEEK) by EPD (Boccaccini et al., 2006a) on Nitinol wires. PEEK is a semicrystalline thermoplastic polymer (T_m=340°C) with excellent performance also for biomedical applications. A suspension containing PEEK and Bioglass in ethanol was used for EPD co-deposition. Sintering was performed at 340°C for 20 min, just to melt the polymer and therefore to embed the ceramic particles. After sintering, the surface of coatings became smooth due to viscous flow of melted polymer. The PEEK/Bioglass coating deposited on Nitinol showed to be adherent to substrate also when it was bended. Moreover, the composite coating demonstrated to have two functions: to protect the NiTi substrate from corrosion when it was in contact with body fluids, and to improve the bonding of bone or soft tissue to an implant, being Bioglass a bioactive material.

Another example of application of composite coatings based on Bioglass concerns composites containing carbon nanotubes (Charlotte Schausten et al., 2010). The EPD parameters were optimised in order to obtain both co-deposition and sequential deposition of Bioglass particles and multi-walled CNTs on stainless steel substrates. Co-deposition produced homogenous and dense coatings with well-dispersed CNTs placed between the Bioglass particles. So, the network of CNTs acted as a reinforcement and contributed to the improvement of the mechanical stability of the coatings. The coatings obtained by sequential deposition formed a two-dimensional nanostructured mesh of CNTs on the Bioglass layer and produced a surface nanotopography, with a great potential in the formation of bone-like nanosized HA crystal with the presence of body fluids.

The combined use of Bioglass and CNTs is motivated also by recent results on differentiation of ostheoblast cells during their growth based on the substrate nanostructure. Moreover, as CNTs have a similar dimension to proteins, they demonstrated a high reactivity for interactions involved in the cell attachment mechanism. An interesting formation of CNTs/Bioglass composites by EPD was described by Boccaccini (Boccaccini et al., 2007) who referred to CNTs deposited on porous Bioglass-based scaffolds for bone tissue engineering. Concentrated CNTs suspensions (0.5-5 g/l) were used to create a deposit on interconnetted and macroporous structure of Bioglass, placed between two stainless steel electrodes. The final structure showed a scaffold with substantially unchanged porosity, with nanostructured pore walls. Moreover, the presence of CNTs conferred biosensing properties to the scaffold by adding an electrical conduction function, potentially useful for stimulation of cell growth and tissue regeneration by a physioelectrical signal transfer.

Alumina and ultrahigh molecular weight polyethylene (UHMWPE) are used as biomaterials for prostheses and joint replacement, but they do not bond with live bone. Coatings of bioactive ceramics can improve the osteoconductivity of these materials. Wollastonite (Yamaguchi et al., 2009) was used to form a composite bioactive material based on Alumina and UHMWPE. Wollastonite powder was prepared by calcination of silicon dioxide and calcium carbonate in an equimolar ratio. Substrates of porous alumina (10 μm average pore size) and porous UHMWPE (30 μm average pore size) were placed between two gold plates, acting as electrodes. After EPD, the subtractes were placed in simulated body fluid (SBF), then after 14 days an apatite layer was observed on the cathode-side of the porous substrates, induced by deposited wollastonite. The adhesive strength of the apatite layer to the porous substrates was higher than commercially available coated alumina and UHMWPE.

As a biomaterial, titania is normally applied as a coating on metallic substrates in order to improve the integration of orthopaedic implants. Bacterial colonisation of implanted materials occurring on the coating surface represents major complications in orthopaedic surgery. Therefore, in order to reduce the risks of bacterial infection after implantation, inorganic antibacterial materials were used with better results than those organic in terms of durability, toxicity and selectivity. Silver is one of the preferred elements as antibacterial agent, so Ag-TiO_2 coatings were developed with both bioactive and antibacterial properties (Santillán et al., 2010). Ag nano-particles (4 nm) were grown directly on the surface of commercial TiO_2 nano-particles (23 nm) from nucleophilic reaction. In such a way, the Ag nano-particles were uniformly distributed on TiO_2 nano-particles, contributing to control the release of Ag, in the presence of bacterial colonies. An aqueous stable suspension containing Ag-TiO_2 particles was used to deposit a composite coating on Ti sheets. EPD parameters (3V for 90 s) were optimised referring to macroscopic homogeneity of coatings and to absence of cracks. In vitro bioactivity tests in SBF showed an increasing formation of HA at increasing time and decreasing silver content in the coatings.

4.2 Nanomaterials

Among nanomaterials, enormous attention is devoted to CNTs for their extraordinary properties due to structure, aspect ratio and size. Their applications range from microelectronics to structural composites and to biomedical materials, but particular arrangements of CNTs need to enhance their positive contribution in specific applications. If CNTs are used in combination with other materials to form a composite, appropriate processing methods have to be adopted in order to obtain homogeneous distribution of CNTs in a matrix. EPD is a very convenient technique for manipulating CNTs and producing ordered CNTs arrays (Boccaccini et al., 2006b). As for all EPD experiments, the composition of suspensions and EPD parameters (electric field, deposition time and electrode material) are critical for the performance of CNTs based coatings. The functionalisation of CNTs is also very important, which obviously influences the interaction between CNTs and other components both in EPD suspension and in a formed deposit.

Boccaccini (Boccaccini et al., 2010) reported on two alternative methods to produce CNTs based composite coatings by EPD. In the first method, a CNTs layer was deposited with a more or less isotropic planar distribution of CNTs on a substrate. Then the CNTs coating was employed as an electrode to deposit ceramic or metallic nanoparticles, again by EPD, with the aim of infiltrating the porous CNTs structure. The second method consisted of a co-deposition of ceramic particles with CNTs. Three are the alternative cases of the components arrangement in the suspension: i) self-assembly of nano-particles coating individual CNT, ii) heterocoagulation of CNTs on individual larger particles, and iii) simultaneous deposition of CNTs and particles (ceramic or metallic) with the same charge polarity in suspension. As an example, the first approach was used to obtain a CNT-TiO_2 coating. The presence of CNTs together with titania enhanced the photocatalytic effectiveness of titania and improved the mechanical properties of titania coatings.

CNTs have been also used to make more effective the interface between fibres and matrix in SiC/SiC composites for fusion reactor applications (König et al., 2010). EPD was used to deposit multi-walled CNTs onto SiC fibres and to infiltrate the CNTs coated fibres mats with SiC powder. The CNTs coating obtained on fibres was firm and homogeneous, with uniformly distributed nanotubes on the surface of the fibres. The fibre mats were then placed in contact with an electrode of the EPD cell and the migration of SiC particles under the electric field allowed to gradually fill the spaces between the fibres with a high grade of infiltration. After sintering at 1300°C in air, a denser composite than the one without CNTs on the surface of the fibres was obtained.

An improved planar-gate triode with CNTs field emitters was successfully realised by combining photolithography, screen printing, and EPD (Zhang et al., 2011). In order to deposit selectively CNTs acting as field emitters onto cathode electrodes, an EPD suspension containing CNTs was used. Previously, an acid activating treatment was performed in order to functionalise the CNTs' surface. During the EPD process, gate electrodes of a planar-gate triode were used as an anode electrode of the EPD cell and cathode electrodes of the triode were used as a cathode electrode of the EPD cell. In such a way, positively charged CNTs were selectively deposited onto the cathode of the triode. The field emission performance of the realised devise was so good that practical applications of dynamic back light units and field emission displays could be implemented.

To improve further the field emission performance of CNTs, they should be vertically aligned so as to obtain a well-defined structural anisotropy and a maximal packing density with the characteristics of a 1D nanomaterial. CNTs forests could be grown in situ by vapour, liquid or solid mechanisms, or could be aligned by a post-growth method. EPD was the method utilised successfully by Santhanagopalan (Santhanagopalan et al., 2010) to control CNTs' orientation and to obtain an alignment in direction of the applied electric field. A high-voltage electrophoretic deposition (HVEPD) process was optimised through three key elements: i) high deposition voltage, to align of nanomaterial with electric field, ii) low concentration of nanomaterial in a suspension to prevent aggregation before deposition and to avoid bundle formation, and iii) simultaneous formation of an holding layer to keep the nanomaterial deposit resistant to manipulating.

This method was also successfully used to deposit aligned manganese oxide nanorods on stainless steel plates, demonstrating the versatility of HVEPD that potentially can be utilised to obtain forests from any nanomaterial that can be charged suitably in a solvent.

Porous template were often used to obtain aligned nanomaterials, which were then removed after the growth of aligned nanostructures. Also in this subject, EPD demonstrated its effectiveness. Deposition of diamond nanoparticles (4-12 nm) was performed (Tsai et al., 2011) on a surface of porous anodic alumina (PAA) with pore size of 20 nm. These diamond nanoparticles acted as nucleation sites for the following growth of diamond nano-tips by HFCVD process. EPD allowed to obtain a continuous diamond HFCVD film with a very fine microstructure. On the contrary, when no EPD of nucleating diamond nanoparticles was performed, diamond film grown up by HFCVD showed a fragmentary distribution of large diamond particles (200 nm).

In order to control the structure of an EPD coatings, other stratagems can be used. Uchikoshi (Uchikoshi et al., 2004, 2010) and Kawakita (Kawakita et al., 2009) demonstrated diffusely that the use of a magnetic field during an electrophoretic deposition induced a preferential orientation of particles with an anisotropic magnetic susceptivity. For example, titania particles in the form of anatase were deposited with *c*-axis aligned with a strong external magnetic field. The deposit was dense, uniform and crack-free, and therefore suitable for energy conversion applications, where the efficiency depends strongly on the crystals orientation.

An AC or a pulsed DC electric field can modify the structure of the EPD deposit. Riahifar (Riahifar et al., 2011) varied time, frequency, voltage, and particles concentration in suspension and obtained titania nanoparticles coatings with different deposition patterns. They used two gold planar electrodes, with a spacing of 150 μm, realised by scratching a continuous gold film on glass substrate. The AC field applied between the two electrodes produced a deposit on the electrodes edge at a low voltage, in a short time, with low particle concentration and high frequencies. On the other hand, at high voltages, a longer time, a higher particle concentration, and lower frequencies, the deposited particles filled the gap between the electrode, demonstrating how it was possible to control the deposition pattern by changing appropriately the process parameters.

The use of EPD is a relatively new technique in the field of polymer electronics. One of the most useful features is the use of dilute solutions, not suitable for other conventional casting techniques, such as spin coating. Particularly interesting results were obtained by Tada (Tada & Onoda, 2011) who prepared composite films of conjugated polymers and fullerene by EPD, starting with the optimisation of the suspension composition. Therefore dense composite coatings, suitable for heterojunction systems, were deposited. Moreover, they found different distributions of fullerene crystals depending on conjugated polymer, with spontaneous stratification in presence of an inhomogeneous suspension. This result confirmed the possibility to control the deposit morphology by EPD.

Other examples of the use of EPD in the energy conversion field are represented by the deposition of nanomaterials such as supercapacitors, photofuntional compounds, or photo anodes in dye-sensitised solar cell (DSSC). As a supercapacitor, manganese oxide is a low cost raw material and therefore it is an alternative material to the conventional oxides, that are RuO_2 and IrO_2. Coatings of needle-like MgO_2 powders with a diameter of 10 nm and a length ranging from 50 to 400 nm, were deposited by EPD by Chen (Chen et al., 2009). The deposited coating showed a porous microstructure and a slight decrease of specific capacitance, from 200 to 190 F/g after 300 cycles, attributed to reduction reaction from Mn^{4+} to Mn^{3+} during EPD process and recovered during the cyclic voltammetry tests.

Titania nanosheets (TN) are promising nanomaterials for the design of UV-visible light sensitive energy conversion systems. EPD was used to deposit TN from an aqueous suspension on ITO substrates (Yui et al., 2005). The addition of poly (vinyl alcohol) (PVA) made the TN coating more adherent to substrates. Then intercalation of methyl viologen (MV^{2+}) was achieved by soaking the EPD TN deposit in an aqueous MV^{2+} solution. The photocatalytic activity was demonstrated under irradiation at the absorption band of TN for the TN/MV^{2+} thin films, that maintained stable for a longer time than those in absence of TN.

Electrodes for flexible DSSCs were prepared by EPD depositing commercial zinc oxide powders on a conducting plastic substrate (Chen et al., 2011). After deposition under an electric field of 200 V/cm for 1 min, ZnO coatings showed some cracks and porosity. Then, in order to improve the films quality, a mechanical compression was applied at different pressures (25-100 MPa). The adhesion of ZnO coatings to the plastic substrates was clearly improved, so that they could not be peeled off even after bending the substrate. Experimental tests on DSSC prepared with this ZnO electrode demonstrated that the photocurrent density and the light-to-electricity conversion energy correlated with the applied pressure during the compression. EPD confirmed again to be a versatile and efficient coating technique.

5. Conclusions

By using EPD it is possible to obtain coatings with an excellent performance in a very large range of applications. However, it needs to control process parameters and to design suitable suspensions in order to obtain outstanding results. Moreover, the properties of the coatings can be tailored through the tuning the applied electric field and the choice of appropriate starting materials, which also influence the eventual densifying post-deposition treatment.

In spite of its numerous advantages and the wide range of applications, efforts have to be devoted to develop theories and models valid for the electrophoretic deposition of nanoscaled materials. In fact, it is expected that the field in which EPD will expand its applications will be those related to nanotechnology, especially for fabrication of nanostructured and hybrid composite materials, both in the form of dense and porous materials.

6. References

Baldisserri, C.; Gardini, D. & Galassi, C. (2010). An Analysis Of Current Transients During Electrophoretic Deposition (EPD) From Colloidal TiO_2 Suspensions, *Journal of Colloid and Interface Science,* Vol. 347, No. 1 (2010), pp. 102–111, ISSN 0021-9797

Baufeld, B., Van der Biest, O. & Rätzer-Scheibe, H.-J. (2008). Lowering The Sintering Temperature For EPD Coatings By Applying Reaction Bonding, *Journal Of The European Ceramic Society,* Vol. 28, No. 9 (2008), pp. 1793–1799, ISSN 0955-2219

Besra, L. & Liu, M. (2007). A Review On Fundamental And Application Of Electrophoretic Deposition (EPD), *Progress In Materials Science,* Vol. 52, No. 1 (2007), pp. 1-61, ISSN 0079-6425

Boccaccini, A.R.; Trusty, P.A. & Telle, R. (1996). Mullite Fabrication From Fumed Silica And Bohemite Sol Precursors, *Materials Letters,* Vol. 29, No. 1-3 (1996), pp. 171-176, ISSN 0167-577X

Boccaccini, A.R.; MacLauren, I.; Lewis, M.H. & Ponton, C.B. (1997). Electrophoretic Deposition Infiltration Of 2-D Woven SiC Fibre Mats With Mixed Sols Of Mullite Composition, *Journal of The European Ceramic Society,* Vol. 17, No.13 (1997), pp. 1545-1550, ISSN 0955-2219

Boccaccini, A.R. & Zhitomirsky, I. (2002). Applications of Electrophoretic Deposition Techniques in Ceramic Processing, *Current Opinion in Solid State & Materials Science*, Vol. 6, No. 3 (2002), pp. 251-260, ISSN 1359-0286

Boccaccini, A.R.; Roether, J. A.; Thomas, B.J.C., Shaffer, M.S.P.; Chavez, E.; Stoll, E. & Minay, E.J. (2003). The Electrophoretic Deposition of Inorganic Nanoscaled Materials, *Journal of The Ceramic Society of Japan*, Vol. 114, No. 1 (2003), pp. 1-14

Boccaccini, A. R.; Peters, C., Roether, J. A.; Eifler, D.; Misra, S. K. & Minay, E. J. (2006a). Electrophoretic Deposition Of Polyetheretherketone (PEEK) And PEEK/Bioglass Coatings On NiTi Shape Memory Alloy Wires, *Journal of Materials Science*, Vol. 41, No. 24 (2006), pp. 8152-8159, ISSN 0022-2461

Boccaccini, A.R.; Cho, J.; Roether, J. A.; Thomas, B.J.C.; Minay, E.J. & Shaffer, M.S.P. (2006b). Electrophoretic Deposition Of Carbon Nanotubes, *Carbon*, Vol. 44, No. 15 (2006), pp. 3149-3160, ISSN 0008-6223

Boccaccini, A.R.; Chicatun, F., Cho, J.; Bretcanu, O., Roether, J.A.; Novak, S. & Chen Q. (2007). Carbon Nanotubes Coatings On Bioglass-Based Tissue Engineering Scaffolds, *Advanced Functional Materials*, Vol. 17, No. 15 (2007), pp. 2815-2822, ISSN 1616-301X

Boccaccini, A.R.; Cho, J.; Subhani, T.; Kaya, C. & Kaya, F. (2010). Electrophoretic Deposition Of Carbon NanoTube-Ceramic Nanocomposites, *Journal of The European Ceramic Society*, Vol. 30, No. 5 (2010), p. 1115-1129, ISSN 0955-2219

Castro, Y.; Ferrari, B.; Durán, A. & Moreno, R. (2004) Effect Of Rheology And Processing Parameters On The EPD Coatings Of Basic Sol-Gel Particulate Sol, *Journal of Materials Science*, Vol. 39, No. 3 (2004), pp. 845-849, ISSN 0022-2461

Charlotte Schausten M.; Meng, D.; Telle, R. & Boccaccini, A.R. (2010). Electrophoretic Deposition Of Carbon nanotubes And Bioactive Glass particles For Bioactive Composite Coatings, *Ceramics International*, Vol. 36, No. 1 (2010), p. 307-312, ISSN 0272-8842

Chen, C.-Y.; Wang, S. C.; Lin C.-Y.; Chen, F.-S. & Lin C.-K. (2009). Electrophoretically Deposited Manganese Oxide Coatings For Supercapacitor Application, *Ceramics International*, Vol. 35, No. 8 (2009), pp. 3469–3474, ISSN 0272-8842

Chen, H.-W., Lin, C.-Y., Lai, Y.-H., Chen, J.-G., Wang, C.-C., Hub, C.-W., Hsu, C.-Y., Vittal, R. & Ho, K.-C. (2011). Electrophoretic Deposition Of ZnO Film And Its Compression For A Plastic Based Flexible Dye-Sensitized Solar Cell, *Journal of Power Sources*, Vol. 196, No. 10 (2011), pp. 4859–4864

Corni, I.; Ryan, M.P. & Boccaccini, A.R. (2008). Electrophoretic deposition: From traditional ceramics to nanotechnology, *Journal of The American Ceramic Society*, Vol. 28, No. 7 (2008), pp. 1353-1367, ISSN 0002-7820

De, D. & Nicholson P. S. (1999). Role Of Ionic Depletion In Deposition During Electrophoretic Deposition, *Journal of The American Ceramic Society*, Vol. 82, No. 11, (1999), pp. 3031-3036, ISSN 0002-7820

De Riccardis, M.F.; Carbone, D. & Rizzo, A. (2007). A Novel Method For Preparing And Characterizing Alcoholic EPD Suspensions, *Journal of Colloid and Interface Science*, Vol. 307, No. 1 (2007), pp. 109-115, ISSN 0021-9797

De Riccardis, M.F.; Carbone, D.; Piscopiello, E. &. Vittori Antisari, M. (2008). Electron Beam Treatments Of Electrophoretic Ceramic Coatings, *Applied Surface Science*, Vol. 254, No. 6 (2008), pp. 1830-1836, ISSN 0169-4332

Ferrari, B. & Moreno, R. (1996). The Conductivity Of Aqueous Al_2O_3 Slips For Electrophoretic Deposition, *Materials Letters*, Vol. 28, No. 4-6 (1996), pp. 353-355, ISSN 0167-577X

Ferrari, B.; Moreno, R. & Cuesta, J.A. (2006). A Resistivity Model For Electrophoretic Deposition, *Key Engineering Materials*, Vol. 314 (2006), pp. 175-180, ISSN 1013-9826

Friedrich, C.; Gadow, R. & Speicher, M. (2002). Protective Multilayer Coatings For Carbon Carbon Composites, *Surface & Coatings Technology*, Vol. 151-152 (2002), pp. 405-411, ISSN 0257-8972

Fukada, Y.; Nagarajan, N.; Mekky, W.; Bao, Y.; Kim, H.-S. & Nicholson, P.S. (2004). Electrophoretic Deposition - Mechanisms, Myths, And Materials, *Journal of Materials Science*, Vol. 39, No. 3 (2004), pp. 787-801, ISSN 0022-2461

Grenci, G.; Denis, S.; Dusoulier, L.; Pavese, F. & Penazzi, N. (2006). Preparation And Characterization Of $YBa_2Cu_3O_{7-x}$ Thick Films Deposited On Silver Substrates By The Electrophoretic Deposition Technique For Magnetic Screening Applications, Superconductor Science and Technology, Vol. 19, No. 4 (2006), pp. 249-255, ISSN 0953-2048

Guelcher, S.A., Solomentsev, Y. & Anderson, J.L. (2000). Aggregation Of Pairs Of Particles On Electrodes During Electrophoretic Deposition, *Powder Technology*, Vol. 110, No. 1-2 (2000), pp. 90-97, ISSN 0032-5910

Hamaker, H.C. (1940). Formation Of A Deposit By Electrophoresis, *Transactions of The Faraday Society*, Vol. 35 (1940), pp. 279-287

Hatton, B. & Nicholson, P.S. (2001). Design and Fracture of Layered Al2O3/TZ3Y Composites Produced by Electrophoretic Deposition, *Journal of The American Ceramic Society*, Vol. 84, No. 3, (2001), pp. 571-576, ISSN 0002-7820

Hosomi, T.; Matsuda, M. & Miyake, M. (2007). Electrophoretic Deposition For Fabrication Of YSZ Electrolyte Film On Non-Conducting Porous NiO-YSZ Composite Substrate For Intermediate Temperature SOFC, *Journal of The European Ceramic Society*, Vol. 27, No. 1 (2007), pp. 173-178, ISSN 0002-7820

Huang, S.-L.; Schoenwaelder, B.; Dew-Hughes, D. & Grovenor, C.R.M. (1995). Thermal Shock Resistance And Bending Strain Tolerance Of Electrophoretically Deposited $Bi_2Sr_2CaCu_2O_y$/Ag Tapes, *Materials Letters*, Vol. 24 (August 1995), pp. 271-274, ISSN 0167-577X

Huang, J.-C.; Ni, Y.-J. & Wang, Z.-C. (2010). Preparation Of Hydroxyapatite Functionally Gradient Coating On Titanium Substrate Using A Combination Of Electrophoretic Deposition And Reaction Bonding Process, *Surface & Coatings Technology*, Vol. 204, No. 21-22, (2010), pp. 3387-3392, ISSN 0257-8972

Kawakita, M.; Uchikoshi, T.; Kawakita, J. & Sakka, Y. (2009) Preparation of Crystalline-Oriented Titania Photoelectrodes on ITO Glasses from a 2-Propanol-2,4-Pentanedione Solvent by Electrophoretic Deposition in a Strong Magnetic Field, *Journal of The American Ceramic Society*, Vol. 92, No. 5 (2009), pp. 984–989, ISSN 0002-7820

Kaya, C.; Boccaccini, A. R. & Trusty, P.A. (1999). Processing And Characterization Of 2-D Woven Metal Fibre-Reinforced Multiplayer Silica Matrix Composites Using Electrophoretic Deposition And Pressure Filtration, *Journal of The European Ceramic Society*, Vol. 19, No. 16 (1999), pp. 2859-2866, ISSN 0955-2219

Kaya, C.; Boccaccini, A. R. & Chawla, K.K. (2000). Electrophoretic Deposition Forming of Nickel - Coated - Carbon - Fiber - Reinforced Borosilicate - Glass - Matrix Composites, *Journal of The American Ceramic Society*, Vol. 83, No. 8 (2000), pp. 1885-1888, ISSN 0002-7820

Kaya, C.; Kaya, F.; Boccaccini, A. R. & Chawla, K.K. (2001). Fabrication And Characterisation Of Ni-Coated Carbon Fibre-Reinforced Alumina Ceramic Matrix Composites Using Electrophoretic Deposition, *Acta Materialia*, Vol. 49, No. 7 (2001), pp. 1189-1197, ISSN 1359-6454

König, K.; Novak, S.; Iverkovic, A.; Rade, K.; Mang, D.; Boccaccini A.R. & Kobe, S. (2010). Fabrication of CNT/SiC/SiC Composites By Electrophoretic Deposition. *Journal of The European Ceramic Society*, Vol. 30 (2010), pp. 1131-1137, ISSN 0955-2219

Kooner, S.; Westby, W.S.; Watson, C.M.A. & Farries, P.M. (2000). Processing of Nextel™ 720/Mullite Composition Composite Using Electrophoretic Deposition, *Journal of The European Ceramic Society*, Vol. 20, No. 5 (2000), pp. 631-638, ISSN 0955-2219

Krause, D., Thomas, B.; Lienenbach, C.; Eifler, D.; Minay, E J. & Boccaccini, A.R. (2006) The Electrophoretic Deposition Of Bioglass Particles On Stainless Steel And Nitinol Substrates, *Surface & Coatings Technology*, Vol. 200, No. 16-17 (2006), p. 4835-4845, ISSN 0257-8972

Laubersheimer, J.; Ritzhaupt-Kleissl, H.-J.; Haußelt, J. & Emig, G. (1998). Electrophoretic Deposition of Sol-Gel Ceramic Microcomponents Using UV-curable Alkoxide precursors, *Journal of The European Ceramic Society*, Vol. 18, No. 3 (1998), pp. 255-260, ISSN 0955-2219

Lessing, P.A., Erickson, A.W. & Kunerth, D.C. (2000). Electrophoretic Deposition [EPD] Applied To Reaction Joining Of Silicon Carbide And Silicon Nitride Ceramics, *Journal of Materials Science*, Vol. 35, No. 35 (2000), pp. 2913-2925, ISSN0022-2461

Lewis, J.A. (2000). Colloidal Processing Of Ceramics, *Journal of The American Ceramic Society*, Vol. 83, No. 10 (2000), pp. 2341-2359, ISSN 0002-7820

Ma, J. & Cheng, W. (2002). Electrophoretic Deposition Of Lead Zirconate Titanate Ceramics, Journal of The American Ceramic Society, Vol. 85, No. 7 (2002), pp. 1735-1737, ISSN 0002-7820

Meng, X.; Kwon, T.-Y.; Yang, Y.; Ong, J.L. & Kim, K.-H. (2005). Effects Of Applied Voltages On Hydroxyapatite Coating Of Titanium By Electrophoretic Deposition, *Journal of Biomedical Materials Research Part B: Applied Biomaterials*, Published online 16 December 2005 in Wiley InterScience (www.interscience.wiley.com). DOI: 10.1002/jbm.b.30497

Moreno, R. & Ferrari, B. (2000). Effect Of The Slurry Properties On The Homogeneity Of Alumina Deposits Obtained By Aqueous Electrophoretic Deposition, *Materials Research Bulletin*, Vol. 35, No. 6 (2000), pp. 887-897

Put, S.; Anné, G.; Vleugels, J. & Van der Biest, O. (2004), Advanced symmetrically graded ceramic and ceramic-metal composites, *Journal of Materials Science*, Vol. 39, No. 3 (2004), pp. 881-888, ISSN 0022-2461

Ren, C.; He, Y.D. & Wang, D.R. (2010). Al_2O_3/YSZ Composite Coatings Prepared By A Novel Sol-Gel Process And Their High-Temperature Oxidation Resistance, *Oxidation of Metals*, Vol. 74, No. 5-6 (2010), p. 275-285, ISSN 0030-770X

Riahifar, R.; Raissi, B.; Marzbanrad, E. & Zamani C. (2011) Effect Of Parameters On Deposition Pattern Of Ceramic Nanoparticles In Non-Uniform AC Electric Field,

Journal of Materials Science: Materials in Electronics, Vol. 22, No. 1 (2011), pp. 40–46, ISSN 0957-4522

Ristenpart, W.D.; Aksay, I.A. & Saville, D.A. (2007). Electrically Driven Flow Near A Colloidal Particles Close To An Electrode With A Faradaic Current, *Langmuir*, Vol. 23, No. 7 (2007), pp. 4071-4080, ISSN 0743-7463

Santhanagopalan, S.; Teng, F. & Meng, D.D. (2010). High Voltage Electrophoretic Deposition For Vertically Aligned Forests Of One-Dimentional Nanoparticles, *Langmuir*, Vol. 27, No. 2, (2010), p. 561-569, ISSN 0743-7463

Santillán, M.J.; Quaranta, N.E. & Boccaccini, A.R. (2010). Titania And Titania-Silver Nanocomposite Coatings Grown By Electrophoretic Deposition From Aqueous Suspensions, *Surface & Coatings Technology*, Vol. 205, No. 7 (2010), p. 2562-2571, ISSN 0257-8972

Sarkar, P. & Nicholson, P. S. (1996). Electrophoretic Deposition (EPD): Mechanisms, Kinetics, and Application to Ceramics, *Journal of The American Ceramic Society*, Vol. 79, No. 8,(1996), pp. 1987-2002, ISSN 0002-7820

Shacham, R.; Mandler, D. & Avnir, D. (2004). Electrochemically Induced Sol-Gel Deposition Of Zirconia Thin Films, *Chemistry - A European Journal*, Vol. 10, No. 8 (2004), pp. 1936-1943, ISSN 0947-6539

Scherer, G.W. (1990). Theory of Drying, *Journal of The American Ceramic Society*, Vol. 73, No. 1 (1990), pp. 3-14, ISSN 0002-7820

Streckert, H.H.; Norton, K.P.; Katz, J.D. & Freim, J.O. (1997). Microwave Densification Of Electrophoretically Infiltrated Silicon Carbide Composite, *Journal of Materials Science*, Vol. 32, No. 24 (1997), pp. 6429-6433, ISSN 0079-6425

Tada, K. & Onoda M. (2011). Spontaneous Stratification In Composite Films Consisting Of Conjugated Polymers And Neat C_{60} Prepared By Electrophoretic Deposition, *Materials Letters*, Vol. 65, No. 9 (2011), pp. 1367–1370, ISSN0167-577X

Tsai, H.-Y; Liu, H.-C.; Chen, J.-H. & Yeh, C.-C. (2011) Low Cost Fabrication Of Diamond Nano-Tips On Porous Anodic Alumina By Hot Filament Chemical Vapour Deposition And The Field Emission Effects, *Nanotechnology*, Vol. 22, No. 23 (2011), art. no. 235301 (5p), ISSN 0957-4484

Uchikoshi, T.; Suzuki, T.S.; Okuyama, H. & Sakka, Y. (2004). Fabrication Of Textured Alumina By Electrophoretic Deposition In A Strong Magnetic Field, *Journal of Materials Science*, Vol. 39, No. 3 (2004), pp. 861-865, ISSN 0022-2461

Uchikoshi, T.; Suzuki, T. S. & Sakka, Y. (2010) Fabrication Of C-Axis Oriented Zinc Oxide By Electrophoretic Deposition In A Rotating Magnetic Field, *Journal of the European Ceramic Society*, Vol. 30, No. 5 (2010), pp. 1171–1175

Van der Biest, O. & Vandeperre, L.J. (1999). Electrophoretic Deposition of Materials, *Annual Review Material Science*, Vol. 29, (1999), pp.327-352, ISSN 0084-6600

Van der Biest, O.; Put, S.; Anné G. & Vleugels, J. (2004). Electrophoretic Deposition For Coatings and Free Standing Objects, *Journal of Materials Science*, Vol. 39, No. 3 (2004), pp. 779-785, ISSN 0022-2461

Vandeperre L.J. & Van der Biest, O. (1998). Electric Current And Electric Field for electrophoretic deposition from non-aqueous suspensions, In: *Innovative Processing and synthesis of ceramics, glasses, and composites*, Ed. Bansal N.P., Logan, K.V., Singh J.P., *Ceramic Transaction*, Vol 85 (1998), Westerville, OH, ISSN: 1042-1122

Wang, Z.; Shemilt J. & Xiao P. (2000a). Novel Fabrication Technique For The Production Of Ceramic/Ceramic And Metal/Ceramic Composite Coatings, *Scripta Materialia*, vol. 42, No. 7 (2000), pp. 653-659, ISSN 1359-6462

Wang, Z.; Xiao, P. & Shemilt J. (2000b). Fabrication Of Composite Coatings Using A Combination Of Electrochemical Methods And Reaction Bonding Process, *Journal of the European Ceramic Society*, Vol. 20, No. 10 (2000), pp. 1469-1473, ISSN 0955-2219

Wang, Z.; Shemilt J. & Xiao P. (2002). Fabrication of Ceramic Composite Coatings Using Electrophoretic Deposition, Reaction Bonding And Low Temperature Sintering, *Journal of The European Ceramic Society*, Vol. 22, No. 2 (2002), pp. 183-189, ISSN 0955-2219

Westby, W.A; Kooner, S.; Farries, P. M.; Boother, P. & Shatwell, R. A. (1999). Processing Of Nextel 720/Mullite Composition Composite Using Electrophoretic Deposition, *Journal of Materials Science*, Vol. 34, No. 20 (1999), pp. 5021-5031, ISSN 0022-2461

Xiao, X. F. & Liu, R.F. (2006). Effect Of Suspension Stability On Electrophoretic Deposition Of Hydroxyapatite Coatings, *Materials Letters*, Vol. 60, No. 21-22 (2006), pp. 2627-2632, ISSN 0167-577X

Yamaguchi, S.; Yabutsuka, T.; Hibino, M. & Yao, T. (2009). Development Of Novel Bioactive Composites By Electrophoretic Deposition, *Materials Science and Engineering C*, Vol. 29, No. 5 (2009), p. 1584-1588, ISSN 0928-4931

Yau, J.K.F. & Sorrell, C.C. (1997). High-J_c $(Bi,Pb)_2Sr_2Ca_2Cu3O_{10+x}$ Tapes Fabricates By Electrophoretic Deposition, *Physica C*, Vol. 282-287, Part 4 (1997), pp. 2563-2564, ISSN 0921-4534

You, C.; Jiang, D.; Tan, S. (2004) Deposition of Silicon Carbide/Titanium Carbide Laminar Ceramics by Electrophoresis and Densification by Spark Plasma Sintering, *Journal of The American Ceramic Society*, Vol. 87, No. 4,(2004), pp. 759-761, ISSN 0002-7820

Yui, T.; Mori,T.; Tsuchino,T.; Itoh,T.; Hattori, T.; Fukushima, Y. & Takagi, K. (2005) Synthesis of Photofunctional Titania Nanosheets by Electrophoretic Deposition, *Chemistry of Materials*, Vol. 17, No. 1 (2005), pp. 206-211, ISSN 0897-4756

Zhang, Y.A; Wu, C.X.; Lin, J.Y.; Lin, Z.X. & Guo, T.L. (2011). An Improved Planar-Gate Triode With CNTs Field Emitters By Electrophoretic Deposition, Applied Surface Science, Vol. 257 (2011), pp. 3259-3264, ISSN 0169-4332

Zhitomirsky, I. & Gal-Or, L. (1997). Electrophoretic Deposition Of Hydroxyapatite, *Journal of Materials Science: Materials in Medicine*, Vol. 8, No. 4 (1997), pp. 213-219, ISSN 0957-4530

Zhitomirsky, I. & Petric, A. (2000). Electrophoretic Deposition Of Ceramic Materials For Fuel Cell Applications, *Journal of the European Ceramic Society*, Vol. 20, No. 12 (2000), pp. 2055-2061, ISSN 0955-2219

Zhitomirsky, D.; Roether, J.A.; Boccaccini, A.R. & Zhitomirsky, I. (2009). Electrophoretic Deposition Of Bioactive Glass/Polymer Composite Coatings With And Without HA Nanoparticles Inclusions For Biomedical Applications, *Journal of Materials Processing Technology*, Vol. 209, No. 4 (2009), p. 1853-1860, ISSN 0924-0136

Part 2

Physical Deposition Process

3

Magnetron Sputtered BG Thin Films: An Alternative Biofunctionalization Approach – Peculiarities of Bioglass Sputtering and Bioactivity Behaviour

George E. Stan[1] and José M.F. Ferreira[2]
[1]*Nanoscale Condensed Matter Physics Department, National Institute of Materials Physics, Bucharest-Magurele*
[2]*Department of Ceramics and Glass Engineering, CICECO, University of Aveiro, Aveiro*
[1]*Romania*
[2]*Portugal*

1. Introduction

Nowadays orthopaedic and dental metallic prostheses are widely used in the medical field, the most common being 316L stainless steel, Co-Cr alloys, titanium (Ti) or Ti superalloys. These metallic materials were preferred due to their good mechanical performance, adequate stiffness, non-magnetic properties and to their intrinsic property of promoting on their surface in contact with air or biological media a very thin and biologically inert oxide film (Cr_2O_3 in case of stainless steel and Co-Cr alloy or TiO_2 for (Ti) and Ti alloys), which could act as a metallic ions diffusion barrier layer. However, due to corrosion in the aggressive biological media, this thin protective layer could easily be shattered locally and metallic ions could enter the biological environment causing adverse reactions. Allergies, bone necrosis and the accumulation of metal particles in organs were detected in some cases (C. Brown et al., 2006; C. Brown et al., 2007).

The new generation of orthopaedic and dental implants aims towards the increase of biocompatibility by replacing the biotolerated metallic surfaces with bioactive ones. For increasing the bioactivity of prostheses and implants, there were designed devices coated with biologically active materials such as hydroxyapatite, simple or doped with different metallic ions or functional groups, various calcium phosphates, and more recently, bioglasses and glass-ceramics. Therefore, considerable attention has been given to the use of implants with bioactive fixation in the past decade (L.L. Hench & J. Wilson, 2003).

The commercial solution currently applied worldwide is the orthopaedic and dental titanium implants biofunctionalized with thick (>50 μm) bioactive coatings of hydroxyapatite [HA, $Ca_{10}(PO_4)_6(OH)_2$] prepared by plasma spraying.

Although this type of implant structure proved to be successful clinically, there are still significant deficiencies hard to ignore (e.g. low mechanical strength, difficulty in controlling the solubility in vivo, etc.), both in the preparation of biofunctional coatings and in their long-term *in situ* functional operation (Batchelor & Chandrasekaran, 2004; Epinette et al., 2003). Albeit their biological properties are excellent, the large thickness of the HA films synthesized by plasma spray determines the susceptibility to cracking and/or delamination due to poor adherence, phenomena that will allow the diffusion of implant's metal ions into the surrounding tissues and can lead to malfunctioning of the medical device in question. To address these shortcomings, currently a variety of alternative coating methods are being studied, from wet sol-gel technology, electrophoretic deposition and pulsed laser ablation, to magnetron sputtering for producing thinner adherent films of hydroxyapatite, calcium phosphates or bioactive glasses.

Bioactive glasses or bioglasses (BG) are osteoproductive-type inorganic materials far from proving their fully operative potential yet. Since the discovery of Bioglass® (45S5) by Larry Hench (L.L. Hench & J. Wilson, 2003), many bioglass compositional systems have been proposed and proved their suitability to form a bond with the living bone tissue and enhance the osteosynthesis at the implant site due to the favourable chemical interactions with the body fluid in the tissue rehabilitation process.

The behaviour of bioactive glasses in the formation of new bone tissue depends on the chemical composition and textural properties (Saravanapavan & Hench, 2001; Sepulveda et al., 2002). Glasses of the Na_2O-CaO-P_2O_5-SiO_2 system can be either formed from the traditional melt-quenching (Wu et al., 2011) or by the modern sol-gel method (Balamurugan et al., 2007). It has also been proved that an increase in the growth rate of apatite-like layer as well as the wider bioactivity were observed depending on the compositional range used for the preparation of bioglass (Rámila & Vallet-Regí, 2001; Vallet-Regí et al., 2003). The recent progresses made on the synthesis and processing of bioglasses that allowed the formulation of new compositional systems with lower thermal expansion coefficients (CTE) and enhanced bioactivity (Agathopoulos et al., 2006; Balamurugan et al., 2007; Tulyaganov et al., 2011) reopened the issue of bioglass coatings as a viable implantologic solution for load-bearing applications.

Generally bioactive glasses and glass-ceramics have been extensively developed and investigated for non-loading applications as bone grafts, fillers or auricular implants owing to their ability to form a bond with the living bone and put into clinical use following many years of animal testing in a variety of experimental models (Hench, 1991; Ratner et al., 2004).

To the best of our knowledge there are still no commercial titanium (Ti) implants functionalized with bioactive glass (BG) coatings, due to their poor adhesion to the metallic substrate determined by their native friability and to the significant mismatch of the CTEs for the BG coating (12–17 x 10-6/°C) and Ti-based substrate (~9.2–9.6 x 10-6/°C). Pull-out adherence values higher than 40 MPa are accepted for such implant-type coatings (ASTM, 2009; FDA, 1997; ISO/DIS, 1999).

The research on implants with thick coatings made of bioglasses prepared by using an enamelling process has shown that, in time, cracks appear in the coatings, allowing metallic ions to spread inside the human body, and producing finally their delamination. Moreover, in comparison with hydroxyapatite films, the control of composition and adhesion to metallic substrates seems to be more difficult to accomplish in the case of the BG ones.

The implants must simultaneously satisfy requirements such as biocompatibility, strength, corrosion resistance and sometimes aesthetics. It is widely accepted that both mechanical properties and chemical composition are important factors in the preliminary physiological bond of such implants to living tissues. Low mechanical properties are the major problem that prevented the use of BG/Ti structures for load-bearing applications.

This chapter aims to introduce magnetron sputtering technique as a solid alternative for bioactive implants' functionalization, taking a new step in the research of implant-type structures based on bioactive glasses. The chapter will present our recent findings on the correlation between bioactive powder targets/RF-MS deposition parameters versus composition tailoring of BG thin films, and their mechanical and in vitro behaviour in simulated body fluids (SBF). Understanding these correlations could be important for fundamental physics, materials science and prosthetic medicine as well as from a technological point of view.

Radio Frequency - Magnetron Sputtering (RF-MS) deposition is nowadays one of the most popular techniques to grow thin films in research and in decorative and semiconductor industry. In this method the plasma is used as a source of energetic ions (within the energy range 10–500 eV) that are accelerated towards the cathode target. When energetic ions reach the target surface with energy above the surface binding energy (the minimum threshold is typically somewhere in the range 10–100 eV), an atom can be ejected. This way free atoms and clusters are produced by sputtering, which are subsequently deposited on a substrate as well as the reactor chamber walls (Palmero et al., 2007).

Recently, Radio Frequency - Magnetron Sputtering (RF–MS) has emerged a promising alternative for preparing adherent bioactive glass films (G.E. Stan et al., 2009; G.E. Stan et al., 2010a, 2010b, 2010c, 2010d) due to its tailoring possibilities and due to some advantages: low pressure operation, low substrate temperature, high purity of the films, ease of automation, and excellent uniformity on large area substrates (Wasa et al., 2004).

In this chapter we present recent findings on the adherence and bioactivity of bioglass coatings prepared by magnetron sputtering technique. The study will indicate how features such as composition, structure, adherence and bioactivity of bioglass films can be tailored simply by altering the magnetron sputtering working conditions, proving that this less explored technique is a promising alternative for preparing implant-type coatings. Extensive multi-parametrical structural, compositional, morphological and mechanical characterizations were employed by FTIR, GIXRD, SEM, and pull-out tests.

2. Solutions for increasing the adherence at the titanium substrate / glass coating interface

It is widely accepted that the integrity of the substrate/coating interface is always critical in determining the performance and the reliability of any implant-type coating. Generally, low values of adhesion for bioglass coatings were published (Mardare et al., 2003; Goller, 2004; Peddi et al., 2008). Among the deposition techniques available for producing bioglass coatings, magnetron sputtering is the less explored. Only three papers have been published by other groups on this topic to the best of our knowledge (Mardare et al., 2003; Wolke et al., 2008; Saino et al., 2010). The main impediment in using bioglass coatings as implant applications is the high thermal expansion coefficients of bioglass, about 12–17 x 10^{-6}/°C, relative to that of medical-grade titanium ~9.2 x 10^{-6}/°C (L.L. Hench & J. Wilson, 2003; Goller, 2004).

The adherence of the substrate/coating interface is always considered when estimating the implant-type coating reliability in medical practice. The mechanical quality of the interface can be evaluated by pull-out strength measurements. The pull-out measurements were carried out using an adhesion tester – DFD Instruments PAT MICRO adhesion tester AT101 (maximum pull force = 1 kN) equipped with Φ 2.8 mm stainless steel test elements. The test elements were glued to the film's surface with a cyanoacrylate one-component Epoxy adhesive, E1100S. The stub surface was first polished, ultrasonically degreased in acetone and ethanol and dried in a nitrogen flow. After gluing, the samples were placed in an oven for thermal curing (130°C/1 h). Each test element was pulled-out vertically with a calibrated hydraulic pump until detachment. The experimental procedure was conducted in accordance with the ASTM D4541 and ISO 4624 standards. The adhesion strength was determined from the recorded failure value divided by the quantified detached surface area. Mean values and standard deviations were computed. The statistical significance was determined using an unpaired Student's t–test. The differences were considered significant when $p < 0.05$.

In a first step, a bioglass (BG1) mild-pressed powder having the following composition (wt %): SiO_2 - 55, CaO - 15, P_2O_5 - 10, K_2O - 10, MgO - 5, Na_2O – 5, having a low CTE (~ 10.2 x 10^{-6}/°C) was used as cathode target. The BG1 samples preparation details are presented in *Table 1*.

Bioglass target	Sputtering pressure (Pa)	Working atmosphere	Substrate type	Film thickness	Post-dep. heat-treatments
BG1	0.3	Ar	Ti6Al7Nb	750 nm	550°C/2h in air 750°C/2h in air

Table 1. BG1 sputtering deposition conditions and additional sample preparation details

The as-deposited structures as well as the heat-treated ones were investigated. *Figure 1* presents the GIXRD patterns of the BG1 structures before and after the devitrification heat-treatments. The structure of the as-deposited film is amorphous at the sensitivity limit of the measurement. The amorphous component is probably based on amorphous silica, as it can be deduced from the main hump centred on $2\theta \approx 22°$. After heat treatment at 550°C, several weak TiO_2-rutile - ICDD: 12-1276 lines appear, whose intensity increases by annealing at 750°C. During heat treatment at 750°C, crystallization occurs in the layer, but the amorphous component only slightly diminishes. The few rather intensive peaks which appear, were assigned to: $Na_2Mg(PO_3)_4$ – ICDD: 22-477 and TiO_2-rutile. There are also several weak lines that might be associated with small percents of perovskite ($CaTiO_3$ – ICDD: 78-1013) and oxygen deficient titanium oxides (Ti_2O – ICDD: 11-0218 and TiO – ICDD: 72-0020). The formation of calcium titanate and, at least partly, of titanium oxides in the heat-treated samples, is due to the inter-diffusion at the film-substrate interface. The presence of perovskite in heat-treated hydroxyapatite films deposited onto titanium had already been observed by other authors (Wei et al., 2005; Berezhnaya et al., 2008).

In case of the as-deposited and 550°C annealed structures the failure always occurs in glue's volume without damaging the film integrity at an average value of 85 MPa. This represents the bonding limit of the resin, as confirmed by the manufacturer. For the structures annealed at 750°C the adherence was estimated at 72.9±7.1 MPa, the film being detached each time. However, the measured film-substrate adhesion values are much higher than those reported in literature (Mardare et al., 2003; Goller, 2004; Peddi et al., 2008).

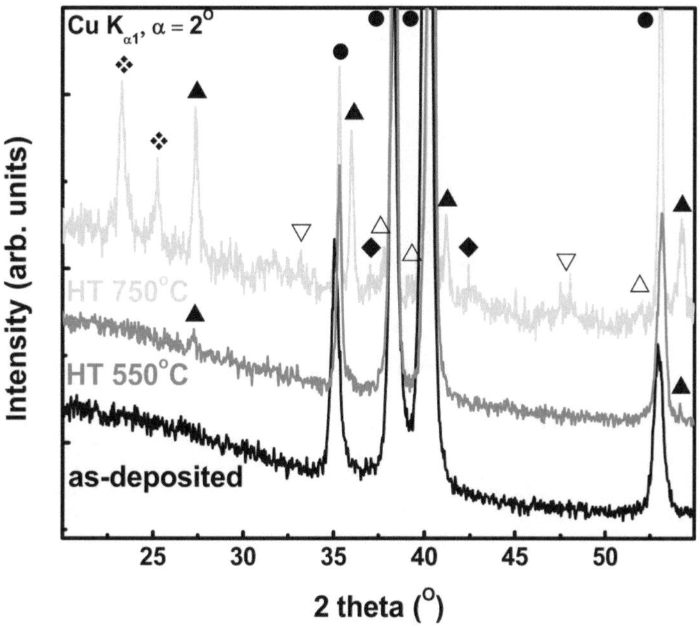

Fig. 1. GIXRD patterns collected for the BG1 films before and after the post-deposition heat-treatments: ●- titanium alloy substrate; ❖- $Na_2Mg(PO_3)_4$; ◆- $CaTiO_3$; ▲- TiO_2-rutile; △- Ti_2O; ▽- TiO

These high values of pull-out strength for the as-deposited coatings, and those annealed at 550°C, are attributed to the sputter cleaning and ion bombarding processes of the substrate, done before deposition (G.E. Stan et al., 2009). As it is known, the poor adhesion of the coatings could be ascribed to the natural oxide layers present on the titanium alloy surface prior deposition. An optimal solution for removing these contaminants in order to increase the adherence is an argon plasma etching pre-treatment of the titanium substrates. The sputter cleaning process largely removes not only the native oxide layers, but also the adsorbed gas molecules, to produce a clean, highly active surface (Lacefield, 1988). The argon ion bombarding process during sputtering might enhance the atomic diffusion and mixing in the near-interface region. The etching process has been optimized by tuning plasma power (~200 W), surface DC bias (0.4 kV) and etching time (10 min), based on own experience and literature (Mattox, 1994). The sputtering time offers a maximum of the adhesion properties after 10 min, as a result of the removal of surface oxides and contaminants, which leads to good bonding at the coating-substrate interface (G.E. Stan et al., 2009).

The decrease of the adherence after heat-treatment at 750°C, is due the BG layers/titanium substrate dilatation coefficients misfit at high temperatures and to the strong oxidation phenomena which leaded to the of needle-shaped rutile crystals agglomerates which penetrate through the film (*Fig. 2*).

On the other hand, the formation during the heat-treatments of a perovskite-type phase ($CaTiO_3$) suggested an inter-diffusion phenomenon between the coating and the substrate during heat-treatment, which appears to be the predominant factor in determining a film adhesion even in case of high temperature annealing treatments. The formation of various types of titanium oxides and sub-oxides is also in agreement with this hypothesis. This finding could be further exploited by designing special heat-treatments which would lead to nucleation at the interface of inter-mix BG-Ti phases with role in strengthening the coating adherence.

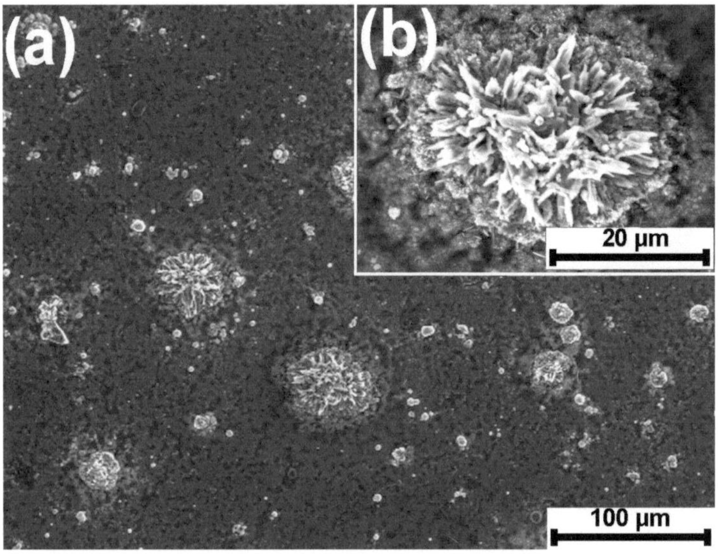

Fig. 2. SEM images of a BG1 film heat-treated in air at 750°C/2h

The classical 45S5 composition system (SiO_2 - 45 wt%, CaO - 24.5 wt%, P_2O_5 - 6 wt%, Na_2O - 24.5 wt%), patented by Hench a couple of decades ago, has a significantly higher thermal expansion coefficient (15 - 17 x 10^{-6}/°C) than the titanium and titanium alloys materials. 45S5 commercial powders have been mild-pressed to prepared cathode targets (BG2). The BG2 samples preparation details are presented in *Table 2*.

Bioglass target	Sample type	Sputtering pressure	Working atmosphere	Substrate type	Film thickness	Post-dep. heat-treatments
BG2	Simple BG2-A (BG/Ti)	0.3 Pa	Ar	cp-Ti grade 4	~ 700 nm	650°C/2h in air
BG2	Graded BG2-G ($BG/BG_{1-x}Ti_x/Ti$)	0.3 Pa	Ar	cp-Ti grade 4	~ 770 nm	650°C/2h in air

Table 2. BG2 sputtering deposition conditions and additional sample preparation details

In such a case, in the attempt to increase the adherence properties of the BG2 film to the titanium substrate, a buffer layer with chemical gradient of composition was introduced. The graded $BG_{1-x}Ti_x/Ti$ (x=0-1) structure was prepared by a slow continuous shifting of the rotating substrate holder during the co-sputtering deposition, from the Ti target towards the BG one (G.E. Stan et al., 2010c). This way, a functionally graded transition zone with a variable chemical composition was forming between the BG biofunctional coating and the Ti substrate (*Fig. 3*). This process lasted for 5 minutes, the graded layer thickness being estimated at ~70 nm. Next, the formed graded structure was placed in front of the BG target and the sputtering continued for 1 hour in order to deposit the functional BG layer with a thickness of ~ 700 nm (BG2-G).

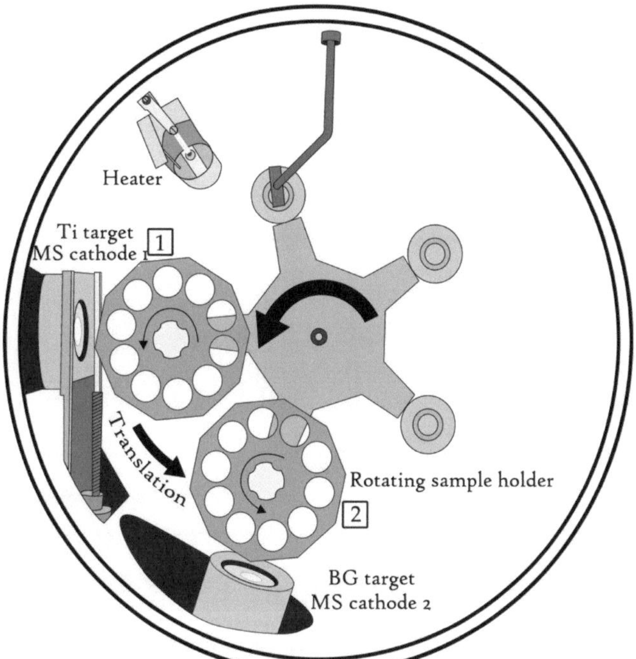

Fig. 3. Schematic diagram of the graded $BG_{1-x}Ti_x/Ti$ (x=0-1) structure deposition by magnetron co-sputtering

For bonding strength comparison, we synthesized by RF-MS, under identical experimental conditions, 700 nm BG/Ti "abrupt coatings" (BG2-A) without the intermediate buffer layer between the substrate and the film.

As an improvement of the mechanical and biological properties of films is known to be achieved by the transformation of BG into glass-ceramics via heat-treatments (Peitl Filho et al., 1996; El Batal et al., 2003), we have chosen for our study an annealing temperature of 650°C/2h in order to induce the partial crystallization of the combeite ($Na_2CaSi_2O_6$) phase. As known (Lefebvre et al., 2007), the crystallization of the combeite is reached within the range (610 - 700°C) and the larger the temperature, the higher the crystallization degree. Low heating and cooling rates (1°C/min) have been applied in order to minimize the residual mechanical stress in films at the end of the thermal cycle (Berbecaru et al., 2010).

The GIXRD analysis of the BG2-G coating before heat treatment confirmed the amorphous nature of the deposited films (*Fig. 4*). The titanium sub-oxide phase, Ti_3O (ICDD: 73-1583) was present in the structure of the untreated sample. The structure was crystallized after heat-treatment with combeite - ICDD: 75-1687, $CaSiO_3$ (wollastonite) - ICDD: 42-550, and Na_3PO_4 - ICDD: 76-202 as main crystalline phases. There was noticed a strong signal originating from the Ti substrate - ICDD: 44-1294. After the heat-treatment, Ti lines were shifted towards lower angles, simultaneously with a strong broadening. Moreover, the Ti sub-oxide completely disappeared and was replaced by two well crystallized Ti dioxide phases, anatase - ICDD: 89-4203 and rutile - ICDD: 12-1276 (G.E. Stan et al., 2010c).

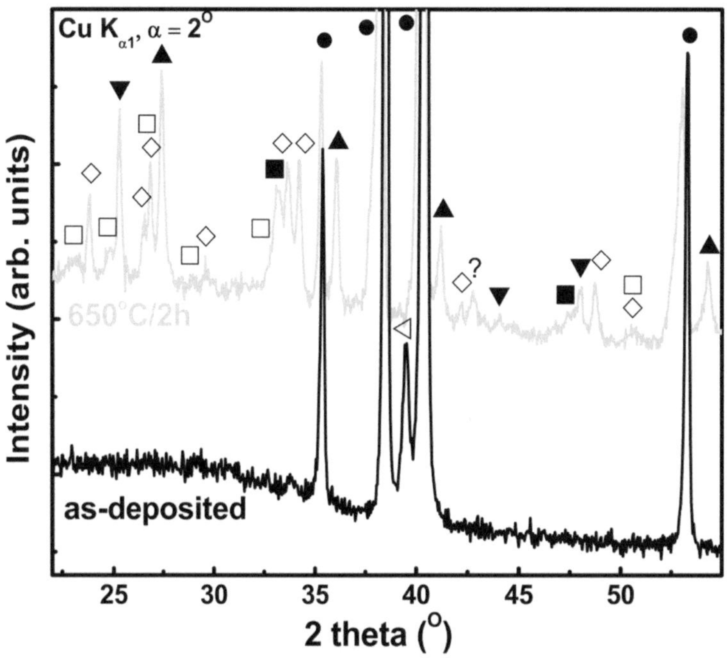

Fig. 4. GIXRD patterns collected for the BG2 films before and after the post-deposition heat-treatments: ●- titanium substrate; ◇- combeite ($Na_2CaSi_2O_6$); □- wollastonite ($CaSiO_3$); ■- Na_3PO_4;▼- TiO_2-anatase; ▲- TiO_2-rutile; ◁- Ti_3O

The partial crystallization of the BG structure may prove advantageous for biomedical applications as it results in improved bioactivity due to formation on their surface of calcium phosphate (CaP) rich layers in contact with simulated body media, as would be demonstrated in the Subsection 3.3. The formation of such an apatite type layer is very important for bone growth and bonding ability.

A statistical analysis was performed based on the average value for ten different BG samples in case of each type of structure. For the BG2-A structure, the bonding failure occurred at a mean value of 29.2 ± 7 MPa (G.E. Stan et al., 2010c). This is a rather low bonding strength value, but similar to those reported in literature (Mardare et al., 2003; Goller, 2004; Peddi et al., 2008). This effect is mainly due to the significant difference between the thermal expansion coefficients of the BG film and the titanium substrate at high temperatures.

A 1.7 times higher bonding strength (50.3 ± 5.8 MPa) was obtained in case of the BG2-G structure (G.E. Stan et al., 2010c). The bonding strength to Ti substrates characteristic to heat-treated BG2-G samples is larger than current values reported in literature (50 MPa vs. 30 MPa). The BG2-G design eliminates the material interface discontinuity due to the formation of a BG_xTi_{1-x} (x=0-1) functionally graded buffer layer that improves the bonding strength. The result indicated that the synthesis of BG structures with graded buffer layers is a feasible solution for preparing adherent BG coatings, even in case of bioglasses with high thermal expansion coefficients (G.E. Stan et al., 2010c).

In the following paragraphs results will be presented depicting the influence of typical sputtering variables (deposition pressure, working gas composition) on the BG thin films adherence. For these studies a novel complex bioglass powder composition (BG3) was chosen as cathode target (wt %): SiO_2 - 40.08, CaO - 29.1, MgO - 8.96, P_2O_5 - 6.32, CaF_2 - 5.79, B_2O_3 - 5.16, and Na_2O - 4.59. The deposition conditions are presented in *Table 3*.

Bioglass target	Sample denomination	Sputtering pressure	Working atmosphere composition	Substrate type	Deposition time	Film thickness
BG3	BG3-1	0.2 Pa	100%Ar	cp-Ti grade 1	70 min	646 nm
BG3	BG3-2	0.3 Pa	100%Ar	cp-Ti grade 1	70 min	510 nm
BG3	BG3-3	0.4 Pa	100%Ar	cp-Ti grade 1	70 min	480 nm
BG3	BG3-4	0.3 Pa	93%Ar+7%O_2	cp-Ti grade 1	70 min	380 nm

Table 3. BG3 sputtering deposition conditions and additional sample preparation details

The GIXRD measurements evidenced the amorphous state of all as-deposited BG3 thin films. *Figure 5* displays the SEM micrographs of the as-deposited BG films. One can see important modifications of the morphology at a sub-micrometric level when varying the deposition conditions. The SEM micrographs revealed well-adhered films with a homogeneous surface microstructure for all the as-deposited samples. No signs of micro-cracks or delaminations were noticed. The rough microstructure, consisting in parallel alternant, stripe-like tall regions, delimited by narrow depressions was probably induced by titanium substrate, while the fine structure consists of nano-sized merged-granules (*Fig. 5 - inset*). One can not see significant influence of pressure upon the microstructure (BG3-1, BG3-2 and BG3-3). In case of reactive atmosphere (BG3-4) the films' surfaces presented interesting features. The coating is uniformly covered by tower-shaped nano-formations with an average diameter of ~70 nm. The tower-shaped nano-aggregates seem to be ingrowths nucleated on a matrix which is similar to BG3-1, BG3-2 and BG3-3 (G.E. Stan et al., 2011).

The physics of the magnetron sputtering process at different working pressures and compositions of the working atmosphere determines the thickness and morphology of the as-deposited BG films. The films' thicknesses varied between 380 and 646 nm, the thicker for the lowest pressure non-reactive deposition atmosphere. The decrease of the film thickness with the increase of Ar pressure from 0.2 to 0.4 Pa might be assigned to a decreased fraction of sputtered particles reaching the substrate, due to the increased probability of collision with other particles when running from target to substrate. But, despite the masses of the sputtered atoms (Ms) in case of BG material are comparable to that of the background Ar gas (M_{Ar}/M_s= 1-1.8), thus favouring the kinetic energy loss by collision, the thermalization or removal while running towards the substrate is unlike because of the short target-to-substrate distance (only 30 mm) (Palmero et al., 2007, G.E. Stan et al., 2010d).

Under these deposition conditions it is highly probable that the main phenomenon leading to decreasing growth rate of BG films is the occurrence of charge transfer reactions in Ar (van Hattum et al., 2007). These processes lead to possible modifications of the energy and extent of the argon ion and neutral bombardment during the deposition in the considered pressure region. The occurrence of resonant charge transfer reactions is known to lower the energy of bombarding ions, determining significant variations of the sputter yield.

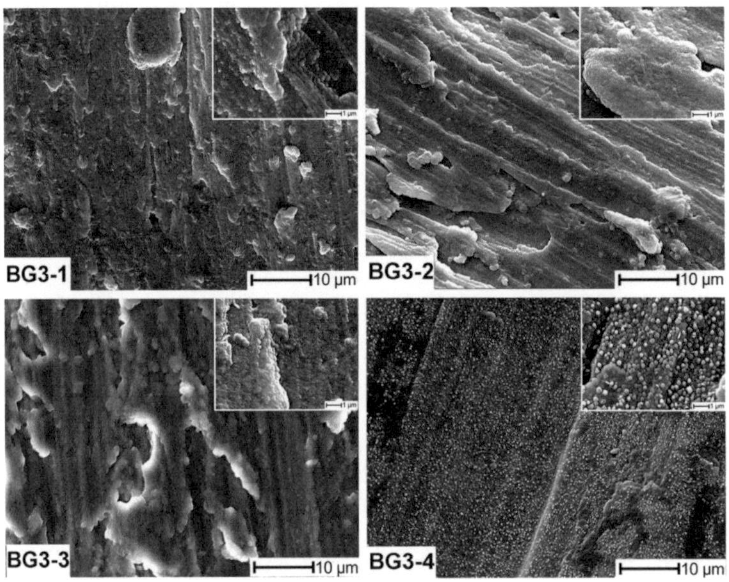

Fig. 5. SEM images of as-deposited BG3 sputtered films

The decreased deposition rate observed in the presence of oxygen in the working atmosphere can be attributed to "target poisoning" induced by chemisorption and oxygen ion implantation (Berg & Nyberg, 2005). During the sputtering process the BG target is bombarded by ions from the plasma, including reactive oxygen ions. This leads to the formation of a compound film not only on the substrate as desired, but also on the sputtering target. This results in a significantly reduced sputter yield, and thereby, a reduced deposition rate (G.E. Stan et al., 2010a).

Significant pull-out adherence differences were found function of sputtering deposition regime. In the case of BG3-1 and BG3-2 films, the failure occurred in the epoxy adhesive's volume at 84.8 ± 1.5 MPa without damaging the film integrity (G.E. Stan et al., 2011). As this value represents the bonding limit of the epoxy adhesive as confirmed by the manufacturer, the true BG coating - Ti substrate bonding strength could be even higher. This adhesion value is much higher than the usual ones reported in literature (Mardare et al., 2003; Goller,

2004; Peddi et al., 2008). For the films deposited at higher argon pressure (BG3-3), the adherence dramatically decreased to a mean value of 34.2 ± 12.0 MPa, the bonding failure occurring at the film - substrate interface. When using a reactive atmosphere (BG3-4), but keeping the total pressure constant at 0.3 Pa, the adhesion strength of monolithic BG coating declined down to 44.0 ± 6.8 MPa (G.E. Stan et al., 2011). Similar adherence values have been generally reported in literature (Mardare et al., 2003; Goller, 2004; Peddi et al., 2008). The two tailed t-testing, assuming unequal variances, showed statistically significant differences between the results obtained for the different coatings ($p<0.05$).

The excellent adherence value of BG3-1 and BG3-2 films must be emphasized. The adherence values are significantly higher than those reported in literature for this kind of implant coatings, thereby opening new perspectives in implantology. The high adherence is related to the processes characteristic of magnetron plasma sputtering. At a lower pressure the sputtered atoms collide with the substrate with higher kinetic energy, creating the possibility of forming chemical bonds with atoms from the substrate or being implanted into substrate. Such phenomena could lead to an increased adherence in case of the BG3-1 and BG3-2 films. Using a higher pressure (BG3-3) results in a spatial density variation of the background gas and affects the magnetron sputtering discharge as well as the transport of the particles towards the substrate. The energy flux at the substrate is thus affected, which in turn affects the properties of the growing film such as density, grain size, columnar structure, stoichiometry, coverage and adhesion (G.E. Stan et al., 2011). Moreover, their surface mobility is dramatically reduced, causing possible film inhomogeneities, such as voids or clustering hillocks (G.E. Stan et al., 2010d). In case of BG3-4 films, the presence of the tower shaped glassy nano-aggregates on the BG surface could be the underlying reason for the decreased adherence. Usually when dealing with tensile forces, rupture involves only a few molecules in the material causing the whole specimen to fracture in a "domino effect" (G.E. Stan et al., 2011).

3. Bioactivity tailoring of bioglass and glass-ceramic sputtered thin films

3.1 As-deposited films analysis

An implant-type ideal coating should constitute a proper mechanical support while exhibiting an enhanced bioactivity.

The bioglass structure is very complex; there is only short and medium range order determined by chemical bounding and steric hindrance. The silica-based glass structure is generally viewed as a matrix composed of SiO_4 tetrahedra connected at the corners to form a continuous tri-dimensional network with all bridging oxygen (BOs). The SiO_4 tetrahedra network is slightly distorted due to variations in the bond angles and the torsion angles. The network modifiers (alkali and alkali-earth ions typical to bioglasses) enter the structure as singly or double charged cations and occupy interstitial sites. Their charge is compensated by non-bridging oxygen bonds (NBOs), created by breaking bridges between adjacent SiO_4 tetrahedra. The increase of modifier content generates the creation of large NBOs concentrations, reducing the connectivity of the BG network, with direct effect upon electrical conduction, the thermal expansion coefficient, glass transition temperature, chemical corrosion in aqueous media and reactivity (Serra et al., 2002; Liste et al., 2004).

The biomineralization activity of a bioglass is influenced by the concentrations of bridging and non-bridging oxygen atoms per silicon oxygen tetrahedron as a function of the alkali oxide concentration. The Q^n notation expresses the concentration of bridging oxygen atoms per tetrahedron, where the value of n is equal to the number of bridging oxygen atoms (Elgayar et al., 2005). The Q^n species are detected in IR spectra in the 850–1200 cm^{-1} region by broad bands developing in accordance with the glass alkali and alkali-earth composition. The depolymerization of silicate network is defined by bands present at lower wave numbers in the absorption envelope.

RF-MS is known as a non-equilibrium preparation method which generally produces non-stoichimetric films relative to the sputtering target composition. Therefore it is possible to prepare nanostructured BG films with stoichiometries which can not be obtained by the classical equilibrium bulk synthesis methods. However, the preliminary studies have revealed that by varying the deposition pressure and/or working atmosphere composition it is possible to obtain a stoichiometric transfer even for complex compositional systems such as BGs (G.E. Stan et al., 2010a, 2010b, 2010c, 2010d). When selecting the RF-MS deposition parameters for the BG films preparation, one should take into account the standard free energy of the oxidation reactions of the different elements involved (Alcock, 2001):

$$2\,Ca + O_2 \rightarrow 2\,CaO \qquad \Delta G^\circ = -1\,267\,600 + 206.2\,T\ (298\text{–}1124\ K),\ (Jmol^{-1})$$

$$2\,Mg + O_2 \rightarrow 2\,MgO \qquad \Delta G^\circ = -1\,206\,300 + 273.7\,T\ (300\text{–}900\ K),\ (Jmol^{-1})$$

$$Si + O_2 \rightarrow SiO_2 \qquad \Delta G^\circ = -907\,030 + 175.7\,T\ (300\text{–}1700\ K),\ (Jmol^{-1}).$$

$$4\,Na + O_2 \rightarrow 2\,Na_2O \qquad \Delta G^\circ = -830\,180 + 260.3\,T\ (300\text{–}350\ K),\ (Jmol^{-1})$$

$$4\,K + O_2 \rightarrow 2\,K_2O \qquad \Delta G^\circ = -729\,100 + 287.3\,T\ (400\text{–}370\ K),\ (Jmol^{-1})$$

Table 4 presents the BG films atomic concentration with respect to the cathode target original composition. As can be seen Ca is among the most reactive elements towards oxygen, and this explains why its concentration decreased in the BG3 films. The reaction between Ca and the impinging oxygen ions might account for target poisoning by binding calcium ions at the target's surface. According to this reasoning, one would expect almost similar concentration changes for Mg since its standard free energy of oxidation is close to that of Ca. However, *Table 4* shows that higher concentrations of Mg were determined for all the films in comparison to the starting bioglass powder. This suggests that the atomic weight might also play a role, with heavier elements diffusing more slowly. Na enjoys of these two characteristics, i.e., it is lighter and its standard free energy of oxidation is lower, being more easily sputtered. The formation of P_4O_{10}, and especially of PO requires higher partial pressure of oxygen. This might explain why a so small amount of P was found in the films. On the other hand, silicon also forms two possible phases, SiO_2 or SiO upon oxidation with the first being favored at the temperature of the experiments. The formation of these species also requires higher partial pressure of oxygen in comparison to the formation of oxides of Na, Ca, Mg (G.E. Stan et al., 2011).

Sample type	Concentration (at %)					
	Si	Ca	P	Na	Mg	K
BG1 powder target	50.28	14.68	7.73	8.85	6.81	11.65
BG1 as-deposited film	50	16	10	19	2.6	2.4
BG2 powder	36.34	21.2	4.1	38.36	-	-
BG2 as-deposited film	35	20	2	43	-	-
BG3 powder target	38.43	34.18	5.9	8.5	12.88	-
BG3-1 as-deposited film	30	27	3	20	20	-
BG3-2 as-deposited film	30	32	3	16	19	-
BG3-3 as-deposited film	30	33	1	20	16	-
BG3-4 as-deposited film	30	20	3	27	20	-

Table 4. Chemical compositions in at % for the BG target powders and for the BG films deposited onto titanium substrates. The values were determined by EDS for the films, and were calculated on the basis of the nominal oxides composition for the target.

The *in vitro* bioactivity of these BG samples, reflected in their capability of inducing HA-formation onto their surfaces, was investigated by immersion in SBF at 37°C for various periods of time up to 30 days. The SBF had the following ionic concentrations (in mM) of 142.0 Na^+, 5.0 K^+, 2.5 Ca^{2+}, 1.5 Mg^{2+}, 147.8 Cl^-, 4.2 HCO^{3-}, 1.0 HPO_4^{2-} and 0.5 SO_4^{2-}, buffered at pH=7.4 with tris-hydroxymethyl-amminomethane (Tris, 50 mM) and hydrochloric acid solutions according to Kokubo (Kokubo & Takadama, 2006). A surface area to volume ratio of 0.1 cm^{-1} was maintained for all immersions. The biomineralization processes at the BG–SBF interface were monitored, on the surface and in the volume, by FTIR, GIXRD and SEM. Hench's theory states that the first stage of a bioglass mineralization upon immersion in SBF involves the rapid exchange of Na^+ and Ca^{2+} ions from the glass for H^+ and H_3O^+ ions from the solution, which will initialize the hydrolysis of the Si–O–Si bonds of the glass structure and the forming of silanol groups. The dissolution of the glass network, leading to the formation of silica-rich gel layer, the supersaturation of SBF solution with respect to hydroxyapatite and the subsequent deposition of an apatite-like layer on the glass surface, were found to be essential steps in the bioactivity evaluation both *in vivo* and *in vitro* studies (Hench, 1991). The formation of stable, mechanically strong interface with both bone and soft connective tissues is essential for the clinical success.

FTIR spectroscopy is a powerful method to obtain useful information concerning the short-range order for the as-deposited amorphous films as well as for SBF tested films, allowing the identification of specific features in the IR vibrational spectrum, such as those related to silicate or phosphate groups.

Figure 6 shows the FTIR spectra of the as-deposited and heat-treated BG films together with the spectra of the cathode target powder. Similar absorption envelopes for the target and as-deposited BG films have been recorded, exhibiting the same broad IR bands typical to an amorphous structure. However all the as-deposited BG films presents a shift to higher wave numbers of the maximum absorption peak with respect to the powder spectrum which indicate a certain degree of modification of the silicate tetrahedron network during sputtering, in good correlation with the compositional modification.

In case of BG1 structures (*Fig. 6-a*) we noted the presence of three intense vibration bands:

- 1001-1030 cm^{-1} assigned to asymmetric stretching vibrations of Si-O-Si in the Q^2 and Q^3 units;
- 1102-1130 cm^{-1}, more intense for the as-deposited samples could be attributed to the anti-symmetric stretching mode of Si–O–Si groups;
- 794 cm^{-1} correspond to the bending motion of Si–O–Si links (Socrates, 2001; Agathopoulos et al., 2006).

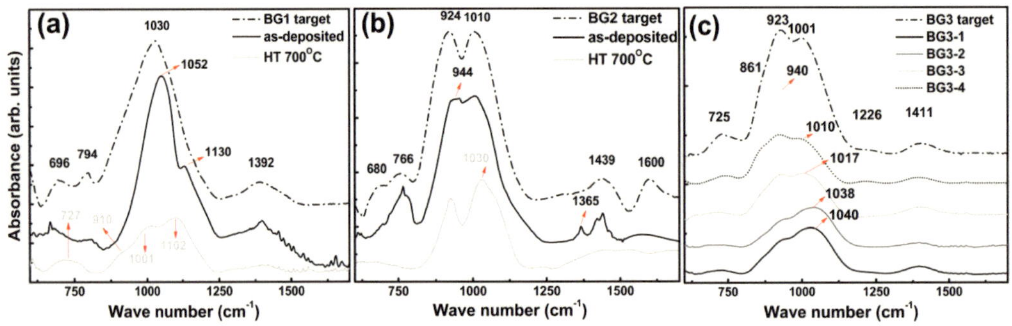

Fig. 6. Comparative FTIR spectra for the target powders and the studied films: a) BG1; b) BG2; c) BG3

The peak at 1383 cm^{-1} might be assigned to a shifted $(CO_3)^{2-}$ stretching band. The presence of the weaker band at 696 cm^{-1} corresponds to symmetric stretching bands of PØP Q^3 in Q^2 and Q^1 units. After the heat-treatment at 700°C/2h one can observe the splitting of the envelope in two shoulders, indicating a crystallization of the BG structure, along with the appearance of new vibration band at 910 cm^{-1} (the stretching of the Si-O-3NBO and Si-O-2NBO) groups). A strong shift to lower wave numbers of the bands positioned at 1030 and 1120 cm^{-1} was also noticed.

The IR spectra of BG2 structures (*Fig. 6-b*) revealed three strong vibration bands:

- 924-944 cm^{-1} - attributed to the stretching vibration of the SiO_4 units with three and two non-bridging oxygen atoms, the Q^1 (Si-O-3NBO), and Q^2 (Si-O-2NBO) groups;
- 1010-1022 cm^{-1} assigned to the coexistence of various Q^2 and Q^3 Si-O-Si asymmetric stretching vibration;
- 766 cm^{-1} correspond to the bending motion of Si–O–Si links (Socrates, 2001; Agathopoulos et al., 2006).

The weaker band at 680 cm^{-1} might correspond to symmetric stretching bands of PØP Q^3 in Q^2 and Q^1 units. Other vibrations of phosphate groups present in the bioglass are difficult to emphasize because of the superimposition of the strong bands of SiO_4 units. Previous IR studies noticed also the presence of Q^2, Q^1, and Q^0 phosphate units in the 1400–400 cm^{-1} IR spectra range (Socrates, 2001; Agathopoulos et al., 2006). The weak broad shoulder present at 1600 cm^{-1} can be assigned to water bending vibrations, indicating that BG1 material is highly hygroscopic, absorbing water vapour when in air. The presence of absorption band at 1439 cm^{-1} is attributed to the stretching vibrations of carbonate $(CO_3)^{2-}$ structures incorporated during the deposition process. After the post-deposition annealing a clear splitting of the bands was noticed, pointing a strong crystallization of the BG coating, in agreement with the GIXRD measurements (*Figs. 1* and *4*).

All the BG3 films FTIR spectra displayed as weal broad absorption bands: two dominant peak maxima at ~1030 cm^{-1} (Si–O stretching in Q^2 and Q^3 units) and at ~940 cm^{-1} (Si–O stretching in Q^1 and Q^2 units). The weak shoulder at ~725 cm^{-1} corresponds to Si-O-Si bending motion. Broad vibration bands centred at 1218 and 1411 cm^{-1} were evidenced, and are related to B–O stretching of BO_3 units in borates with bridging oxygen, vibrations of metaborate triangles and B–Ø stretching of $BØ_4$ and $BØO_2^-$ units (Agathopoulos et al., 2006).

In case of BG3 films the FTIR analysis revealed a dependence between argon deposition pressure and the short range order of the sputtered glass structure. The displacement of the asymmetric stretching vibration present at 1040 cm^{-1} (BG3-1) to lower wave numbers (*Fig. 6-c*) indicates the Si-O-Si linkages perturbation by a continuous formation of non-bridging oxygen type linkages with the increase of argon sputtering pressure, and the weakening of the structural bonds sustained by the bridging oxygen atoms (*Fig. 6-c*). The FTIR spectra displayed indeed an increase of the intensity ratio of the bands: 923–944 cm^{-1} / 1010–1040 cm^{-1}, which denote the enrichment in Q^1 and Q^2 structural groups, with increasing the deposition pressure. This suggests that as the deposition pressure increases a higher concentration of alkali and alkali-earth oxides might be incorporated in the glass structure, the Q^3 groups being progressively converted to Q^2 groups (Socrates, 2001; Serra et al., 2002). A direct correlation between the glass thin films' composition and their structure that reflects directly in their biomineralization capability was reported in literature (Serra et al., 2002).

3.2 Bioactivity tests: Observations & considerations

Figure 7 displays a comparison of the BG films IR spectra before and after immersion in simulated body fluid up to 30 days.

For the as-deposited and annealed BG1 structures, the FTIR measurements (*Fig. 7-a,b*) showed no changes in intensity and position of the original vibration bands with immersion time, suggesting the inert character of this material.

In case of as-deposited BG2 samples (*Fig. 7-c*) after only 24 hours of immersion, dramatic changes were observed. The spectrum revealed the disappearance of the initial BG film vibration bands and the emergence of two new and strong bands positioned at 816 and 1107 cm^{-1}, which could be assigned to bending and stretching bands of the amorphous silica phase (Bal et al., 2001). Weaker bands at 736, 892 and 968 cm^{-1}, were noticed, belonging most probably to the original BG layer. The intensity of these bands decreased with the immersion time suggesting a continuously leaching of BG ions into the SBF solution. After 30 days of immersion the BG films is completely dissolved, no IR bands could be emphasized, indicating the resorbability of this material.

The FTIR results of the annealed BG2 samples after in-vitro testing in SBF are presented in *Figure 3-d*. After 24 hours soaking in SBF solution the amplitude of the peak at 924 cm^{-1} present increased values and displaced at ~ 940 cm^{-1}. This maximum peak at 940 cm^{-1} is also present related to the Si-OH groups (Berbecaru et al., 2010). In comparison the peak positioned at 1030 cm^{-1} is decreased in amplitude. The process could be related to the continuous leaching of Na, Ca and Si ions in the SBF solution. The exchange of the ions between the surface layer and SBF solution lead to a rearranging of the bonds in the sample surfaces. At the same time the diffusion of the H^+ ions, because of the increase of the electronegativity of the surface toward the BG films will initiate the formation of the Si-OH bonds on their surface (L.L. Hench & J. Wilson, 2003; Berbecaru et al., 2010). The small shoulder at 895 cm^{-1} revealed the presence of the stretching vibrations of the Si-O bonds with two non-bridging oxygens (Q^1 and Q^0 units) from the original BG layer. After 72 hours of immersion in SBF solution the peaks at 776 and 1030 cm^{-1} increased in amplitude suggesting the enrichment of the Si-O-Si bonds in the SiO_4 tetrahedra (Q^4 units). This suggests that after a certain period of time the polymerization reaction begins at the film surface. Also, the peak at 940 cm^{-1} strongly decreases in amplitude, revealing three new weak bands at 960 cm^{-1}, 1021 and 1087 cm^{-1} respectively. These absorption bands are assigned to v_1 symmetric stretching mode of $(PO_4)^{3-}$ (960 cm^{-1}) and to v_3 asymmetric stretching of the phosphate groups, respectively (1021 and 1087 cm^{-1} bands) (Socrates, 2001).

This phenomena could be related to simultaneous phenomena such: formation of the silicic acid $Si(OH)_4$ and subsequent polycondensation reaction, along with the beginning of the precipitation of calcium phosphate (Ca-P) phases on the surface from the SBF solution suprasaturated in Ca-P ions (Berbecaru et al., 2010). The released water in the polycondensation reaction is known to remain physically bonded with the Si-O-Si surface forming the hydrated silica rich layer (Hench, 1991; L.L. Hench & J. Wilson, 2003). This is leading to an increase of the pH at the surface level which will be favourable to the absorption of the cations and anions on the surface, the precipitation processes of Ca-P rich phases being under these auspicious conditions.

The process of precipitation seems to continue up to 30 days, the relative integral area of the bands at 960 cm^{-1}, 1021 and 1087 cm^{-1} monotonously increasing with immersion time. Thus, it is suggested that on top of the annealed BG1 coatings chemically develops in vitro a Ca-P type layer. The broad aspect of the bands suggests the amorphous nature of this chemically grown layer. For comparison the spectrum of a crystalline synthetic hydroxyapatite powder is displayed. No crystallization processes were observed in XRD after 30 days of immersion.

Figure 7-e,f,g displays the evolution of the FTIR spectra for the BG3 structures after SBF immersions for 3, 15 and 30 days. Similar IR spectral evolutions were observed for all BG3 films. A change of the BG3 spectra envelope was noticed after 3 days of immersion in SBF (*Fig. 7-e*). There are some perceptible differences in shape and amplitude of the IR spectra between these three investigated glass films. The two dominant maxima (~1090 and ~1120 cm^{-1}, respectively) are assigned to the $(PO_4)^{3-}$ (v_3) asymmetric stretching, hinting that after three days of immersion the precipitation of a Ca–P layer had started (G.E. Stan et al., 2011). A splitting of the phosphate stretching band, more clearly in case of BG3-4, could be observed, suggesting a more advanced stage of amorphous calcium phosphates phases' precipitation. Thus, at this point, one can deduce that partial dissolution of BG3 structures occurred up to 3 days, as demonstrated by the diminishing overall intensity of the spectrum and the lower contribution of silicates bands overlapped by the emergence of dominant phosphate bands due to Ca–P precipitation. The leaching of BG ions along with the dissolution-diffusion of silicon atoms from the glass structure resulting in a supersaturation at the film-SBF interface is a prerequisite condition for the starting of precipitation process (Hench, 1991). One can hypothesize that the peak positioned at ~1125 cm^{-1} could obscure also the presence of Si-O-Si asymmetric stretching vibrations (Hench, 1991; Socrates, 2001), owned to the silica-rich layer formed at the film-solution interface at an early stage (Hench, 1991). After 15 days (*Fig. 7-f*), all the BG3 samples displayed similar IR spectra, defined by the phosphate's two intense shoulders positioned around 1028 and 1121 cm^{-1}, respectively. Increases in intensity of water and O–H bands along with the appearance of $(CO_3)^{2-}$ stretching vibration bands suggest a continuous incorporation of various molecular structures at the liquid-solid interface. After 30 days of immersion the FTIR absorbance spectra (*Fig. 7-g*) revealed strong vibrations at the following wave numbers: 876, 960, 1019, 1107, 1413, 1465, and 1635 cm^{-1}, corresponding to crystalline carbonated apatite (CHA). The sharp bands at 1019 cm^{-1} and 1107 cm^{-1} correspond to (v_3) asymmetric stretching of phosphate groups. The splitting of the stretching and bending IR absorption bands in two narrow components suggests a crystalline apatitic growth. The weak band centred at 960 cm^{-1} is assigned to $(PO_4)^{3-}$ (v_1) symmetric stretching mode. All apatitic layers obtained were hydroxylated and carbonated as demonstrated by the presence of strong O–H bands (the bending mode centred at ~1635 cm^{-1}) and the sharp C–O bending (v_2) and stretching (v_3) lines at (876, 1413 and 1465 cm^{-1}). Considering the hydroxyapatite structure, the carbonate group can substitute both the hydroxyl and the phosphate ions, giving rise to the A-type and B-type carbonation, respectively. The positions of the carbonate bands indicate that $(CO_3)^{2-}$ groups are substituting $(PO_4)^{3-}$ ions, suggesting the predominance of B-type CHA (Markovic et al, 2004). The B-type is the preferential substitution in the human bone and is known to have better bioactivity and osteoinductivity (Spence et al., 2008).

The GIXRD patterns are presented in *Fig. 8* revealed after 30 days immersion in SBF the characteristic lines of hydroxyapatite (HA) as large, overlapping peaks (ICDD: 9–432 , for instance) for all the samples. There are deviations of relative intensities with respect to the reference ICDD card, due probably to imperfect stoichiometry. The increased relative intensity of the (002) line is often reported for chemically grown HA layers and it is assigned to the preferred orientation of crystallites and is typical for the biological apatite. The HA peaks are broad due either to the small crystallite size or to lattice disorder or to both (G.E. Stan et al., 2011). The mean crystallite size along the c axis of the hexagonal HA structure, estimated from the (002) line broadening by using the Scherrer formula, is about 25-30 nm (strain-broadening neglected). Because of the large line broadening one can not distinguish between different HA types, for instance between HA and carbonated HA.

After three days of immersion in SBF a strong and narrow diffraction peak appeared in the GIXRD patterns of BG3-1, BG3-2 and BG3-3 at $2\theta = 26.5°$, whose intensity is correlated with that of a weak line at $2\theta = 54.6°$. The most plausible phase is calcium phosphate hydrate type (i.e. $Ca(H_2P_2O_7)$ ICDD: 70-6384), whose role in the subsequent HA formation has been the subject of several detailed studies (Liu et al., 2008).

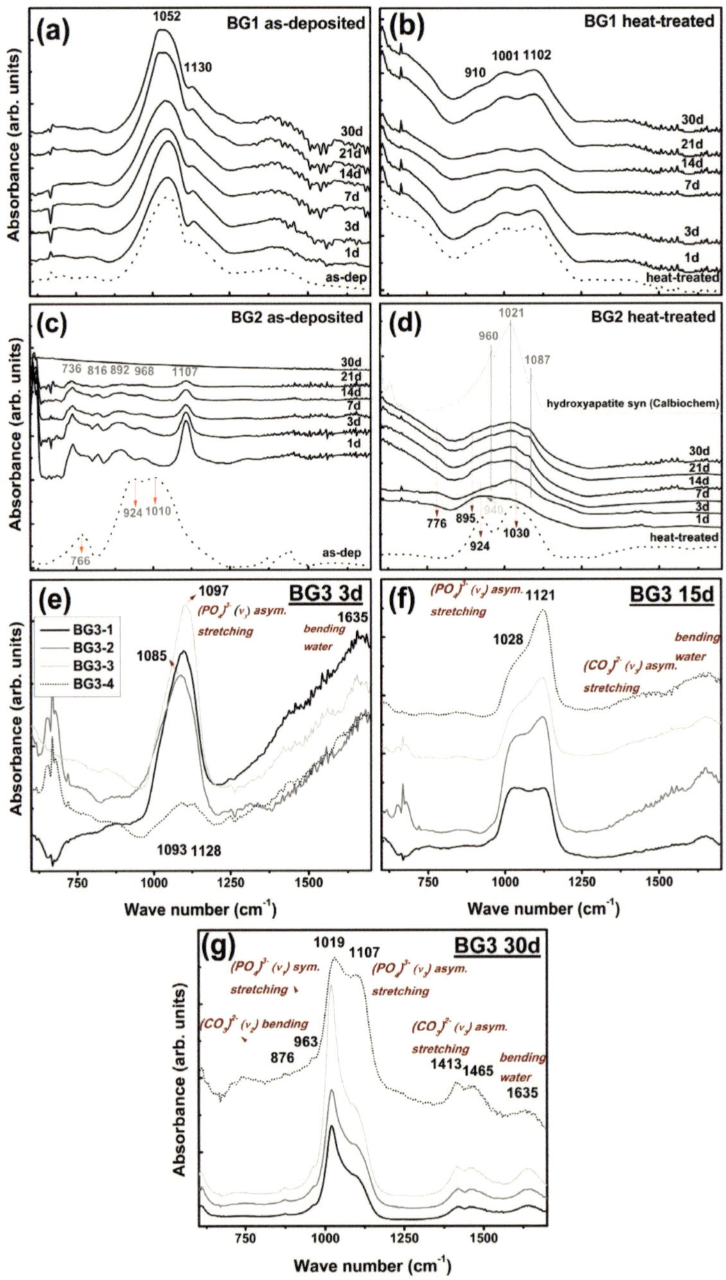

Fig. 7. Evolution of FTIR spectra of the BG films with increasing immersion time in SBF (1 - 30 days)

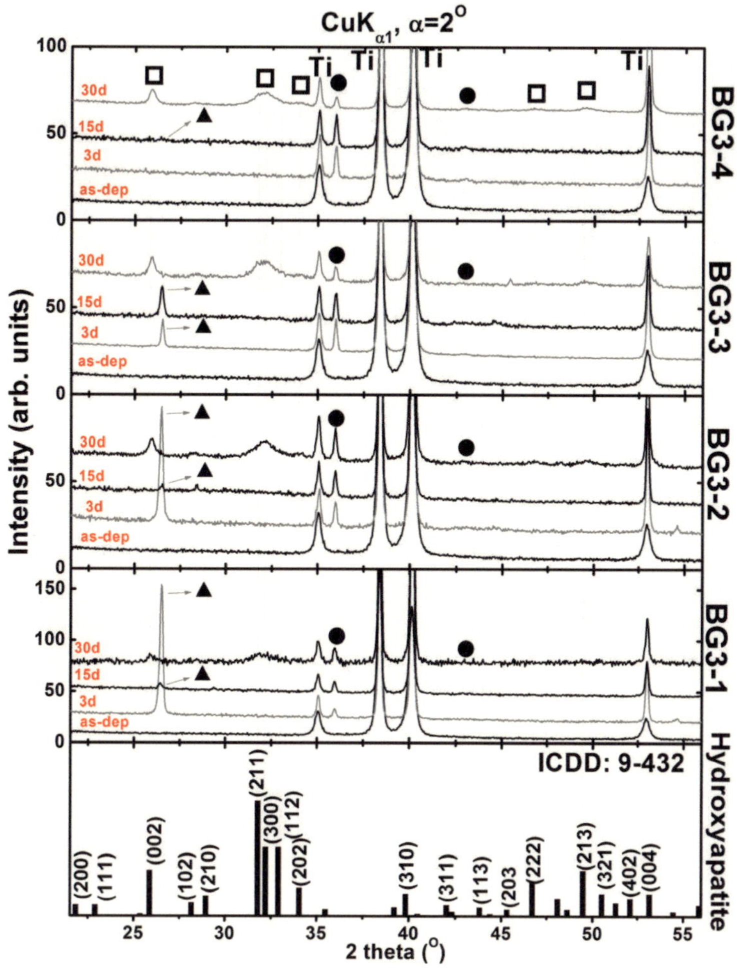

Fig. 8. Comparative representation of the GIXRD patterns of the as-deposited BG films before and after 3, 15 and 30 days immersion in SBF. □= Hydroxylapatite (ICDD:9-432); ●= $TiH_{1.7}$ (ICDD:40-1244); ▲= calcium phosphate hydrate (ICDD: 70-6384)

The intensity of the line at $2\theta = 26.5°$ start to diminish until the 15th day of immersion in SBF, and almost disappears after 30 days for all the samples. It is interesting to note that this phase has a strong intensity only in the structures deposited in pure Ar (BG3-1, BG3-2 and BG3-3) and it is very weak for the BG3-4 samples deposited in reactive atmosphere (7% oxygen). Another quite intensive line in the GIXRD patterns of the samples after SBF immersion appears at $2\theta = 36.0°$. The best assignment for this line is TiH (ICDD: 40-1244). This peak is present in all the structures that were immersed in SBF and it is expected that its presence is related to the uncovered regions of the titanium substrate.

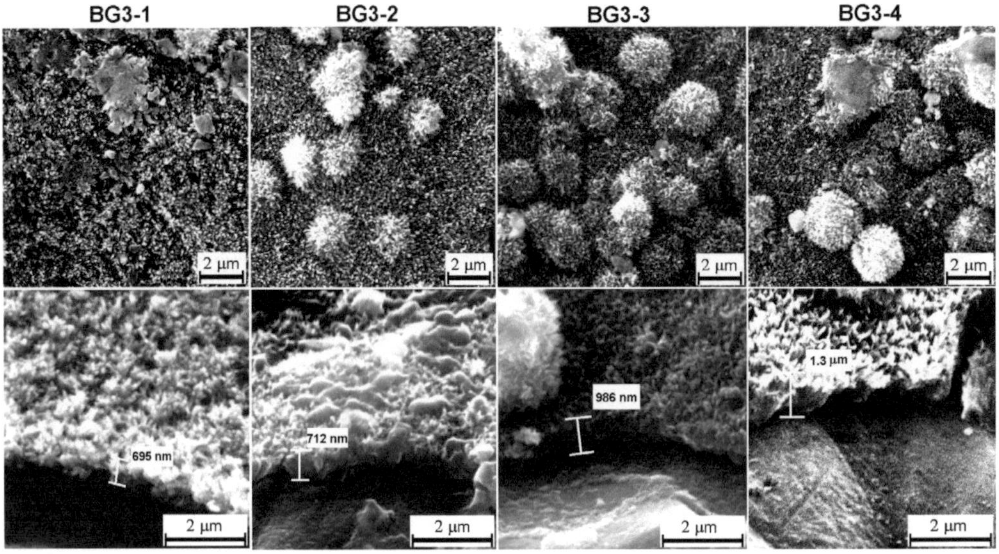

Fig. 9. SEM micrographs of BG3 films after immersion in SBF for 30 days. Left side, SEM-top view; Right side, ESEM-cross view

The SEM microstructures of BG3 films immersed for 30 days clearly asserted the growth of thick and rough coatings with randomly distributed irregular spherulitic shaped micro-sized agglomerates on top of all three types of samples, evidencing the good biomineralization capability of all coatings (*Fig. 9-top row*). The thicknesses of the chemical grown HA layers were determined by tilt-SEM (*Fig. 9-bottom row*). One can observe that the BG3-3 and BG3-4 coatings led to the thickest chemical growths: total film thickness ~1 µm, and ~1.3 µm, respectively, compared to the as-deposited film thickness of ~ 0.4 µm. The HA chemically grown layers had in case of BG3-1 and BG3-2 structures a lower average thickness of ~0.7 µm. Thus, the best biomineralization, correlated with the thickest HA layer growth, was obtained for the BG3-4 coating (G.E. Stan et al., 2011).

3.3 Discussion

Thus, the IR spectra of the samples immersed in SBF for 1 month showed different behaviour *in vitro* as a function of composition and crystallization state.

The inertness of BG1 structures might be determined by their chemical composition. High contents of SiO_2 have influences both in bioactivity and the thermal expansion coefficient of the glass. In general, SiO_2-rich glasses with silica contents higher than 60 wt.% have a better mechanical stability and adhesion to the metallic substrate, but are not soluble in body fluids (Lopez-Esteban et al., 2003). Decreasing the silica content is therefore mandatory for improving the bioactivity and partial dissolution of the glass surface. For the BG1 structures one can notice an important silicon (~50 at.%) and phosphorous content (~10 at.%). In silica-based glasses phosphorous acts as a network former producing structural rearrangement where the silicate network is more interconnected by creating P-O-Si cross-links. At higher content of phosphorus, the glass is increasing in connectivity. The glass compositions with higher content of network formers, do not bond either to bone or to soft tissues and elicit formation of a non-adherent fibrous interfacial capsule. No surface changes during SBF tests were evidenced for the annealed BG1 structure. Unlike in case of BG2, the crystallization had no effect on BG1's surface reactivity. The nucleated $Na_2Mg(PO_3)_4$ phase proved to be inert in SBF. It can be concluded that the chemical composition, bonding configuration and crystallinity could play important roles in BG films' reactivity in SBF.

In case of as-deposited BG2 samples a continuous dissolution process occurs, after 30 days the entire film is dissolved into the surrounding fluid. One can notice the high content of Ca (~20 at.%) and Na atoms (~43 at.%) in the BG2 film (*Table 3*). It can be speculated that when very high numbers of alkali and alkali-earth ions are released in SBF, the pH will rapidly and drastically change at the film-fluid interface, producing a chemical unbalance which will suppress the polymerization of silanols and the further formation of the SiO_2-rich surface layer. The bioactive process is thus disrupted, leading to continuous dissolution of the BG amorphous film. On the other hand the annealing of BG1 films followed by slow cooling promotes formation of combeite, wollastonite and Na_3PO_4, which have smaller solubility, and consequently a decreased leaching rate of Na^+ and Ca^{2+} into the SBF solution. This will determine the conditions at the surface of the film to be more stable, therefore allowing polymerisation of soluble $Si(OH)_4$ in a SiO_2-rich layer which will act as a nucleation site for apatite (L.L. Hench & J. Wilson, 2003). During 30 days of SBF soaking, dissolution-reprecipitation processes took place, the annealed BG2 layer is partially dissolved and finally we obtain a multilayer structure containing a bottom BG layer coated by an amorphous SiO_2-rich thin film and at the top a Ca-P type layer. Thus in case of BG2 annealed samples our FTIR results are consistent with Hench's theory. Previous studies have also reported that both the combeite (Chen et al., 2006) and wollastonite (Xue et al., 2005) can generate on their surface Ca/P rich layers when in contact with SBF. The growth of the bioapatite layer is essential for bone generation and bonding ability. However, the combeite was proved to rapidly transform into amorphous calcium phosphate phase in contact with SBF, but it delays the process of crystallization into a hydroxyapatite phase (L.L. Hench & J. Wilson, 2003). Besides biocompatibility, these structures could actively improve proteins and osteocytes adhesion, significantly shortening the osteointegration time.

Regarding the higher biomineralization rates of BG3-3 and BG3-4 films one can hypothesize that the chemical processes involved in the bioactivity mechanism are accelerated, because of the increased sodium content of these films and a propitious bridging oxygen/non-bridging oxygen ratio (*Table 4*), speeding up the ionic exchange and the chemical growth of HA. Hench's theory states that the first stage of SBF immersion of a bioglass involves the rapid exchange of Na^+ ions from the glass for H^+ and H_3O^+ ions from the solution, which shall initialize the formation of silica-rich layer known to favour the nucleation of Ca-P type layers. Therefore, we can suppose that a higher number of sodium ions released into SBF solution will change the chemical equilibrium of the precipitation reaction, thereby catalyzing the CHA chemical growth mechanism.

Therefore, for achieving bioactive properties, an optimal ratio of network formers/network modifiers is required, but not sufficient. The disruption of the Si-O-Si bonds and creation of Si-O-NBO bonds, determined by increased network modifiers content, is mandatory in the first steps of Ca-P chemical growth mechanism. The disruption of the Si-O-Si bonds evidenced by the increasing content of Si-O-NBO groups plays an important role in the dissolution rate of silica in the first steps of Hench's chemical growth mechanism, favouring the bioactivity processes of the films by a higher rate of biomineralization (Serra et al., 2002; Liste et al., 2004). This hypothesis is in agreement with our results, the stronger biomineralization, expressed by the thickest CHA film grown, was obtained for the BG3 sample. Although BG compositions with very high network modifier content have a native poor stability, we can tailor their in-vitro behaviour by thermal treatments of crystallization to obtain the best bioactive outcome (G.E. Stan et al., 2010b).

The mechanism of calcium phosphates formation onto the surface of bioactive glasses is widely accepted, and involves the dissolution of cations from the surface of bioactive glasses and the consequent increase of the supersaturation degree in the surrounding fluid, with respect to HA components. Hench's theory states that the first stage of a bioglass mineralization upon immersion in SBF involves the rapid exchange of Na^+ ions from the glass for H^+ and H_3O^+ ions from the solution, which shall initialize the breaking of the Si-O-Si bonds of the glass structure and the subsequent formation of silanol groups. In a next step the silanol groups polycondensate forming a silica-rich layer at the bioglass surface, which favours the growth of Ca-P type layers. Homogenous nucleation (precipitation) of biomimetic apatite occurs spontaneously in a solution supersaturated with respect to hydroxyapatite. Heterogeneous precipitation, on the other hand, takes place on the sample's surface. Both, homogenous and heterogeneous nucleation, could be considered in competition during the SBF immersion.

Due to their designed low thickness (lower than 700 nm), even in case of total dissolution, the amount of BG film that could be released in the SBF solution is around few micrograms. Thus, intuitively the bioactivity processes will not be governed in case of thin BG films preponderantly by the supersaturation of SBF solution in Na, Ca and P and further precipitation of Ca-P phases, but by the pH evolution at the sample-solution interface, surface energy and electrostatic interactions between electrically negative charged sample surface and the cations and anions from stagnant solution in the proximity of the BG surface.

A reinterpretation of the classical biomineralization process based on the electrical double layer effect will be presented. The electrical double layer refers to two parallel layers differing in charge surrounding an object (BG sample). The increased number of surface hydroxyl groups could act as nucleation sites for HA. Thus the first layer, with negative surface charge, comprises $Si\text{–}OH^-$ groups primarily formed during chemical interactions with the fluid. Ca^{2+} ions and with them also anions, such as $(HPO_4)^{2-}$, are present in the electrolyte space charge facing the sample surface enriched in polymerising silanol groups. Thus, the second layer is composed of calcium ions attracted to the surface charge via the Coulomb force, electrically screening the first layer. This second layer could be seen in the first stages as a diffuse layer being loosely connected with the silica-rich one, as it consists of free calcium ions which move in the fluid under the influence of electric attraction and thermal motion. As the $(HPO_4)^{2-}$ anions interact and react with this second layer, the heterogeneous nucleation of $CaO\text{–}P_2O_5$ layer starts, which is anchoring progressively into the sample's surface, resulting in a hydrated precursor cluster consisting of calcium hydrogen phosphate, and in time by consuming and incorporating various ions [$(OH)^-$, F^- or $(CO_3)^{2-}$] from the surrounding media crystallize into a carbonated-flour-hydroxyapatite mixed layer. Consequently in case of BG thin films one should consider the heterogeneous nucleation as the dominant mechanism of HA growth.

4. Conclusions

The effects of sputtering pressure, composition of working atmosphere and deposition parameters on the compositional, morpho-structural and mechanical properties of the coated implants were emphasized, and their influence on the *in vitro* behaviour of the coatings in simulated body fluid (SBF) was studied.

The adhesion strength at the coating - substrate interface is a critical factor in successful implantation and long-term stability of such implant-type structures. The chapter presented few technological RF-MS derived approaches to surpass the low adherence drawback when dealing with BG having high CTE: by introducing intermediate buffer layers [$BG_{1-x}Ti_x$ (x=0-1)] with compositional gradient, or by strong inter-diffusion phenomena induced via heat-treatments, with the aim in the nucleation of BG-Ti mix phases at the substrate-coating interface. The importance of a proper *in situ* substrate cleaning, performed by argon ion sputtering processes which will enhance the ad-atoms diffusion, should be also emphasized. Also when using BG cathode target with CTE closely matching the Ti substrates, the RF-MS deposition conditions exert great influence on the adherence of films as well as on their composition. The application of the above mentioned technological algorithms conducted to pull-out adhesion values higher than the usual ones reported in literature, opening new perspectives in oral and orthopaedic implantology.

Peculiarities of the *in vitro* behaviour of the BG thin films were discussed in correlation with the widely accepted Hench bioactivity mechanism. Based on our *in vitro* studies it was concluded that the chemical composition, bonding configuration and crystallinity could play important roles in BG films' reactivity in SBF. For achieving bioactive properties an optimal ratio of network formers/network modifiers is a necessary, but not sufficient, condition. BG compositions with very high network modifier content have a poor native stability, and their *in vitro* behaviour can be tailored by thermal treatments of crystallization to obtain the best biological outcome.

By varying the sputtering pressure and working atmosphere we attempted to tailor the films' composition and structure, hinting optimal mechanical performance and bioactive behaviour. Understanding these correlations is important for biomaterials science, implantology, as well as from an applied physics and technological point of view, opening new perspectives in regenerative medicine.

5. Acknowledgments

The financial support of Centre for Research in Ceramics and Composites (CICECO) - University of Aveiro and of the Romanian Ministry of Education, Research, Youth and Sport (Core Program - Contract PN09-45) is gratefully acknowledged. Authors thank Dr. Iuliana Pasuk for the professional assistance with GIXRD measurements.

6. References

Agathopoulos, S.; Tulyaganov, D.U.; Ventura, J.M.G.; Kannan, S.; Karakassides, M.A. & Ferreira J.M.F. (2006). Formation of hydroxyapatite onto glasses of the CaO-MgO-

SiO_2 system with B_2O_3, Na_2O, CaF_2 and P_2O_5 additives. *Biomaterials*, Vol.27, No.9, (March 2006), pp. 1832–1840, ISSN 0142-9612

Alcock, C.B. (2001). *Thermochemical Processes – Principles and Models*, Butterworth-Heinemann, ISBN 0-7506-5155-5, Oxford, UK

American Society for Testing and Materials [ASTM]. (2009). *Standard specification for composition of ceramic hydroxylapatite for surgical implants* F 1185-03, pp. 514-515

Bal, R.; Tope, B.B.; Das, T.K.; Hegde, S.G. & Sivasanker, S. (2001). Alkali-loaded silica, a solid base: Investigation by FTIR Spectroscopy of adsorbed CO_2 and its catalytic activity. *Journal of Catalysis*, Vol.204, No.2, (December 2001), pp. 358–363, ISSN 0021-9517

Balamurugan, A.; Balossier, G.; Kannan, S.; Michel, J.; Rebelo, A.H.S. & Ferreira J.M.F. (2007). Development and in vitro characterization of sol–gel derived $CaO–P_2O_5–SiO_2–ZnO$ bioglass. *Acta Biomaterialia*, Vol.3, No.2, (March 2007), pp. 255–262, ISSN 1742-7061

Batchelor, A.W. & Chandrasekaran, M. (2004). *Service characteristics of biomedical materials and implants*, Imperial College Press, ISBN 1-86094-475-2, London, UK

Berbecaru, C.; Alexandru, H.V.; Stan, G.E.; Marcov, D.A, Pasuk, I. & Ianculescu, A. (2010). First stages of bioactivity of glass-ceramics thin films prepared by magnetron sputtering technique. *Materials Science and Engineering B: Solid-State Materials for Advanced Technology*, Vol.168, No.1-3, (May 2010), pp. 101-105, ISSN 0921-5107

Berezhnaya, A. Yu.; Mittova, V.O.; Kostyuchenko, A.V. & Mittova, I. Ya. (2008). Solid-phase interaction in the hydroxyapatite/titanium heterostructures upon high-temperature annealing in air and argon. *Inorganic Materials*, Vol.44, No.11, (November 2008), pp. 1214–1217, ISSN 0020-1685

Berg, S. & Nyberg, T. (2005). Fundamental understanding and modeling of reactive sputtering processes. *Thin Solid Films*. Vol. 476, No.2, (April 2005), pp. 215-230, ISSN 0040-6090

Brown, C.; Papageorgiou, I.; Fisher, J.; Ingham, E. & Case, C.P. (2006). Investigation of the potential for wear particles generated by metal-on-metal and ceramic-on-metal implants to cause neoplastic changes in primary human fibroblasts. *Journal of Bone & Joint Surgery (British Volume)*, Vol. 88-B, SUPP III, pp. 393, ISSN 0301-620X

Brown, C.; Williams, S.; Tipper, J.L.; Fisher, J. & Ingham E. (2007). Characterisation of wear particles produced by metal on metal and ceramic on metal hip prostheses under standard and microseparation simulation. *Journal of Materials Science: Materials in Medicine*, Vol.18, No.5, (May 2007), pp. 819–827, ISSN 0957-4530

Chen, Q.Z.; Thompson, I.D. & Boccaccini, A.R. (2006). 45S5 Bioglass®-derived glass-ceramic scaffolds for bone tissue engineering. *Biomaterials*, Vol.27, No.11, (April 2006), pp.2414-2425, ISSN 0142-9612

Draft International Standards [ISO/DIS]. (1999). *Implants for surgery – Hydroxyapatite ceramic.* Part 1 and 2, pp. 13779

El Batal, H.A.; Azooz, M.A.; Khalil, E.M.A.; Soltan Monem, A. & Hamdy, Y.M. (2003). Characterization of some bioglass–ceramics. *Materials Chemistry and Physics*, Vol.80, No.3, (June 2003), pp. 599-609, ISSN 0254-0584

Elgayar, I.; Aliev, A.E.; Boccaccini, A.R. & Hill, R.G. (2005). Structural analysis of bioactive glasses. *Journal of Non-Crystalline Solids*, Vol.351, No.2 (January 2005), pp. 173–183, ISSN 0022-3093

Epinette, J.A.; Manley, M.T.; Geesink, R.G.T. (2003). *Fifteen years of clinical experience with hydroxyapatite coatings in joint arthroplasty*, Springer-Verlag, ISBN 2-287-00508-0, Paris, France

Food and Drug Administration [FDA]. (1997). *Calcium phosphate (Ca-P) coating draft guidance for preparation of FDA submissions for orthopedic and dental endooseous implants*, pp. 1-14

Goller, G. (2004). The effect of bond coat on mechanical properties of plasma sprayed bioglass-titanium coatings. *Ceramic International*, Vol.30, No.3, pp. 351–355, ISSN 0272-8842

Hench, L.L. (1991). Bioceramics: From concept to clinic. *Journal of American Ceramic Society*, Vol.74, No.7, (July 1991), pp. 1487–1510, ISSN 0002-7820

Hench, L.L. & Wilson, J. (1993). *Introduction to Bioceramics*, World Scientific, ISBN 981-02-1400-6, Singapore, Singapore

Kokubo, T. & Takadama, H. (2006). How useful is SBF in predicting in vivo bone bioactivity?. *Biomaterials*, Vol.27, No.15, (May 2006), pp. 2907-2915, ISSN 0142-9612

Lacefield, W.R. (1988). Hydroxyapatite coatings. In: *Bioceramics: material characteristics versus in vivo behavior* , Ducheyne. P. & Lemons, J.E., Vol. 523, pp. 72–80, Annals of the New York Academy of Science, ISBN 978-0-89766-437-0, New York, USA

Lefebvre, L.; Chevalier, J.; Gremillard, L.; Zenati, R.; Pollet, G.; Bernache-Assolant, D. & Govin, A. (2007). Structural transformations of bioactive glass 45S5 with thermal treatments. Acta Materialia, Vol.55, No.10, (June 2007), pp. 3305-3313, ISSN 1359-6454

Liste, S.; Serra, J.; González, P.; Borrajo, J.P.; Chiussi, S.; León, B. & Pérez-Amor, M. (2004). The role of the reactive atmosphere in pulsed laser deposition of bioactive glass films. *Thin Solid Films*, Vol.453 -454, (April 2004), pp. 224–228, ISSN 0040-6090

Liu, X.; Zhao, X.;, Li, B.; Cao, C.; Dong, Y.; Ding, C.; Chu, P.K. (2008). UV-irradiation-induced bioactivity on TiO2 coatings with nanostructural surface. *Acta Biomaterialia*. Vol. 4, No.3, (May 2008), pp. 544 - 552, ISSN 1742-7061

Lopez-Esteban, S.; Saiz, E.; Fujino, S.; Oku, T.; Suganuma, K. & Tomsia, A.P. (2003). Bioactive glass coatings for orthopedic metallic implants. *Journal of European Ceramic Society*, Vol.23, No.15, pp. 2921-2930, ISSN 0955-2219

Mardare, C.C.; Mardare, A.I.; Fernandes, J.R.F.; Joannia, E.; Pina, S.C.A.; Fernandes, M.H.V.; & Correia R.B. (2003). Deposition of bioactive glass-ceramic thin-films by RF magnetron sputtering. *Journal of European Ceramic Society*, Vol.23, No.7 (June 2003), pp. 1027–1030, ISSN 0955-2219

Markovic, M.; Fowler, B.O. & Tung, M.S. (2004). Preparation and comprehensive characterization of a calcium hydroxyapatite reference material. *Journal of Research of the National Institute of Standards and Technology*, Vol. 109, No.6, (November 2004), pp. 553-568, ISSN 1044-677X

Mattox, D.M. (1994). Surface Engineering. In: *ASM Handbook*, Reidenbach, F., Vol. 5, pp. 538, ASM International, ISBN 978-0-87170-384-2

Palmero, A.; Rudolph, H. & Habraken, F.H.P.M. (2007). One-dimensional analysis of the rate of plasma-assisted sputter deposition. *Journal of Applied Physics*, Vol.101, No.8, art.no. 083307 (6 pages), ISSN 0021-8979

Peitl Filho, O.; LaTorre, G.P. & Hench, L.L. (1996). Effect of crystallization on apatite-layer formation of bioactive glass 45S5. *Journal of Biomedical Materials Research*, Vol.30, No.4 (April 1996), pp. 509-514, ISSN 1549-3296

Peddi, L; Brow, R.K. & Brown, R.F. (2008). Bioactive borate glass coatings for titanium alloys. *Journal of Materials Science: Materials in Medicine*, Vol.19, No.9, (September 2008), pp. 3145–3152, ISSN 0957-4530

Rámila, A. & Vallet-Regí, RM. (2001). Static and dynamic in vitro study of a sol–gel glass bioactivity. *Biomaterials*, Vol.22, No.16, (August 2001), pp. 2301–2306, ISSN 0142-9612

Ratner, B.D.; Hoffman, A.S.; Schoen, F.J. & Lemons, J.E. (2004). *Biomaterials Science: An Introduction to Materials in Medicine*. Elsevier Academic Press, ISBN 0-12-582463-7, San Diego, USA

Saino, E.; Maliardi, V.; Quartarone, E.; Fassina, L.; Benedetti, L.; De Angelis, M.G.; Mustarelli, P.; Facchini, A. & Visai, L. (2010). In vitro enhancement of SAOS-2 cell calcified matrix deposition onto radio frequency magnetron sputtered bioglass-coated titanium scaffolds. *Tissue Engineering Part A*, Vol.16, No.3, (March 2010), pp. 995–1008, ISSN 1937-3341

Saravanapavan, P. & Hench, L.L. (2001). Low-temperature synthesis, structure, and bioactivity of gel-derived glasses in the binary $CaO-SiO_2$ system. *Journal of Biomedical Materials Research*, Vol.54, No.4, (March 2001), pp. 608–618, ISSN 1549-3296

Sepulveda, P.; Jones, J.R. & Hench L.L. (2002). Bioactive sol-gel foams for tissue repair. *Journal of Biomedical Materials Research*, Vol.59, No.2, (February 2002), pp. 340–348, ISSN 1549-3296

Serra, J.; González, P.; Liste, S.; Chiussi, S.; León, B.; Pérez-Amor, M.; Ylänen, H.O.; Hupa, M. (2002). Influence of the non-bridging oxygen groups on the bioactivity of silicate glasses. *Journal of Materials Science: Materials in Medicine*, Vol.13, No.12, (December 2002), pp. 1221-1225, ISSN 0957-4530

Socrates, G. (2001). *Infrared and Raman Characteristic Group Frequencies – Tables and Charts*, John Wiley & Sons Inc., ISBN 978-0-470-09307-8, Hoboken, USA

Spence, G.; Phillips S.; Campion, C.; Brooks, R. & Rushton, N. (2008). Bone formation in a carbonate-substituted hydroxyapatite implant is inhibited by zoledronate. The importance of bioresorption to osteoconduction. *Journal of Bone & Joint Surgery (British Volume)*, vol. 90B, No.12, pp. 1635-1640, ISSN 0301-620X

Stan, G.E.; Morosanu, C.O.; Marcov, D.A.; Pasuk, I.; Miculescu, F. & Reumont, G. (2009). Effect of annealing upon the structure and adhesion properties of sputtered bio-glass/titanium coatings. *Applied Surface Science*, Vol.255, No.22, (August 2009), pp. 9132–9138, ISSN 0169-4332

Stan, G.E.; Pina. S.; Tulyaganov, D.U.; Ferreira, J.M.F.; Pasuk, I. & Morosanu, C.O. (2010). Biomineralization capability of adherent bio-glass films prepared by magnetron sputtering. *Journal of Materials Science: Materials in Medicine*, Vol.21, No.4, (April 2011), pp. 1047-1055, ISSN 0957-4530

Stan, G.E.; Popa, A.C. & Bojin, D. (2010). Bioreactivity evaluation in simulated body fluid of magnetron sputtered glass and glass-ceramic coatings: A FTIR spectroscopy study. *Digest Journal of Nanomaterials and Biostructures*, Vol.5, No.2, (April-June 2010), pp. 557-566, ISSN 1842-3582

Stan, G.E.; Popescu, A.C.; Mihailescu, I.N.; Marcov, D.A.; Mustata, R.C.; Sima, L.E.; Petrescu, S.M.; Ianculescu, A.; Trusca, R. & Morosanu, C.O. (2010). On the bioactivity of adherent bioglass thin films synthesized by magnetron sputtering techniques. *Thin Solid Films*, Vol.518, No.21, (August 2010), pp. 5955-5964, ISSN 0040-6090

Stan, G.E.; Marcov, D.A.; Pasuk, I; Miculescu, F; Pina, S.; Tulyaganov, D.U. & Ferreira, J.M.F. (2010). Bioactive glass thin films deposited by magnetron sputtering technique: The role of working pressure. *Applied Surface Science*, Vol.256, No.23, (September 2010), pp. 7102-7110, ISSN 0169-4332

Stan, G.E.; Pasuk, I.; Husanu, M.A.; Enculescu, I.; Pina, S.; Lemos, A.F.; Tulyaganov, D.U.; El Mabrouk, K. & Ferreira, J.M.F. (2011). Highly adherent bioactive glass thin films synthetized by magnetron sputtering at low temperature. *Journal of Materials Science: Materials in Medicine*, Vol.22, No.12, (December 2011), pp. 2693-2710, ISSN 0957-4530

Tulyaganov, D.U.; Agathopoulos, S.; Valerio, P.; Balamurugan, A.; Saranti, A.; Karakassides, M.A. & Ferreira, J.M.F. (2011). Synthesis, bioactivity and preliminary biocompatibility studies of glasses in the system $CaO\text{-}MgO\text{-}SiO_2\text{-}Na_2O\text{-}P_2O_5\text{-}CaF_2$. *Journal of Materials Science: Materials in Medicine*, Vol.22, No.2 (February 2011), pp. 217-227, ISSN 0957-4530

Vallet-Regí, R.M.; Ragel, C.V. & Salinas, A,J. (2003). Glasses with medical applications. *European Journal of Inorganic Chemistry*, Vol.2003, No.6, (March 2003), pp. 1029–1042, ISSN 1434-1948

van Hattum, E.D.;Palmero, A.; Arnoldbik, W.M.;Rudolph, H & Habraken, F.H.P.M. (2007). On the ion and neutral atom bombardment of the growth surface in magnetron plasma sputter deposition. *Applied Physics Letters*, Vol. 91, Art.no. 171501 (3 pages), ISSN 0003-6951

Wasa, K.; Kitabatake, M. & Adachi, H. (2004). *Thin Film Materials Technology: Sputtering of Compound Materials*, Springer-Verlag, ISBN 3-540-21118-7, Heidelberg, Germany

Wei, M.; Ruys, A.J.; Swain, M.V.; Milthorpe, B.K. & Sorrell, C.C. (2005). Hydroxyapatite-coated metals: Interfacial reactions during sintering. *Journal of Materials Science: Materials in Medicine*, Vol.16, No.2, (February 2005), pp. 101-106, ISSN 0957-4530

Wolke, J.G.C.; van den Beucken, J.J.J.P. & Jansen, J.A. (2008). Growth behavior of rat bone marrow cells on RF magnetron sputtered bioglass- and calcium phosphate coatings. *Key Engineering Materials*, Vol.361-363, pp. 253-256, ISSN 1662-9795

Wu, Z.Y.; Hill, R.G.; Yue, S.; Nightingale, D.; Lee, P.D. & Jones, J.R. (2011). Melt-derived bioactive glass scaffolds produced by a gel-cast foaming technique. *Acta Biomaterialia*, Vol.7, No.4, (April 2011), pp. 1807–1816, ISSN 1742-7061

Xue, W.; Liu, X.; Zheng, X. & Ding, C. (2005). In vivo evaluation of plasma-sprayed wollastonite coating. *Biomaterials*, Vol.26, No.17, (June 2005), pp. 3455-3460, ISSN 0142-9612

4

Erosion Behavior of Plasma Sprayed Alumina and Calcia-Stabilized Zirconia Coatings on Cast Iron Substrate

N. Krishnamurthy[1], M.S. Murali[2], B. Venkataraman[3] and P.G. Mukunda[4]
[1]Mechanical Engineering Department, K.S. Institute of Technology, Bangalore,
[2]Auden Technology and Management Academy, Bangalore,
[3]Surface Engineering Group, Defence Metallurgical Research Laboratory, Kanchanbagh, Hyderabad,
[4]Mechanical Engineering Department, Nitte Meenakshi Institute of Technology, Bangalore,
India

1. Introduction

In plasma arc spraying process also known as plasma spraying process, the thermal energy of an electric arc (40 kW or 80kW) together with a plasma forming gas, which would be either nitrogen or argon, are utilized in melting and propelling of the deposit material at high velocities (600 mS^{-1}) onto a substrate. This process is capable of generating very high temperature, exceeding 16,000 °C, which can be gainfully employed in the deposition of materials with high melting points. The deposited material is generally in a powder form and requires a carrier gas to feed it into the combustion chamber. The process enables discharging high bond strengths of the coatings due to the very high propulsion velocities of the impinging particles.

In a DC plasma arc process, gas heating is enough to generate core plasma temperatures exceeding 20000 °C depending upon the properties of gas and its electrical break down characteristics. Enthalpy of the gas is an indicator of its heating potential while it is getting translated to plasma state.

1.1 Special features of plasma spraying technique

The following are some of the unique features of the plasma spraying process.

- The technique can be used to deposit a wide range of ceramics and metals and their combinations as well.
- It is possible to deposit alloys and mixed ceramics with components of widely differing vapor pressures without significant changes in composition.
- Homogenous coatings can be formed for any composition while maintaining uniformity in their thickness.

- Fine microstructures with equiaxed grains and without any type of columnar defects are the characteristics of this process.
- High deposition rates are possible without huge investments on capital equipment.
- The process can be carried out virtually in any environment such as air, encoded inert low and high-pressure environments, or underwater.

1.2 Erosion wear

Erosive wear of the solid bodies is caused by the action of sliding or impact of solids, liquids, gases or a combination of these [1]. Erosion can be divided into three basic types: Solid particle erosion, liquid impact erosion and cavitations erosion. Cavitation erosion is the loss of material due to the repeated formation and collapse of bubbles in a liquid. Liquid impact erosion is the damage caused by water droplets. Solid particle erosion is a wear process where they strike against surfaces and promote material loss. It is also caused by the impact of hard particles carried by a fluid stream onto a material surface.

Solid particle erosion is an important material degradation mechanism encountered in a number of engineering systems such as thermal power plants, aircraft gas turbine engines, IC engines, pneumatic bulk transport systems, coal liquefaction/gasification plants and ore or coal slurry pipe lines. At the same time, the erosion process has been used to advantage in number of situations like sand blasting of castings, shot peening of rotating components, cutting of hard and brittle materials by abrasive jets and rock drilling [2, 3].

Manifestations of solid particle erosion in service usually include thinning of components, a macroscopic scooping appearance following the gas/particle flow field, surface roughening and lack of the directional grooving characteristic of abrasion and in some the formation of ripple patterns on metals. Solid particle erosion can occur in a gaseous or liquid medium containing solid particles. In both the cases, particles can be accelerated or decelerated and their directions of motion can be changed by the fluid [4].

Power station boiler-walls and other utility parts of coal-fired plants are subjected to frequent degradation by erosion–corrosion problems relevant to the reliability and economics of these installations. The environment inside the furnaces is characterized by high-temperature conditions together with aggressive atmospheres, leading to corrosive deposits adhering into the walls and to erosion processes caused by the ash particles [5].

In erosion, several forces of different origins may act on a particle in contact with a solid surface. Neighboring particles may exert contact forces and a flowing fluid, if present, will cause the drag. On some situations, gravity may also be important. However, the dominant force on an erosive particle, which is mainly responsible for decelerating it from its initial impact velocity, is usually the contact force exerted by the surface. Erosion of metals usually involves plastic flow, whereas more brittle materials may wear predominantly either by flow or by fracture depending on the impact conditions [6].

Solid particle erosion behavior of most of the materials can be categorized as being either brittle or ductile in nature [7]. The major differentiating characteristic of the two types of mechanism is the dependence of erosion rate on impact angle i.e. the angle between the moving erodent particle and the material surface [8]. There is general agreement that maximum erosion occurs at a low angle (about 30^0) for ductile material and at 90^0 for brittle material. Figure shows the schematic of the expected variation in erosion behavior with impact angles

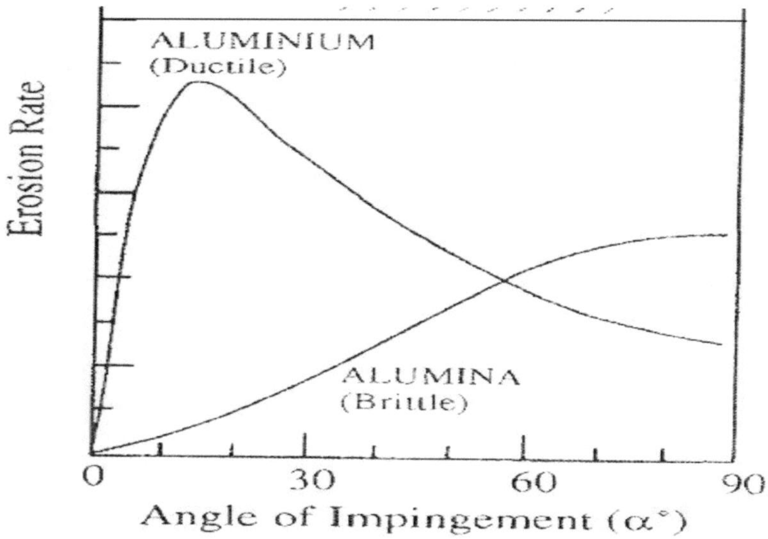

Figure Expected Variation of Erosion Rate with Particle Impact Angle (Ref. 8)

Most applications involve low impact angles at which erosion resistance of ceramics happens to be significant. It should also be noted, that the microstructure of plasma sprayed coatings often differs significantly from that of corresponding bulk material. The structure of plasma sprayed coatings consists of many overlapped lenticular splats which conform more or less either to the morphology of the underlying substrate or to that of previous splats. Although plasma sprayed coatings are anisotropic, their erosion rates tend to exhibit the same dependence on impact angle similar to that of the bulk material of ceramics [9]. On the other both Kingswell [10] and Zhang [11] have noted that the erosion mechanism in plasma sprayed alumina coating is different from those in bulk sintered alumina. Erosion of bulk ceramics generally occurs by a number of fracture mechanisms [12, 13]. During particle impact upon a ceramic surface, median and radial cracks develop at the impact site [14]. Upon rebounding of the particles i.e. unloading of the impact site, lateral cracks develop parallel to the surface and finally follows a curved path before propagating towards the surface, leading to chipping and loss of material. Erosion in plasma sprayed ceramics has been attributed to the failure of the individual splat boundaries.

2. Experimental details

2.1 Plasma spraying and characterization techniques

The surfaces of the substrate materials which are to be plasma coated were examined for dimensional accuracy and surface finish before being degreased in a vapour bath (70- 80^{0}C) of tetra chloro ethylene. The surfaces were then grit blasted by Al_2O_3 (-18+24 mesh) at a pressure of 455 kPa. Plasma spraying process was carried out with the help of proprietary Sulzer Metco Equipment. The composition of cast iron substrate and coating materials is given in Table 1.

Substrate material			
Cast Iron			
C-3.54, Si-2.21, Mn-0.67, Cr-0.025, Cu-0.013, P-0.056, S-0.031,Fe-balance			
Coating material			
Metco105SFP (TC1)	Metco 210NS (TC2)	Metco 452 (BC1)	Metco 410NS (BC2)
99.5 Al_2O_3	ZrO_2 5CaO	Fe 38Ni10Al	$Al_2O_3$30(Ni 20Al)

TC1-Top Coat 1, TC2-Top Coat 2, BC1-Bond Coat 1, BC2-Bond Coat 2

Table 1. Chemical composition of substrate and coating materials

The schematic diagrams of coating layers on cast iron substrate are shown in Fig. 1. The spray parameters for different materials are shown in Table 2.

CI-S1, CI-S2 and CI-S3 CI-S4, CI-S5 and CI-S6

Fig. 1. Schematic diagrams of coating layers with cast iron substrate (number in the bracket indicates the required thickness of each layer)

Materials	Primary gas (Argon) pressure kPa	Secondary gas(H_2) Pressure kPa	Carrier gas (Argon) Flow lpm	Current A	Voltage V	Spray distance mm	Feed rate kg/hr
TC1	700	520	60	600	65	64-125	2.7
TC2	345	345	37	500	75	50-100	5.4
BC1	700	340	37	500	65	100-175	4.1
BC2	700	340	37	500	65	100-175	4.1

Table 2. Plasma spray parameters

Surface texture of the coated samples was examined, employing Mahr Perthometer. The coated plate of 100x100 sq. mm area was selected and it was divided into small segments of 10 x 10 sq. mm. The tracing length was about 5.6 mm for each of the selected segment. Typical parameters describing the surface quality such as Arithmetical Mean Deviation or Average Roughness (Ra), Mean Roughness Depth (Rz), Maximum Roughness Depth (Rmax), Core Roughness Depth (Rk), Reduced Peak Height (Rpk), Distance between the Highest Profile Peak and the Reference Line (Rp), Root Mean Square Deviation (Rq) were recorded on each of the segments subjected to the analysis.

Microstructure analysis and surface morphology studies were carried out on a JOEL-JAPAN JSM-840A Scanning Electron Microscope. Area percentage measurements were done using a Leitz microscope fitted with a Biovis image Analyzer on the polished section of the coating. Care was taken to minimize the pull out of bond coat and top coat particles during polishing of the coated samples. The mounted samples were polished using emery papers of 240, 300, 400, 600 grit sizes and subsequently on 1/0, 2/0, 3/0 and 4/0 grades, successively. Fine polishing was done to obtain a mirror finish using 0.5 µm diamond impregnated cloth. The polished sample was cleaned with acetone before mounting on an optical microscope interfaced to a digital image capture and analysis system. The magnification was chosen such that the coating microstructure image covers the screen and allows the resolution of the voids that contributes significantly to the total porosity area percentage. The process of selecting the appropriate range of grey values was done to ensure that only voids were sampled. About ten separate fields of view were selected to ensure consistency in the analysis.

XRD analysis of the coated test samples were carried out on a Philips X-Ray Diffractometer (Model: PW 1840) using Cu-Kα radiation over a 2θ range of 20 to 100^0. The scanning speed was taken as 2^0 per min.

For Adhesion test cast iron cylindrical substrates were prepared according to ASTM C633-79 standards. The circular face of the test sample was coated according to the procedure explained in first paragraph of this section. One more sample with the same geometry but without coating was then joined to this coated surface employing an adhesive, Epoxy Polymer 15 (EP 15) with the application of the contact pressure varying from 2-3bar. The sample was then heated to 170 ^{0}C and maintained at this temperature for more than 60 minutes, before cooling it to room temperature. The specimens thus prepared were tested in a UTM of 60 tones capacity. The maximum load, maximum tensile strength and the stress strain diagram for the specimen were displayed by the computer connected to UTM. On each sample, five tests were conducted.

The microhardness of the test samples was determined using Leica Vickers Microhardness Tester (Model: VHMT Auto) as per ASTM E384 [15] standards. The test parameters are; 300 g load, 25 µg S^{-1} loading rate, 15 seconds dwell time with a Vickers Pyramid indenter. The measurement of hardness was done along the total thickness of the coating including substrate. An average of ten measurements taken at different locations on the transverse section of the coating was reported.

2.2 Solid particle erosion test

Erosion tests on coated and uncoated test samples were carried out according to ASTM G76-02 [16] standards. The test parameters are shown in Table 3. The sample was first cleaned in acetone using an ultrasonic cleaner, dried and then weighed using an electronic balance having a resolution of 0.01 mg. The sample was then fixed to the sample holder of the erosion rig and eroded with silica sand at the predetermined particle feed rate, impact velocity and impact angle for a period of about 5 min. The sample was then removed, cleaned in acetone and dried and weighed to determine the weight loss. This weight loss normalized by the mass of the silica particles causing it (i.e. testing time x particle feed rate) is then computed as the dimensionless incremental erosion rate. The above procedure was repeated till the incremental erosion rate attained a constant value independent of the mass of the erodent particles or, equivalently, of testing time. This constant value of the incremental erosion rate is defined as the steady-state erosion rate. On each coating system, three tests were conducted. The test parameters are given in Table 3.

Erodent Material	Silica Sand (Angular)
Erodent Size (μm)	150-300
Particle Velocity (m/s)	40
Erodent Feed Rate (g/min)	4.3
Impact Angle (°)	15, 45 and 90
Test Temperature	Room Temperature
Test Time (min)	5 minutes Cycles
Sample Size (mm)	30 x 30 x 5
Nozzle Diameter (mm)	4.5
Stand-off Distance (mm)	10

Table 3. Erosion Test Parameters

3. Results and discussion

3.1 Surface texture of coatings

Average roughness of different coatings is indicated in Table 4.

It is evident that the average roughness values of alumina coatings cast iron substrates vary between 3.5 and 5.5 μm and it is between 4.5 and 7.2 μm in case of ZrO_25CaO coatings. Top coat of test samples such as CI-S4, CI-S5, CI-S6 possess mounds of molten and unmolten particles contribute to the increase in roughness. Flowability of ZrO_25CaO is less compared to alumina and this contributes to the formation of mounds and affects the quality of the surface texture of the coating. Increase in porosity as well as the coating thickness enhances the roughness of top coat. Coating roughness also increases with enhanced coating thickness. Similar observations regarding to the effect of coating thickness on roughness were reported by O. Sarikaya [17].

Fig.2 shows the roughness profiles of few coating systems (CI-S2 and CI-S6) with average roughness (Ra) as the main parameter.

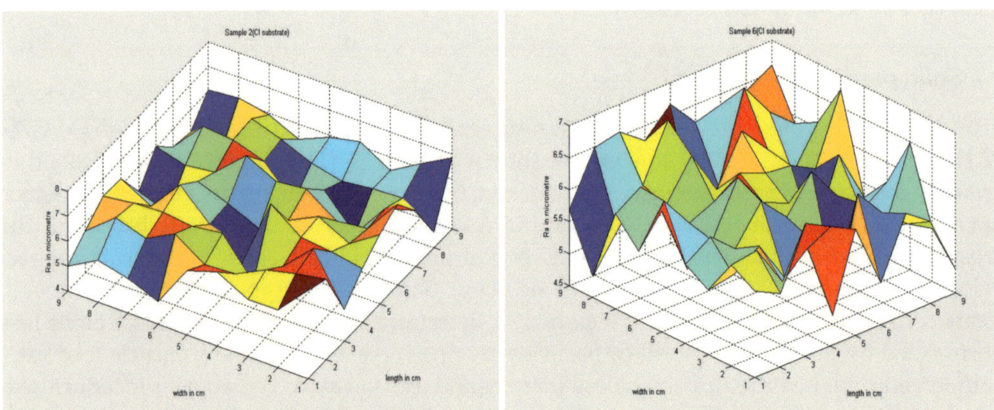

Fig. 2. Roughness profiles of coating systems

3.2 Morphology of coatings

Al_2O_3 coated test samples viz., CI-S1, CI-S2, CI-S3 are characterized by their disc shaped grains (Fig.3). These grains are found to be the flattened solidified droplets of the coating material. The molten particles are found to be distributed more or less evenly producing a smooth coating surface. Enlarged view of marked region of CI-S1 sample (Fig.3) indicates a network of microcracks. Cracks are also observed on the surface of flattened droplets. This may be possibly due to the presence of residual stresses introduced by thermal shocks resulted during the spraying process. Changing the thickness of top coat appears to have no significance on the microstructure as in the case with test samples CI-S2, CI-S3 (Fig.3).

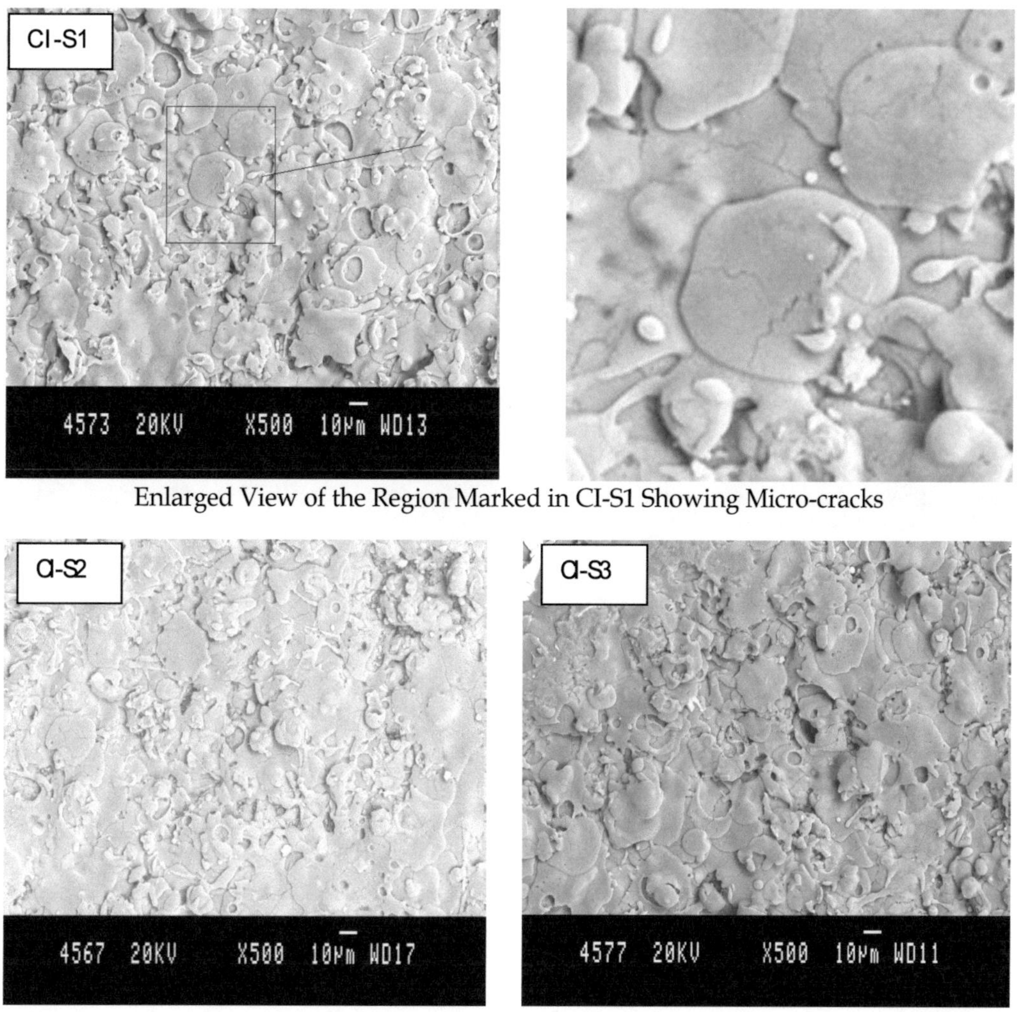

Fig. 3. Topology of Al_2O_3 Coatings Cast Iron

$ZrO_2 5CaO$ coated test samples such as CI-S4, CI-S5, CI-S6 exhibit a dense undulated structure (Fig.4). The enlarged view of marked region of CI-S6 sample indicates a network of microcracks (Fig. 4). The sizes of these microcracks appear to be slightly larger than that observed with Al_2O_3 coated test samples, possibly due to the large difference in the magnitude of thermal conductivity between the substrate and coated material. The thermal conductivity of $ZrO_2 5CaO$ is found between 2 to 4 $Wm^{-1}K^{-1}$, where as it varies from 33 to 37 $Wm^{-1}K^{-1}$ for alumina. Since the difference in the magnitude of thermal conductivity between cast iron (50 to 55 $Wm^{-1}K^{-1}$) substrate and Al_2O_3 coating is less, heat is transferred more or less effectively through the coating system, resulting lower level of thermal stresses which in turn producing smaller size microcracks. On the other hand, the difference in thermal conductivity of alumina and cast iron substrate and $ZrO_2 5CaO$ is large, higher level of thermal stresses will be developed resulting larger size microcracks. Further, the splats in the coatings are separated by inter-lamellar pores resulting from rapid solidification of the lamellae and very fine void are being formed due to incomplete inter-splat contact in and around un-melted particles.

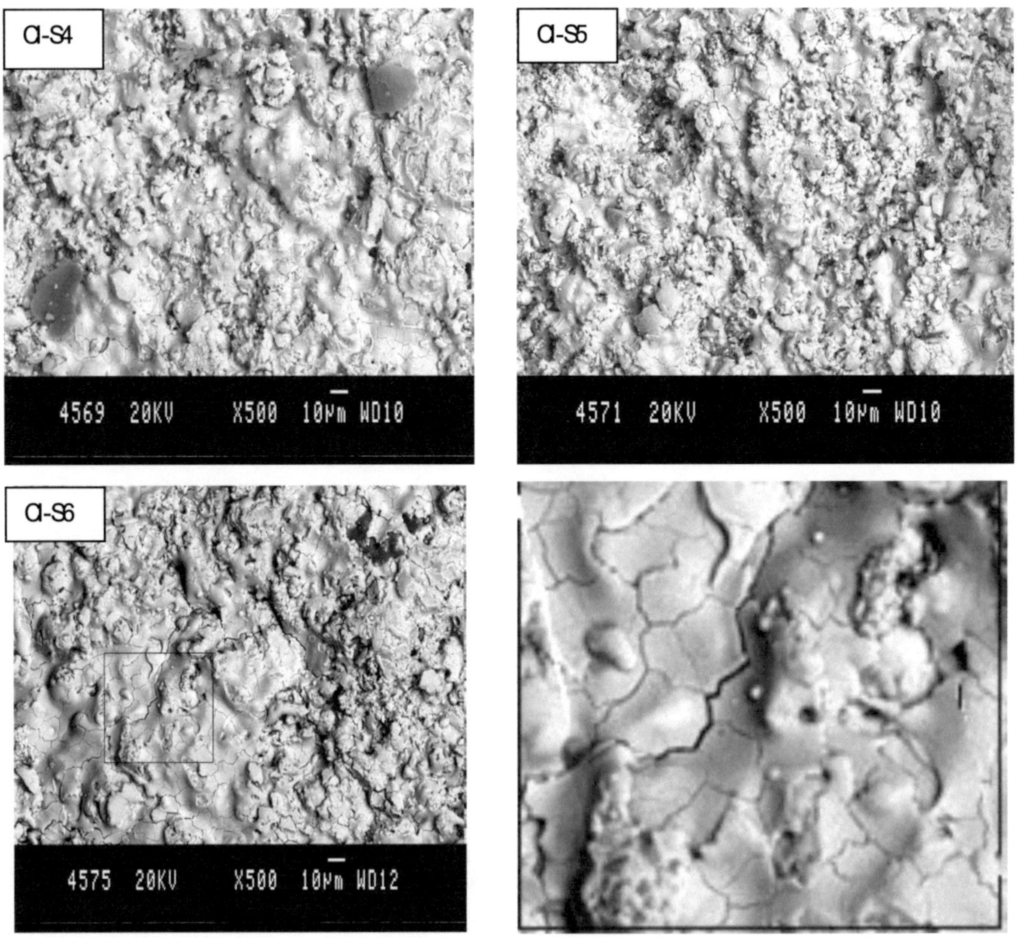

Enalarged View of the Region Marked in CI-S6 Showing Micro-cracks

Fig. 4. Topology of $ZrO_2 5CaO$ Coatings on Cast Iron Substrate

3.3 Coating thicknesses and porosity

It is observed that the variation in coating thickness (Fig.5) is about ±25 μm from the actual required thickness. This is attributable to the variations in speed of the gun during plasma spraying process. This variation can be minimized by applying Robotic Plasma spraying. Sample polishing technique is also believed to contribute to the variations in the thickness of the coating.

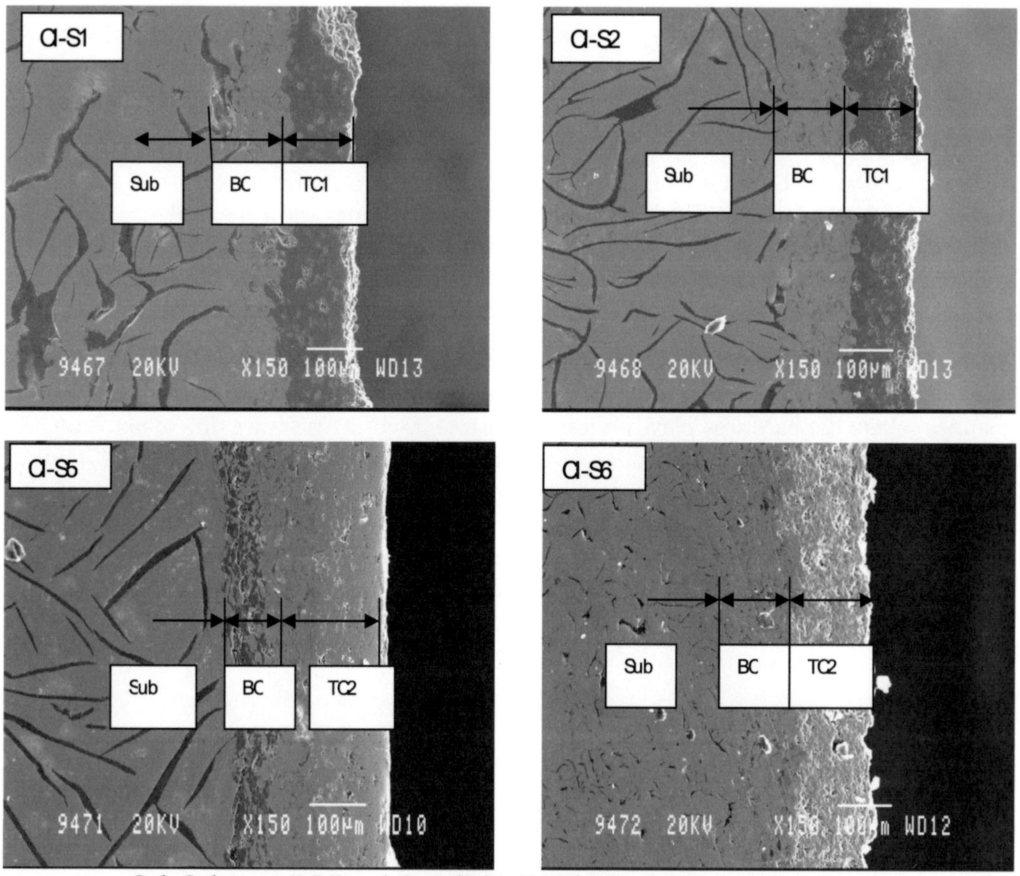

Sub-Substrate BC-Bond Coat (BC1+BC2/ BC3+BC2), TC1-Top Coat 1

Fig. 5. SEM Cross-sections of Al_2O_3 and ZrO_25CaO Coatings

The porosity of Al_2O_3 (CI-S1, CI-S2 and CI-S3) coatings are in the range of 5.7 to 6.4% in case of bond coat and it varies between 6.4 to 7.1% in case of top coat. ZrO_25CaO coated samples such as CI-S4, CI-S5, CI-S6 also shows pores. The porosity of these coatings varies between 6.3 and 6.8% in case of bond coat and 8.2 to 9.4% in case of top coat. Porosity is high, due to formation of rounded pores which are produced by unmelted particles, splats stacking faults and gas entrapment. Porosity of coatings is found to increase with increase in the thickness of top coat.

Porosity formation is due to residual stresses present in coatings. It is found to influence the tendency of the coating to de-bond from the substrate [18-23]. Residual stresses are introduced into the coatings when the molten particles are quenched upon impact causing a difference in the coefficients of thermal expansion between the coating and the substrate. Residual stresses are also indirectly affected by the pore structure since the stresses depend upon the elastic modulus and magnitude of strain as well. Porosity of ZrO_25CaO coatings is slightly higher than that of Al_2O_3 coatings which is due to the larger difference in thermal conductivity between the substrate and top coat in comparison with that found with Al_2O_3 coatings, for the reasons explained earlier.

In the present work, the porosity of coatings is less than that of coating systems reported by Portinha [24].

3.4 X-ray diffraction analysis of coatings

X-ray diffraction (XRD) patterns for the top surfaces of plasma sprayed Al_2O_3 and ZrO_25CaO coatings are shown in Fig.6. The XRD patterns of Al_2O_3 coatings show the presence of γ-Al_2O_3 as a principal phase and α-Al_2O_3 as minor phase. It shows that oxidation has occurred during spraying by converting hard phase of Al_2O_3 into soft phase of γ-Al_2O_3. ZrO_25CaO coatings possess tetragonal ZrO_2 as a principal phase and $CaZrO_3$ as minor phase.

XRD patterns suggest that Al_2O_3 particles are not completely transferred into soften γ-Al_2O_3 phase after the plasma spray process. This is a good result for tribological behavior of coatings where the hardness plays an important role in wear resistance due to abrasion and erosion. The hardness of Al_2O_3 coating is lower than that of bulk alumina (HV-2045) which is mainly due to the intrinsically lower hardness of γ-Al_2O_3 than α-Al_2O_3. The indentation response of a plasma sprayed material is governed not only by the intrinsic hardness of the material, but also by the lamellar microstructure, with splat boundaries giving off under load to facilitate the indenter accommodation [25].

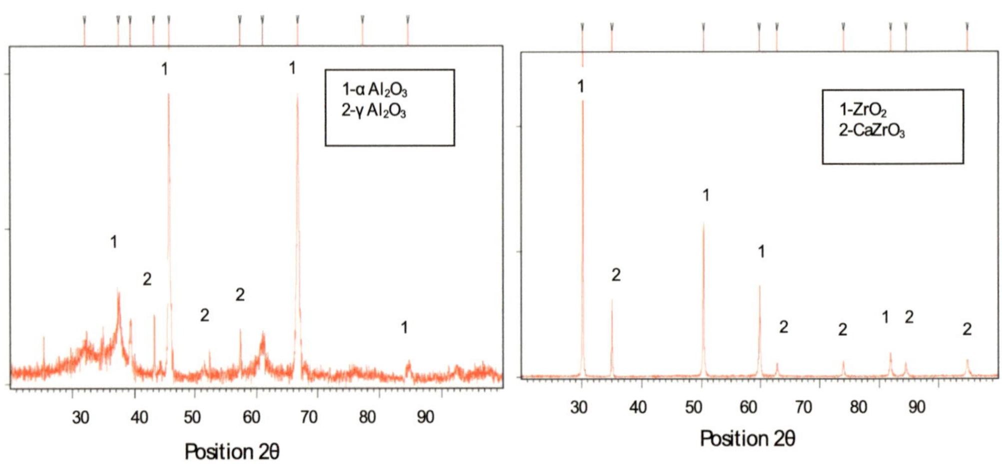

Fig. 6. X-Ray Diffraction Diagrams of Al_2O_3 and (b) ZrO_25CaO Coatings

Coating Type	Average Thickness (μm)		Average Porosity (%)			Avg. Surface Roughness (μm)		
	BC	TC1/TC2	BC1	BC2	TC1/TC2	BC1	BC2	TC1/TC2
CI-S1	80	105	5.7	6.2	6.5	4.8	5.8	3.5
CI-S2	100	275	6.1	6.8	6.9	5.3	5.7	5
CI-S3	95	360	6.4	7.0	7.1	5.1	5.5	5.5
CI-S4	100	100	6.3	----	8.2	5.4	----	6.2
CI-S5	85	270	6.5	----	9.0	5.6	----	6.8
CI-S6	85	380	6.4	----	9.4	5.5	-----	7.2

Table 4. Thicknesses, Porosity and Average Surface Roughness of Coatings

3.5 Adhesion test – Results

The location of coating failures during the test is described in Table.5.

Samples	CI-S1		CI-S2		CI-S3		CI-S4		CI-S5		CI-S6	
	Strength MPa	Failure Location	Strength MPa	Failure Location	Strength MPa	Failure Location	Strength MPa	Failure Location	Strength MPa	Failure Location	Strength MPa	Failure Location
1	20.4	BC3/S	24.8	BC3/S	30.8	BC3/S	37.5	BC3/S	39.8	BC3/S	42.8	BC3/S
2	18.2	BC3/S	20.4	BC3/S	27.4	BC3/S	34.2	BC3/S	34.5	BC3/TC2	46.7	BC3/S
3	22.0	BC3/S	18.8	BC2/TC1	27	BC3/S	32	BC3/TC2	37.6	BC3/S	54	Glue
4	15.8	BC3/S	20.8	BC3/S	29.2	BC3/S	35.4	BC3/S	38.2	BC3/S	45.2	BC3/S
Mean Strength MPa	19.1		21.2		28.7		35.9		38.87		44.2	

BC1/Substrate, BC3/Substrate=Adhesive Failure, BC1/BC2, BC1/TC2, BC2/TC1 and BC3/TC2 = Cohesive Failure, Glue=Failure with in Glue (Poor Test)

Table 5. Adhesion Strength and Failure Location of Coating Systems

The results indicate that the mean values of adhesion for test samples CI-S1to CI-S6. The bond strength is found to increase with the increase in the thickness of the top coat. Analysis of the chemical composition of CI substrate and that of the bond coat layer (BC3) for these samples indicate that the bond coating material consists of as high as 52% of Fe. This will influence the possibility of fusing Fe into the cast iron substrate since the time of exposing the substrate to the plasma spray gun is more in case of samples CI-S2, CI-S3, CI-S5, CI-S6. This probably explains the reason for high adhesion strength of samples coated on cast iron substrate. Fig.7 shows the fracture of samples at substrate/bond coat interface.

Fig. 7. Fractured Surfaces of Coating Systems after ASTM C633 Tensile Test- Bond Coat/ Substrate Interface

But this simple explanation did not represent the exact behavior of each distinctive coating system. A deep analysis can bring up much more information about the behavior of each individual coating system. It is observed further from Table 5, that the location of the coating failure is at the interface between bond coat and substrate. This is called as adhesion failure. It is seen that in the case of sample 4 of CI-S6, that the failure have occurred at 54 MPa along the glue line. It means that a higher value of adhesion strength could probably found for the above specified samples. Sample 3 of CI-S2, 3 of CI-S4 and 2 of CI-S5 have shown the occurrence of fractures at bond coat/ceramic or bond coat/cermet interfaces respectively. This is attributable to the defects at the bond coat/ceramic interface.

Stress-strain relations (Fig. 8) of the samples show two distinct regions. The first region is due to initial slipping of specimen from the fixtures of UTM. The second region is due to the elongation of specimen at bond coat/substrate interface. Neglecting the strain in first region, the strain percentage in the second region is found to vary between 5 to 6%. It clearly indicates that the specimens in tensile test are ductile in nature.

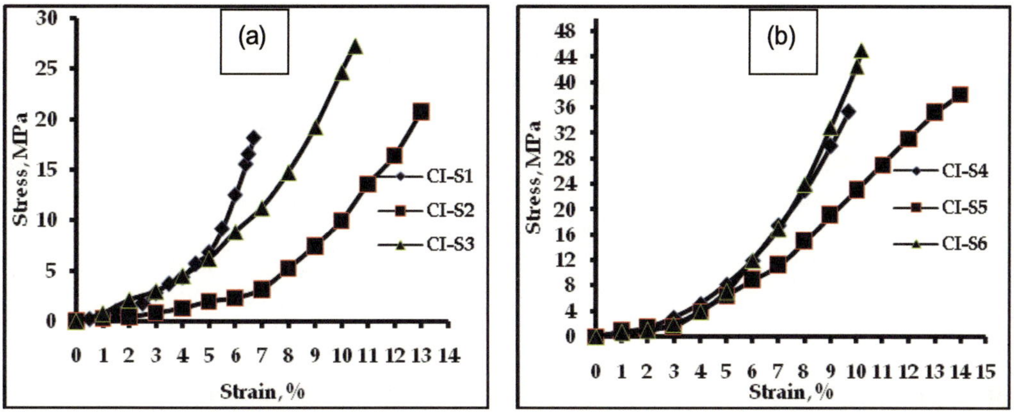

Fig. 8. Stress v/s Strain Diagram for (a)Samples 1, 1, 4 of CI-S1, CI-S2, CI-S3 (b) Samples 3, 3, 4 of CI-S4, CI-S5, CI-S6

3.5.1 Factors affecting adhesion strength of coatings

The amount of adhesion can be evaluated based on degree of coverage [26] of the remaining particles which are bonded after testing of bonding strength. Therefore an adhesion test which presents a singular and partial failure means that true adhesion must be evaluated from that area remaining on substrate after the test which is intact and did not detach or fail. This type of failure can be due to the problems associated with spraying process such as residual stresses, inter splats defects or related to test procedure such as sample alignment or traction speed. It is found that interlocking increases with an increase in the density of coat, the velocity of the impinging droplet and roughness of substrate surface and where as it decreases with the increase in the surface tension at substrate/droplet interface. Rebonding of partially melted particles and stress relaxation from local plastic deformation is found to influence the adhesion strength. On the other, in case of multilayered thermal barrier coatings, adhesion strength mainly depends on the proportion of bond coat, top coat and substrate material. Low bond strength is prevalent when there is a low surface roughness and low mechanical interlocking [27].

Hadad et. al [28] comparing adhesion tests found that interfacial toughness tends to increase with Ra for thin coatings (140 μm) and in their experiments, an opposite trend is seen for thicker coatings (330 μm). Since the crack propagation into a smooth interface is easier than into a rougher one, the interfacial toughness should increase with Ra and then they concluded that the residual stress effect would be dominant for thicker coatings. In the present work, all of the coating systems tested has similar thickness values as reported by Hadad et. al. and they could be considered as thick coatings. The highest roughness value is for CI-S6 (7.2 μm) that also presents one of the highest adhesion mean values (44.2 MPa).

Another issue to be observed is that the roughness just after bond coat application. From Table.4, it is evident that bond coat roughness increases with increasing the thickness. According to Khan et. al. [29] the adhesion of the coating increases with the increase of substrate roughness or bond coat surface roughness up to certain limits (about 5 μm) and then decreases. In case of CI-S2, CI-S3, CI-S5, CI-S6, the adhesion increases with increase in bond coat roughness. With the increase of bond coat roughness there is an increase in interfacial toughness due to high compressive stresses associated with high rough surfaces but further temperature and pressure from the spray process affect the residual stress profile and thus the interfacial toughness of the coatings.

Limarga et al. [27] have carried out investigations on multilayered thermal barrier coatings in which they have obtained adhesion strengths between 5 and 23 MPa depending on the proportion of bond coat, Al_2O_3 and ZrO_2 in the coating systems. In their tests the majority of failures are found to have occurred inside the ceramic layer. They have registered highest values of adhesion strength when the interfacial bond coat/ceramic failure has occurred. According to the authors, the very low bond strength exhibited by some coating systems is due to the low surface roughness of the sprayed ceramics, where as the mechanical interlocking is negligible. Further, they have found that the low surface roughness correspond to the small particle size of materials used in plasma spraying has affected the bond strength. Using the same analogy in the present work and by considering the top ceramic layer only, it can be observed from Table 4 that the highest roughness values are found with CI-S4, CI-S5 and CI-S6 with ZrO_25CaO top coat. These coating systems have highest values of adhesion. The grain size of ZrO_25CaO (-53+11μm) powder is greater than that of Al_2O_3 (-31+3.9μm) powder which is used as top coat in case of CI-S1, CI-S2, CI-S3 coating systems. This is partially in confirming with that observed by earlier investigator [29]. That is, higher the adhesion strength, higher would be grain size as well as surface roughness.

Lima and Trevisan [30] while working with graded TBCs have found that not only by increasing the thickness, coating adhesion can be decreased but also by increasing the number of coating layers for the same thickness. They have reported that increasing the number of layers has indicated a greater interruption time for the spraying of the subsequent layer due to the time required for making necessary arrangements. In the present work, the only difference with the tested coating systems is the greater number of passes required to deposit Al_2O_3 on CI-S2, CI-S3 and ZrO_25CaO on CI-S5, CI-S6 systems. 40 to 50 % higher number of passes as employed in this investigation implies that an increased number of ceramic interfaces as well as more homogeneous ceramic coating with thinner intermediate layers. This would ensure that the specimen would fracture only at ceramic interfaces in tensile adhesion strength giving higher magnitudes of cohesion.

3.6 Microhardness of coatings

Hardness of different layers of coating systems is indicated in Table.6 and their graphical representations are shown in Fig.9. It is evident that there is a marked difference in microhardness on different layers of coatings. The substrate average hardness varies from 328 to 355 HV, whereas the hardness of BC1, BC2 and TC1 in case of CI-S1, CI-S2 and CI-S3 samples varies from 140 to 160 HV, 130 to 140 HV and 1110 to 1190 HV respectively. In case of CI-S4, CI-S5, CI-S6 samples, the hardness of BC1 and TC2 varies from 145 to 160 HV and

780 to 820 HV respectively. It is also observed that for the first three series coating systems, the BC1 thickness is about 50μm and for the next three coatings it is about 100 μm. From this data it can be realized that the microhardness of BC1 decreases as its thickness increases. Similarly, the hardness of TC1 and TC2 decreases with the increase in their thicknesses. Further, the hardness of Al_2O_3 coatings is found to be more than that of ZrO_25CaO coatings. It is also evident that the micro hardness measurements exhibit a wide dispersion. Such dispersion in the microhardness values of the coatings is a typical characteristic of APS ceramic coatings clearly attributable to their microstructural heterogeneity [31].

Samples	Hardness $HV_{0.3}$			
	Substrate	BC1	BC2	TC1/TC2
CI-S1	355	150	135	1190
CI-S2	340	148	130	1158
CI-S3	330	140	140	1110
CI-S4	330	160	-----	820
CI-S5	342	145	-----	800
CI-S6	328	150	-----	780

Table 6. Hardness of Coating Systems

Hardness of the coating is a measure of the resistance to plastic deformation. It is widely recognized that the hardness increases with the increase in the density, i.e. by decreasing the number of pores and micro-cracks. Therefore the hardness of the top coat is a measure of the amount of sintering and integrity and can provide information about the temperature history of the top coat. Bond coat hardness has no effect on life of TBC.

Microhardness measurements of coatings have specific implications with regard to their basic science and technological applications. The effective hardness of a microvolume of a material depends on the cooling rate, phase structure, crack size and distribution, residual stress and strain of the local environment as well. Thus by examining the variation of microhardness within the coatings ensures avenues to understand the processing, structure and property relationships of coatings. Hardness tests may be related to the tensile adhesion tests since both these measurements rely on deformation under stress. Moreover, microhardness studies can give the variation of strength and the flaw distribution throughout the specimen, whereas the strength tests yield the strength of the weakest link of the system.

Portinha [32] has reported about the decrease in microhardness towards the surface in variable porosity samples and slightly increase in microhardness in case of samples with constant porosity. The decrease in microhardness for the graded samples is attributed to the increase in porosity along the cross section. Samples with constant deposition parameters have exhibited only marginal porosity towards the surface with the increase in the surface temperature during deposition process which also contributes to the enhanced hardness. The variation in microhardness within the given thickness of the coating is due to the variation in local structure which is attributable to the pores or lamellar boundaries. In the present investigation, it is observed that the porosity of Al_2O_3 and ZrO_25CaO increases with the increase in coating thickness of the samples. From the graphs (Fig. 9) it is seen that the hardness of top coat decreases with the increase of porosity. It shows that the coating systems used in the present investigation are in good agreement with the results reported by Portinha.

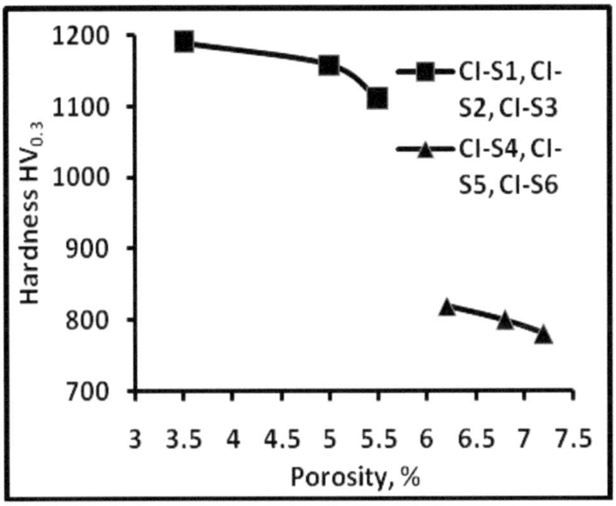

Fig. 9. Variation of hardness with porosity

Microhardness of the coatings increases with the decrease in their porosity. This can be explained based on the principles of microhardness measurements. During the indentation process, a complex elastic-plastic field is formed beneath the indentation. Porosity tends to reduce the effective area supporting the load and is detrimental to strength. When porosity or an equivalent defect is present in a sample, the load bearing area is reduced. It can be safely assumed that the defective region will yield first, thereby inducing strain concentration. However, voids are found to create a multiaxial stress state which can cause local strain concentrations in their vicinity. If all coating systems are considered together, it is obvious that there exists a general tendency that the microhardness decreases with increasing porosity.

3.7 Solid particle erosion test results

3.7.1 Variation of incremental mass loss and cumulative mass loss with time

Fig. 10 shows the photographs of eroded surfaces of coating systems. In erosion test, the samples were allowed to erode by the erodent until a steady state erosion rate was attained. The mass loss of coating systems after every 5 min test is shown in Fig. 11. It is observed that mass loss is suddenly increases in the first 5min of test. After this, mass loss gradually decreases and attains a steady state. From the graphs of cumulative mass loss with time (Fig. 12), it is found that there are two distinct regions under different angles of impact such as 15, 45 and 90^0 for all coating systems on cast iron substrate. The first region is belongs to erosion of top coat ceramic layer. During this period the slope of erosion mass loss is high and it occurs for a period of 10 to 15 minutes from the starting of the experiment. At normal impact this slope is little higher compared to other angles. The main reason for higher slope during this period is removal of top coat ceramic material. Ceramic layer is made by brittle material which under goes brittle fracture and especially the rate of brittle fracture is high at normal angles of impact. The second region of the graph is occurred due to eroding of cermet and metallic bond layers in case of samples CI-S1, CI-S2 and CI-S3 respectively and only metallic layer in case samples CI-S4, CI-S5 and CI-S6. In this region the slope of erosion mass loss is small compared to that of first region. This region is available till the coating system reaches a steady state erosion condition.

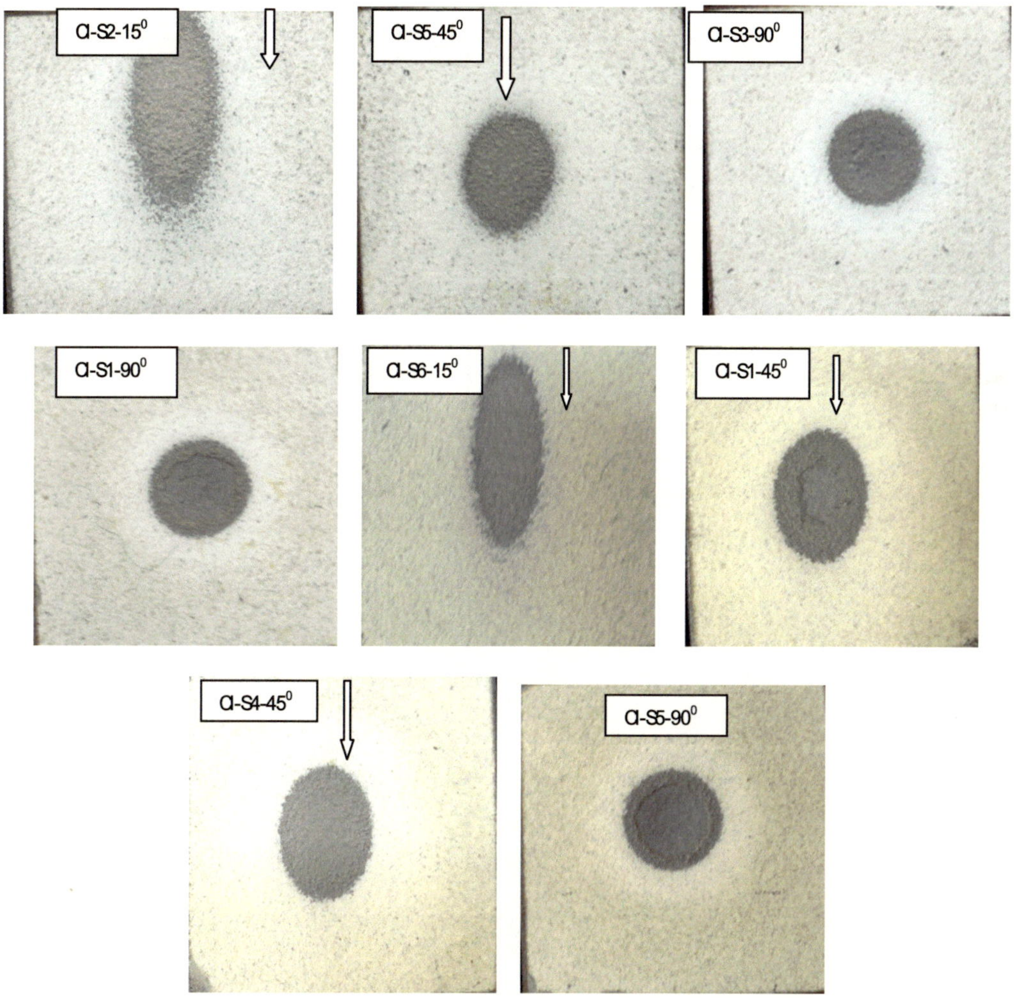

Fig. 10. Photographs of Eroded Surfaces of Al_2O_3 and ZrO_25CaO Coatings on Cast Iron Substrate 15, 45 and 90^0 Angles of Impact (Arrow Mark Indicates the Direction of Silica Sand Jet)

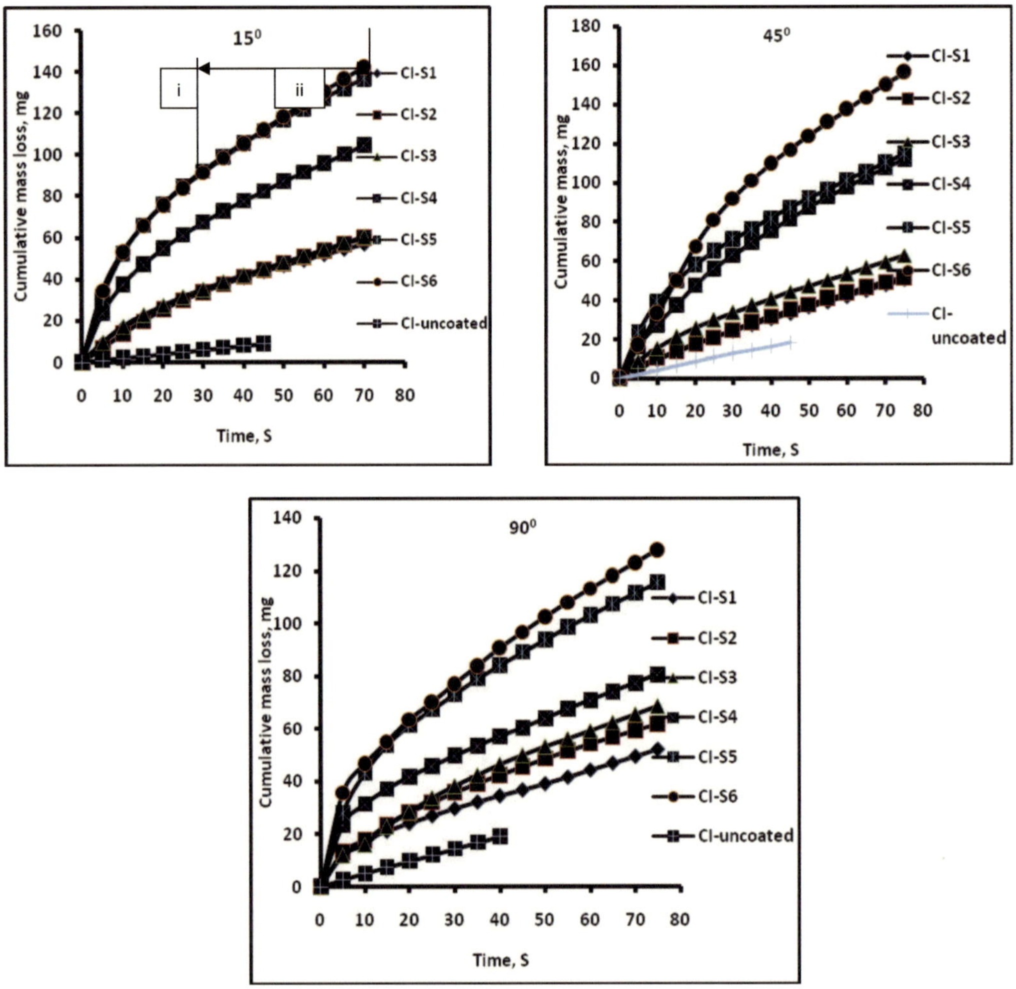

Fig. 11. Cumulative Mass Loss Plots as Function of Time for Alumina and ZrO_25CaO Coatings on Cast Iron Substrates at 15, 45 and 90^0 Angle of Impacts

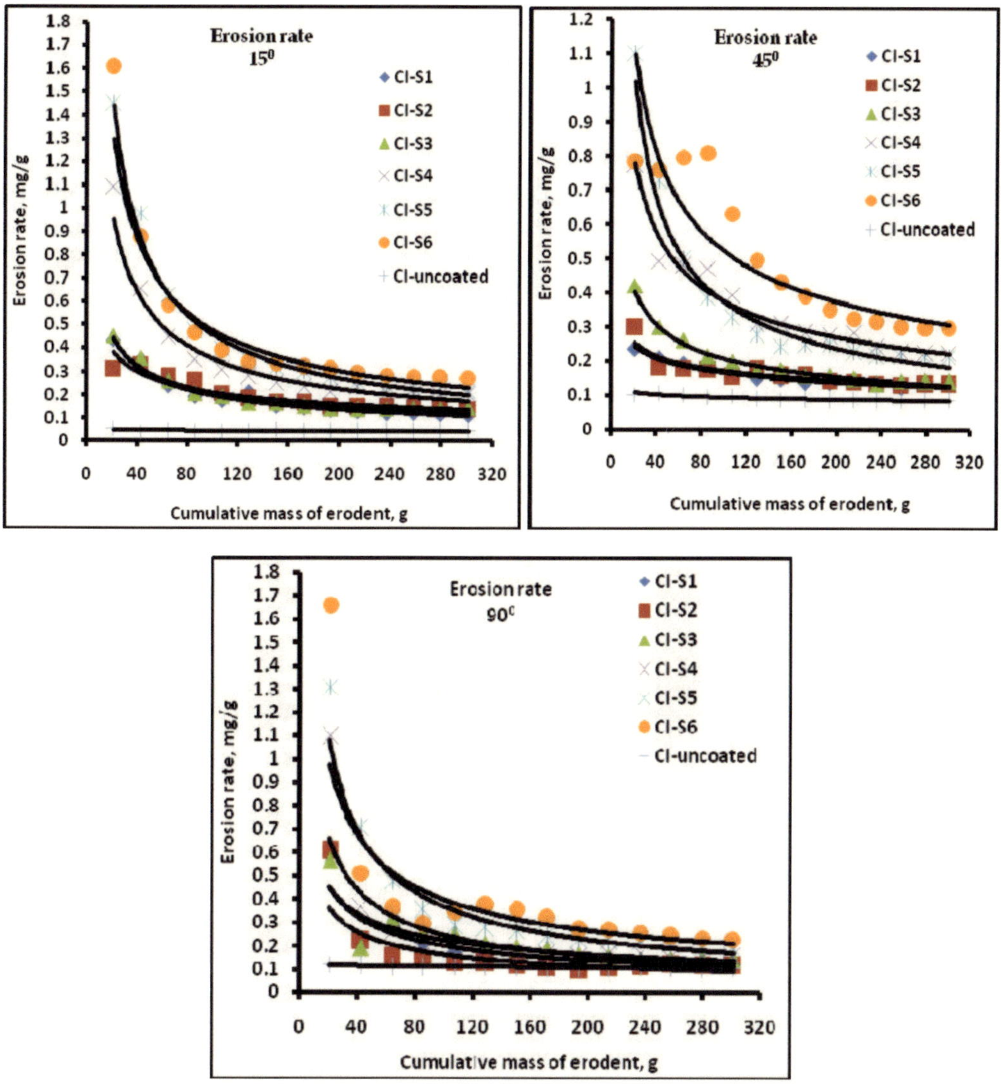

Fig. 12. Erosion Rate Plots as Function of Mass of Erodent for Alumina and ZrO_25CaO Coatings at 15, 45 and 90^0 Angle of Impacts

Cumulative mass loss with time graphs (Fig. 12) shows that the mass loss varies with angle of impinging. According to some engineering model developed so far, for erosive wear, it has been established that the angle at which the stream of solid particles impinges the surface influences the rate at which the material is removed from the surface. This angle determines the relative magnitude of the two velocity components of the impact namely, the component normal to the surface and parallel to the surface. The normal velocity component will determine how long the impact will last and the load. The product of this contact time and the tangential velocity component determine the amount of sliding that takes place. The tangential velocity component also provides a shear loading to the surface, which is in addition to the normal load that the normal velocity component causes. Hence as this angle changes the amount of sliding that takes place also change as does the nature and magnitude of the stress system. Both of these aspects influence the way a material wears. From the graphs (Fig. 12) it can be realized that the erosion mass loss is more at 45^0 angle of impinging. In most of the materials, solid particle erosion behaviour can be categorized as being either brittle or ductile in nature [7]. The major differentiating characteristic of the two types of mechanism is the dependence of erosion rate on impact angle i.e. the angle between the moving erodent particle and the material surface [8]. There is a general agreement that maximum erosion occurs at a low angle (about 30^0) for ductile material and at 90^0 for brittle material. In this investigation, coating systems possess multilayer comprising a ceramic top coat and two intermediate metal and cermet bond coats in case of CI-S1, CI-S2 and CI-S3 and only metallic bond coat in case of CI-S4, CI-S5 and CI-S6. Since the erosion loss is more at 45^0 angle of impact, it can be realized that the coating systems behave neither as purely ductile (where the maximum loss is expected around 15-30^0) nor purely brittle (maximum loss is expected at 90^0) and has a composite behavior. However, the extent of erosion is found to be strongly dependent on impact angle.

3.7.2 SEM micrographs of eroded surfaces

Fig. 13 shows the surface micrographs of worn out region of the coatings at 15, 45 and 90^0 impact angles. At lower impact angles (15^0, 45^0), there are evidences of grooves and ridges (indicated by1) as the material ahead of the erodent is removed by cutting action. Also material removal may occur from the ridges around the grooves by repeated impacts of erodent. The groove formation may predominantly occur within the softer binder region and this may also result in under cutting of the grains, which may get loosened and eventually pulled out. The pull-out of the grains can also be seen in some regions. At 90^0 impact angle, indentation impressions due to impingement of erodent on the surface are clearly seen. In ductile erosion one of the common mechanisms is the removal of material from the lips that are formed around the impact craters due to strain localization. The material removal may occur from the displaced material forming lips around the indentations as a result of repeated impacts of erodent. Thus the surface morphology shown in Fig. 13 indicates that the predominant mechanisms are grooving of binder phase, cratering and particle pull-out that are prevalent in the coatings. These mechanisms are responsible for composite erosion mode. The appearance of eroded surfaces also indicates that cracks tend to follow a variety of weak sites to produce wear debris. From the morphology of as-sprayed coatings (Fig. 3 and 4) it is observed that all coating systems possess cracks. Linkage of these pre-existing cracks with indentation cracks could have aided the material removal process. The thermal cracks normal to the surface, the interfaces between adjacent layers of splats can be identified as structural weakness for both Al_2O_3 and ZrO_25CaO coatings, as described in [33, 34]. The erosion of plasma sprayed coating of lamellar structure occurs through spalling of surface lamella resulting from impact of abrasives. Accordingly the erosion of coating is controlled by the crack propagation along the interface, i.e., the interface bonding between lamellae. Therefore the erosion of the coatings will be dominated by interface bonding condition and lamellar thickness

The eroded surfaces showed evidence of plastic deformation. Ploughing of the surface by the impinging, sharp silica particles, resulting in groove formation, is evident at all angles of impact, but become more pronounced at lower angles as seen from Fig.13. It is apparent that repeated impacts by the hard particles resulted in highly deformed platelets which are removed by subsequent impacts.

It is well established that in bulk brittle materials such as ceramics, the ratio of particle hardness to the target hardness (Hp/Ht) has a controlling influence in the erosion mechanisms [35, 36]. When this ratio is greater than 1, the wear mechanism essentially involves indentation-induced fracture. At lower ratios cracking is suppressed and the material removal occurs by less severe micro-chipping mechanisms. In the present work, the hardness of erodent (silica) is obtained as 12000 HV. The hardness of Al_2O_3 and ZrO_25CaO top coats are 1120-1180 and 830-850 HV respectively. Since Hp/Ht is higher than 1, top coats undergo splat ejection and indentation-induced material removal mechanism. Kingswell et al. [37] have proposed three basic mechanisms of material removal during erosion of thermal spray coatings depending on their microstructure. In poorly bonded thermal sprayed structures, material loss occurs by splat boundary fracture. As splat cohesion is improved, the dominant material removal process becomes splat fracture, micro-chipping and ploughing. Evidently, alumina coatings have a microstructure superior than ZrO_25CaO coatings. Due to this alumina coatings have greater resistance to erosion than ZrO_25CaO coatings.

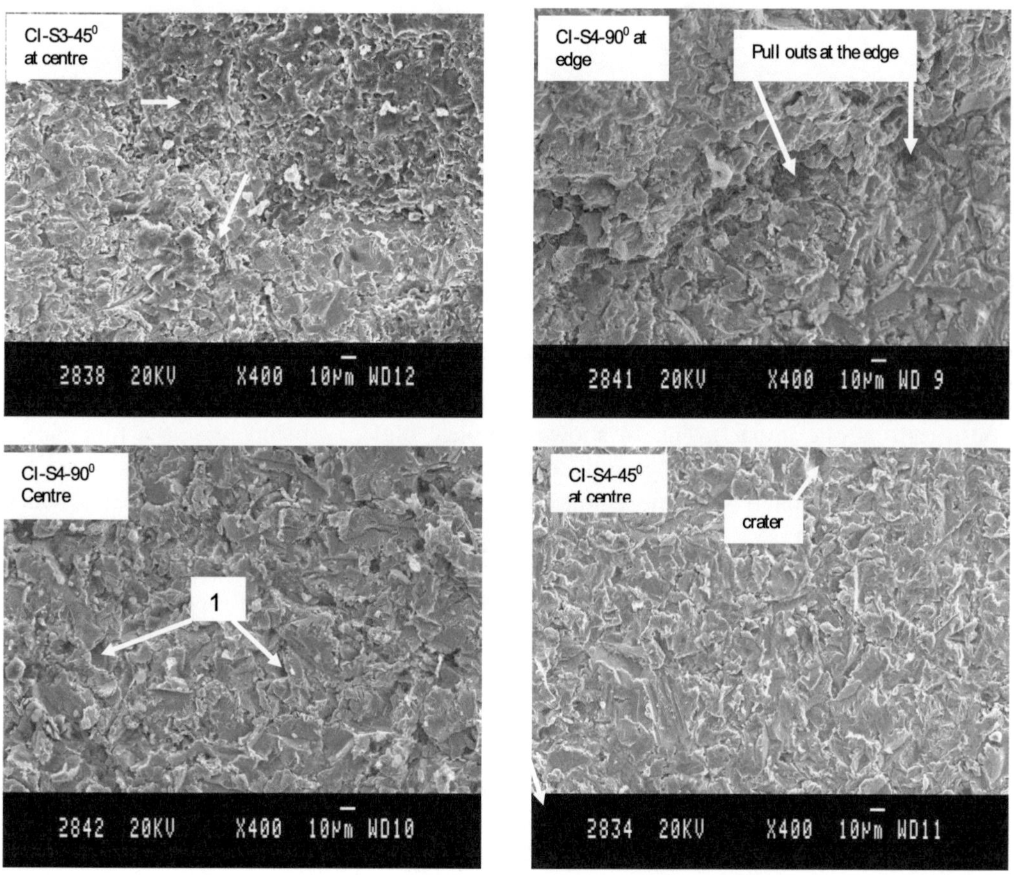

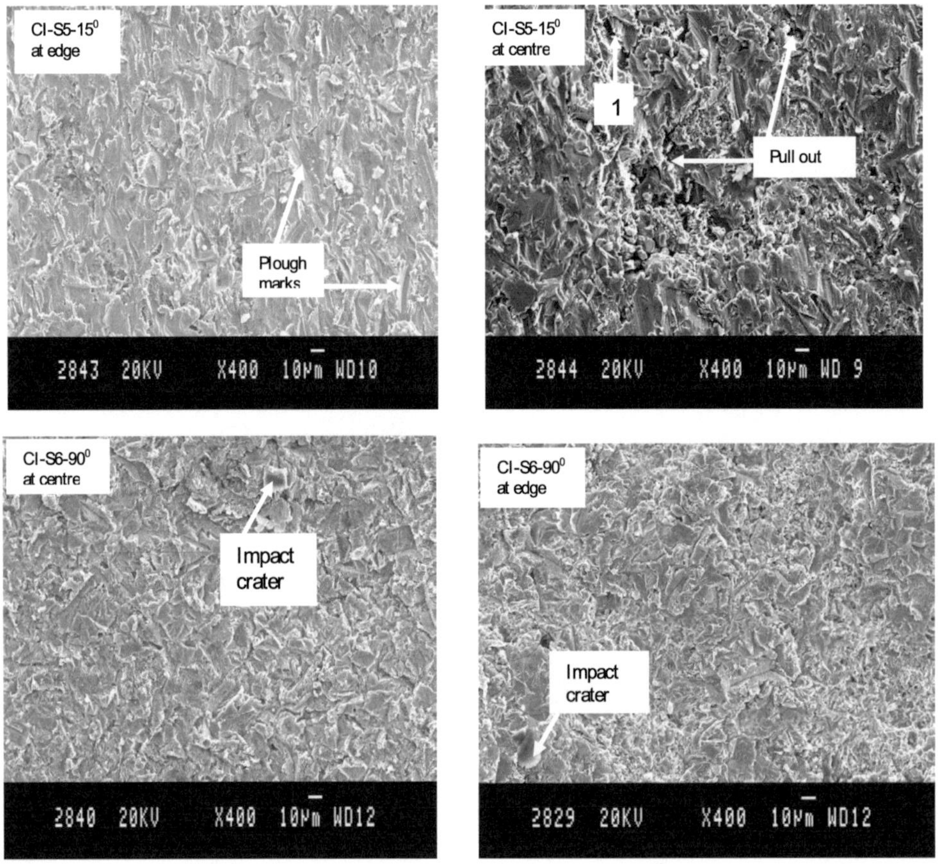

Fig. 13. SEM Micrographs of Eroded Surface of Alumina and $ZrO_2 5CaO$ at Different Angles of Impact.

3.7.3 Incremental erosion rate and volume erosion of coatings

Erosion rate is the ratio of incremental mass loss to mass of erodent per impact (5min. test is considered as one impact). Fig. 14 shows a typical plot of erosion rate as a function of cumulative erodent mass impinging the coating at 15, 45 and 90^0 impact angles. From the graphs it is found that a transient regime has occurred in the erosion process, during which incremental erosion rates decrease monotonically down to a constant steady state value. The starting period of erosion process is also called as incubation period. The decrease in the erosion rate with erosion time or cumulative erodent mass has been reported before [38]. They postulated that, for some brittle materials, initially the target surface is thoroughly cracked with minimum material loss. Then, significant chipping occurs, which leads to a maximum erosion rate. Further particle impact cracking proceeds, with less material removal.

Later, Levy [38] proposed that incremental erosion rate curves of brittle materials start with a high rate at the first measurable amount of erosion and that it then decrease to a much lower steady state value. Another important factor for high initial erosion rate is high surface roughness, where protrusions are easily knocked out from as-sprayed surface. Some insight on the reasons for the solid particle erosion transient as observed in this work can come from the current modeling of brittle erosion. According to it, debris is created due to lateral cracking and intersection between various crack types. The size of these cracks varies with load, or equivalently, impact energy. If one starts a solid particle erosion experiment with a target that has a cracking structure with dimensions lower than expected for the impact energy to be used, the incremental erosion rate should increase as the cracking dimensions increase upto a steady state. If, on the other hand, the cracking dimensions and density are higher than what would be imposed by the experiment impact energy, then the erosion rate should start high and decrease to a steady state value, as the cracking dimensions and density decrease. In the case of plasma sprayed coatings, it is possible that the near surface coating has a defect density higher than the bulk coating. With higher crack density the near surface coating toughness decreases and so does hardness, which according to equation $W_E \sim (C_r^2 h) \alpha (1/K^n H^m)$ [39] where W_E , volume loss per impact, C_r , lateral crack size, K and H, coating toughness and hardness, m and n are constants, should determine a higher erosion rate than the bulk coating. Also, since solid particle impact can promote significant surface heating, it is possible that crack closing happens during erosion. The steady state erosion rate is achieved when bond layer of coating systems is exposed to erodent. The steady state erosion is almost same for the systems CI-S1, CI-S2 and CI-S3 but it is different for CI-S4, CI-S5 and CI-S6 and increases with increasing of top coat thickness. The average mass loss of the coatings under steady state erosion rate conditions is taken for comparing the erosion of coatings.

The steady state volume loss of the coatings as a result of erosion at different angles of impact of the erodent is shown in Fig. 14. From this, it is observed that the volume loss is more at 45^0 angle of impact. The volume erosion loss of cast iron substrate increases with increase of angle of impact showing that the erosion behaviour as brittle. The volume erosion loss of these substrates is less than that of all coating systems. The deference in deformation in uncoated substrates and coating systems can be rationalized based on the deformation response in amorphous and crystalline materials. It is known that amorphous material is prone to shear band formation [40]. Since erosion conditions involve relatively high strain rates, they are quite favorable for shear band formation. The amorphous binder in the plasma sprayed coating systems is expected to form shear bands more easily leading to higher erosion. On the other hand, the deformation in crystalline metal substrates involves strain-hardening leading to higher energy absorption resulting in lower erosion. Cast iron substrate erodes more at 90^0. This clearly shows that cast iron follows brittle erosion behaviour. Again, it is found that volume erosion loss of alumina coatings is more than that of ZrO_25CaO coatings. The volume erosion loss of different coatings at 45^0 impact is 1.203, 1.23 and 1.12 $x10^{-3}cm^3$ for CI-S1, CI-S2 and CI-S3 (alumina coatings) 0.952, 0.9208 and 0.754 $x10^{-3}cm^3$ for CI-S4, CI-S5 and CI-S6 (ZrO_25CaO coatings). Although the cumulative mass loss of ZrO_25CaO coatings is more than that of alumina coatings, the volume erosion loss of these coatings is higher. This is mainly due to higher composite density of ZrO_25CaO coatings (composite density of ZrO_25CaO coatings is about 6.3 to 6.96g cm^{-3} where as density of alumina coatings are about 2.4 to 2.7 g cm^{-3}).

3.7.4 Effect of coating hardness on erosion rate

It is well understood that the erosion rates are affected by various factors [4, 41-44]. These factors can be broadly classified into three types: impingement variables describing the particle flow, particle variables, and material variables. The primary impingement variables are particle velocity, angle of incidence, flux (particle concentration) and target temperature. Particle variables include particle shape, size, hardness, and friability (ease of fracture). Material variables include all the material properties, such as hardness, work hardening behaviour, and microstructure

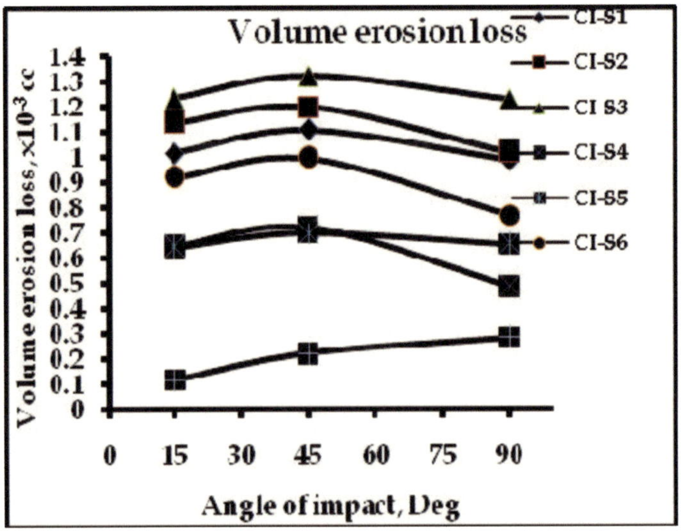

Fig. 14. Volume Erosion Plots as Function of Angle of Impact for Alumina and $ZrO_2$5CaO Coatings

Hardness is one of the most effective factors in predicting erosion behaviour and in modeling erosion processes. The best correlation of hardness with erosion is observed for hardness measured on bulk materials [45]. However, few data concerning ceramic coatings are available till now. Here the effect of hardness of coating systems on erosion rate is studied using Vickers hardness. The relationship between erosion rate and hardness of coatings are shown in Fig. 15. It can be seen that a good correlation of hardness with erosion rate is observed irrespective of the type of ceramic. The higher erosion rate is observed at lower hardness of a ceramic coating.

3.7.5 Effect of coating porosity on erosion rate

The erosion rate of plasma sprayed coatings depends on so many parameters like hardness, rupture strength, etc. However, despite the fact that the coatings have very different mechanical properties at the same porosity content, it is the porosity that dictates the erosion behaviour. From the Fig. 16 it is observed that the erosion rate increases with increase of porosity. This result shows that there must be a strong microstructural feature to be incorporated in erosion models. Porosity is definitely one very important feature, which influences erosion in three ways. Firstly, it decreases the materials strength against plastic deformation or chipping, since the material at the edge of a void lacks mechanical support. Secondly, the concave surface inside a void that is not under the shadow of some void edge will see an impinging particle at an angle higher than the average target surface to impact angle. This will be detrimental for brittle materials and beneficial for ductile ones. Finally pores can act as stress concentrators and decrease the load bearing surface.

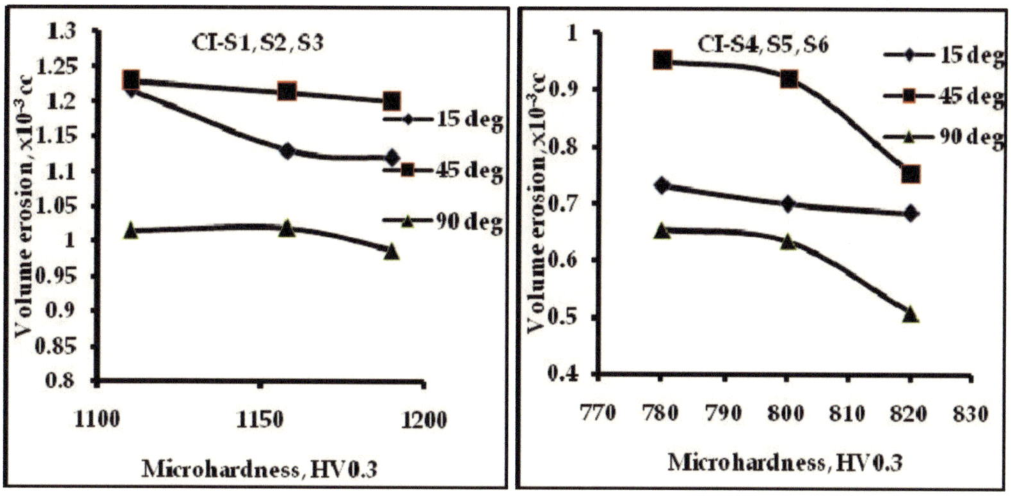

Fig. 15. Effect of Hardness on Volume Erosion of Alumina and ZrO_25CaO Coatings

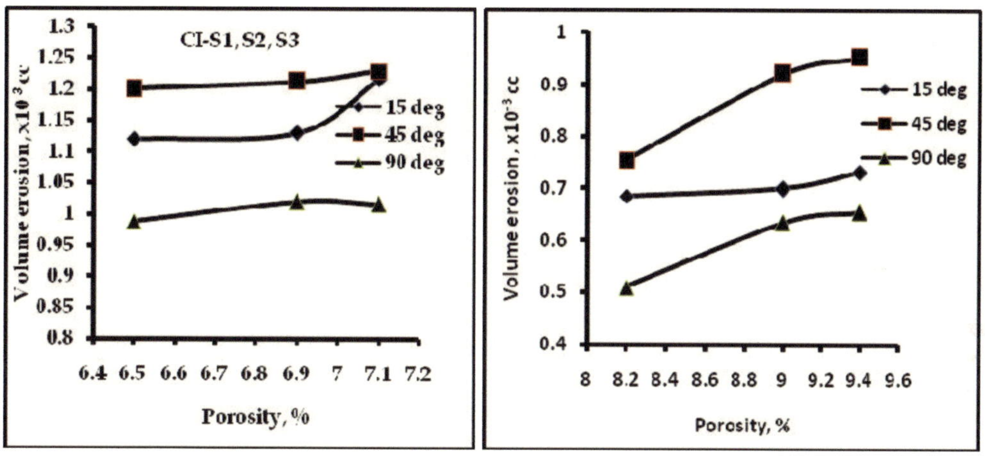

Fig. 16. Effect of Porosity on Volume Erosion of Alumina and ZrO_25CaO Coatings

4. Conclusions

a. From the study of SEM morphology of the coatings it can be concluded that thermal stresses developed during spraying were main causes for presence of micro-cracks in the coatings.
b. From the SEM cross-sections of coating systems it was observed that there is a variation in coating thickness. It is mainly due to variation in speed of the plasma gun during spraying. This variation can be minimized by applying Robot Plasma Spraying technique.
c. For adhesion measurement, ASTM C633 tensile adhesion test was used and proved to be effective for general level of comparison. From the obtained results it can be said that the main failure location was in the bond coat/ceramic interface corresponding to the

lowest adhesion values. For the highest adhesion values, coating failure was mainly located in the bond coat/substrate interface. The highest value of adhesion strength was obtained for ZrO_25CaO coatings.

d. By comparing hardness values it can be concluded that Al_2O_3 coating is harder than ZrO_25CaO coatings. It can be also concluded that microhardness decreases with increase in coating thickness and porosity.

e. It was found that erosion of coating systems occurred through spalling of lamella exposed on coating surface resulting from cracking along the lamellar interface. The material removal may occur from the displaced material forming lips around the indentations as a result of repeated impact of erodent. Erosion wear was more at 45^0 angle of impact showing that erosion behaviour as a composite ductile-brittle. Porosity influences erosion in three ways. Firstly, it decreases the materials strength against plastic deformation or chipping, since the material at the edge of a void lacks mechanical support. Secondly, the concave surface inside a void that is not under the shadow of some void edge will see an impinging particle at an angle higher than the average target surface to impact angle. This will be detrimental for brittle materials and beneficial for ductile ones. Finally pores can act as stress concentrators and decrease the load bearing surface

5. References

[1] Ramesh. C. S, Seshadr. S. K and Iyer, K. J. L, 1991, *A survey of aspects of wear of metals*, Indian Journal of Technology, Vol. 29, pp 179-185.

[2] Sundararajan. G, *The solid particle erosion of metallic materials: The rationalization of the influence of material variables*, Wear, Vol. 186-187, 1995, pp 129-144.

[3] Sundararajan. T. and Roy. M. *Solid particle erosion behavior of metallic materials at room and elevated temperatures*, Tribology International, Vol. 30, 1997, pp 339-359.

[4] Davis. J. R., *Surface engineering for corrosion and wear resistance*, ASM International. 2001.

[5] Fagoaga, I, Viviente, J. L, Gavin, P, Bronte. J. M, Garcia. J and Tagle. J. A, *Multilayer coatings by continuous detonation System spray technique*,Thin Solid Films, Vol. 317, 1998, pp. 259-265.

[6] I. M Hutchings, *Tribology: Friction and wear of engineering materials*, Metallurgy and Material Science Series, Edward Arnold Publishing, England 1992.

[7] G. P. Tilly, Erosion caused by impact of solid particles, Treatise on materials Science technology, 13(1979) 287-319.

[8] I. M. Hutchings, *Recent advances in the understanding of solid particle erosion*, Mecanique Materiaux, 365, 1980, pp 185- 192.

[9] J. G. Murphy, H. W. King, M. L. Taylor, *Pariculate erosion of zirconia-alumina plasma sprayed coatings I: Ceramic coating erosion mechanism*, The Canadian Ceramic Society, 56, 1987, pp 26-32.

[10] R. Kingswell, K. T. Scott, S. Bull, *Erosion behaviour of plasma sprayed alumina coatings*, 2nd Plasma Technik Symposium, 1991, Lucerne, Switzerland, pp 367-377.

[11] X. X. Zhang, I. M. Hutchings, T. W. Clyne, *The effect of deposition conditions on the Erosive Wear resistance of plasma sprayed alumina coatings*, 2nd European Conference on Advanced Materials and Processes, 1991, Euromat 91.

[12] B. R. Lawn, M. V. Swain, *Microfracture Beneath Point Indentations in Brittle Solids*, J. Mat. Sci., 10, 1975, pp 113-122.

[13] B. R. Lawn, D. B. Marshall, *Hardness, Toughness and Brittleness: An Indentation Analysis*, J. Am. Ceram. Soc., 62, 1979, pp 347-350.

[14] B. R. Lawn, R. Wilshaw, Review, Indentation *Fracture: Principles and Applications*, J. Mat. Sci., 10, 1975, pp 1049-1091.

[15] W.P. Dong, P.J. Sullivan, K.J. Stout, *Comprehensive study of parameters for characterising 3D surface topography II: statistical properties of parameter variation*, Wear 167, 1993, pp 9–21.

[16] W. P. Dong, P. J. Sullivan, K.J. Stout, *Comprehensive study of parameters for characterising 3D surface topography III: parameters for characterising amplitude and some functional properties & IV: parameters for characterising spatial and hybrid properties*, Wear 178, 1994, pp 29–60.

[17] O. Sarikaya, *Effect of some parameters on microstructure and hardness of alumina coatings prepared by the air plasma spraying process*, Surf. Coat. & Technol., 190, 2005, pp 388-393.

[18] P. M. Lonardo, H. Trumpold, L. De Chiffre, *Progress in 3D surface microtopography Characterization*, Ann. CIRP 42, 1996, pp 589–598.

[19] R. Groppetti, N. Senin, *A contribution to the development of three-dimensional nano and micro-topography measurement and analysis techniques and systems, in: Eleventh International Metrology Congress*, Toulon, France, 2003.

[20] S. Jahamnir, *Friction and Wear of Ceramics*, Dekker, New York, 1994.

[21] R. Divakar, P. J. Blau, *Wear testing of advanced materials*, ASTM STP 1167, ASTM, 1992.

[22] T. W. Clyne, C. J. Humphreys, *Improvements in Plasma Sprayed Thermal Barrier Coatings for Use in Advanced Gas Turbines*, Department of Materials Science & Metallurgy, University of Cambridge, Pembroke Street, Cambridge CB2 3QZ.

[23] Thompson, J. A., Tsui, Y. C., Reed, R. C., Rickerby, D. S. and Clyne, T. W, *Creep of Plasma Sprayed CoNiCrAlY and NiCrAlY Bond Coats and its Effects on Residual Stresses During Thermal Cycling of Thermal Barrier Coating Systems in High Temperature Surface Engineering*, Nicholls, J. and Rickerby, D.S. (ed.), 2000, IoM, Edinburgh, pp 199-212.

[24] Y. Kubo , Satrou Maezono, Koji Ogura, Toru Iwaio, *Pre-treatment on metal surface of plasma spray with cathode spots of low pressure*, Surf., Coat., Technol., 200, 2005, pp 1168-1172.

[25] J. Alcala, F. Gaudette, S. Suresh, S. Sampath, *Instrumented spherical micro-indentation of plasma sprayed coatings*, Mater. Sci. Eng. A, 369, 2004, pp 124-137.

[26] C. R. C. Lima, J. M. Guilemany, *Adhesion improvements of Thermal Barrier Coatings with HVOF thermally sprayed bond coats*, Surf. Coat. & Technol, 201, 2007 pp 4694-4701.

[27] A. M. Limarga, S. Widjaja, T. H. Yip, Surf. Coat. & Technol. 197, 2005, pp-93.

[28] M. Hadad, G. Marot, J. Lesage, J. Michler, S. Siegmann, in: E. Lugscheider (Ed.), *Thermal Spray Connects: Explore its Surface Potential*, Proceedings of the International Thermal Spray Conference ITSC, Basel, Switzerland, ASM International/DVS, Dusseldorf, Germany, 2005, pp. 759.

[29] A. N. Khan, J. Lu, H. Liao, Surf. Coat. & Technol., 168, 2003 pp 291.

[30] C. R. C. Lima, R. E. Trevisan, J. Therm. Spray Technol. 6(2), 1997, pp 199.

[31] P. P. Psyllaki, M. Jeandin, D. I. Pantelis, *Microstructure and wear mechanism of thermally sprayed alumina coatings*, Materials Letters, 47, 2000, pp 77-82.

[32] A. Portinha, V. Teixeira, J. Carneiro, J. Martins, M.F. Costa, R.Vassen, D. Stoever, *Characterization of thermal barrier coatings with a gradient in porosity*, Surf. Coat. Technol., 195, 2005, pp 245-251.

[33] S. F. Wayne, S. Sampath, *Structure/property relationships in sintered and thermally sprayed WC-Co*, J. Thermal Spray Technol., 1, 1992, pp-307-315.

[34] H. M. Hawthorne, L. C. Erickson, D. Ross, H. Tai, T. Troczynsky, *The microstructural dependence of wear and indentation behaviour of some plasma sprayed alumina coatings*, Wear 203-204, 1997, pp 271-279.

[35] J. L. Routbort, R. O. Scattergood, *Erosion of ceramic materials, Key engineering materials*, Vol.71, Trans. Tech. Publications, Switzerland, 1992, pp 23-50.

[36] S. M. Wierderhorn, B. J. Hockey, J. Mater., Sci. 18, 1983, pp 766-780.

[37] R. Kingswell, D. S. Rickerby, S. J. Bull, K. T. Scott, Thin solid films 198(1991) pp 139-148

[38] Levy .A.V, *The erosion-corrosion behaviour of protective coatings*, Surf. Coat. & Technol., 36, 1998, pp 387-406.

[39] Ives .L. K, Ruff .A. W, Wear 46, 1978, pp 149-162.

[40] A. L. Greer, K.L. Rutherford, I. M. Hutchings, Int. Mat. Rev. 47(2), 2002, pp 87.

[41] Kosel, T. H., *Solid particle erosion, in: Friction," in: Lubrication and Wear Technology*, ASM Handbook, Vol. 18, 1992, pp. 199-213.

[42] Levy. A.V, *Solid Particle Erosion and Erosion-Corrosion of Materials*, ASM International, Materials Park, OH 44073-0002, U.S.A. 1995.

[43] Sundararajan. G., *The effect of temperature on solid particle erosion*," Wear, Vol. 98, 1984, pp 141-149.

[44] Tilly. G. P., *A Two Stage Mechanism of Ductile Erosion*, Wear, Vol. 23, 1973, pp 87-96.

[45] M. Matsumura, Y. Oka, R. Ebara, T. Odohira, T. Wada, M. Hatano in: W.B.Lisagor, T. W. Crooker, B. N. Leis, ASTM STP, 1049, ASTM Publ., Philadelphia, 1990, pp 521.

Part 3

Coating Engineering

5

Investigations of Thermal Barrier Coatings for Turbine Parts

Alexandr Lepeshkin
Central Institute of Aviation Motors
Russia

1. Introduction

Progress in gas-turbine engine (GTE) manufacturing is continuously linked with a rise of operating temperature and stresses of engine gas path elements, especially the turbine parts. More advanced cooling systems, structural materials and thermal barrier coatings (TBC) and other coatings provide the required life and strength reliability of these components. While engines are in use, the necessity arises to repair turbine blades and vanes with a TBC and combustion liners. At the same time, it is difficult to estimate the durability of the turbine blades and combustor components with a TBC and vanes because of the complexity of simulation of the damaging factors acting under service conditions and also because of problems in obtaining the input data required for making such estimations. Therefore, the development of methods for the calculated and experimental investigations thermal barrier coatings, thermophysical and strength properties of TBC and thermal, thermostress state and thermomechanical fatigue engine parts with a TBC is of great importance. While conducting these investigations, the main tasks are the comparative estimation of the design and production (or repair) process solutions and verification of the methods of calculation of the durability of engine parts with a TBC. To provide simulation of loading conditions for the hot engine parts with a TBC under service conditions, the test procedure shall ensure the possibility of cyclic surface heating of the object under testing (simulating its heating in hot gas flow) up to temperatures of 1150 °C and more at heating rates of 150-200 °C/s and subsequent cooling. Also it is desirable to have the possibility for mechanical loading of parts and TBC with a required phase shift between mechanical load and temperature.

2. Ceramic thermal barrier coatings and heating methods

For the purpose of providing the serviceability of high-efficiency aircraft gas turbine engines and gas turbine plants of new generations, it is necessary to improve existing cooling systems, to design new refractory and ceramic high-temperature materials, and to enhance the protection of parts of the high-temperature section of gas turbine engines with the use of heat-resistant and refractory coatings [1-12]. Improvement of the internal heat removal system leads to the transformation of parts into heat exchangers, which is accompanied by an increase in the thermal stress and a decrease in the thermal cycle life. Currently widely used refractory materials based on nickel usually operate in gas turbine engines at maximum allowable temperatures. The gas temperature can be allowed to increase only in the case where care is taken to restrict the passage of heat flow through the wall of the part. The heat flow from the gas to the wall of the base material of the part can be considerably reduced by means of either using a well-organized protective cooling without ejection or depositing thermal barrier coatings on the surface of the most strongly heated regions of the part.

2.1 Ceramic thermal barrier coatings

In recent years, works on introduction and practical use of thermal barrier ceramic coatings on parts of a high-temperature system of gas turbine engines have been carried out especially actively. The protection of the material of the part against the heat flux with heat-barrier coatings is most effective when the ceramic coatings used are based on ZrO_2 [2-4, 8]. The heat-protective effect of the thermal barrier ceramic coating reaches 100-120 °C under operating conditions. The heat-protective effect - decrease of metal temperature is function of thickness and heat conductivity of a thermal barrier ceramic coatings and thermal flows in a wall of a protected detail. The values of thermal flows on workers turbine GTE blades are in a range from 1,0 up to $(2 \div 2,5) \times 10^6$ W/(m·K) in works [1, 2, 3]. In some cases the thermal flow makes 3×10^6 W/(m·K) and more [4]. In the given work with the use of system ANSYS the calculated investigations of influence of the specified factors on decrease of metal temperature of cooled blades were carried out and the estimations of the heat-protective effect of a thermal barrier ceramic coatings for cooled details have been obtained by author. The results of calculated investigations are presented in Fig. 1 and Fig. 2. The values of decrease of metal temperature on a surface of cooled GTE blades depending on thermal flows at thickness h = 0,14 mm of ceramic coatings and different heat conductivity coatings are shown in Fig 1. The values of decrease of metal temperature on a surface of cooled GTE blades depending on thickness of ceramic coatings at gas thermal flow q = $1,8 \cdot 10^6$ W/m^2 and different heat conductivity of coatings in Fig 2. However, the questions regarding the thermal cyclic fatigue life are very problematic because the fracture strength of these coatings under tension is very low and thermal cycling usually leads to the appearance of alternating thermal stresses. Moreover, during operation of turbine blades, oxygen from an oxidizing medium (air, fuel combustion products) penetrates into the "ceramics-metal" interface. The penetration of oxygen through the ceramic layer results in the oxidation of the sublayer. The formation of oxides gives rise to additional stresses and decreases the adhesion of the ceramic layer. Therefore, the above factors must be taken into account in the design of coatings. The efficiency of thermal protection of coatings and their thermal fatigue resistance depend not only on the thermophysical properties (Fig.1 and Fig.2)but also on the technique used for depositing of the coating. Among numerous techniques currently employed for depositing of the coatings, the electron-beam technique provides the best thermal protection with a high thermal fatigue resistance.

2.2 Technique for depositing of ceramic thermal barrier coatings

Development of thermal barrier coatings applied to cooled blades is one of the trends for improving gas turbines. Unlike aluminide protective coatings, the ceramic coatings not only protect blade surfaces from high-temperature oxidation and corrosion but also prevent base material softening at high temperatures. Thermal barrier coating application allows the reduction of the blade temperature and the significant increase in its service life. Under both

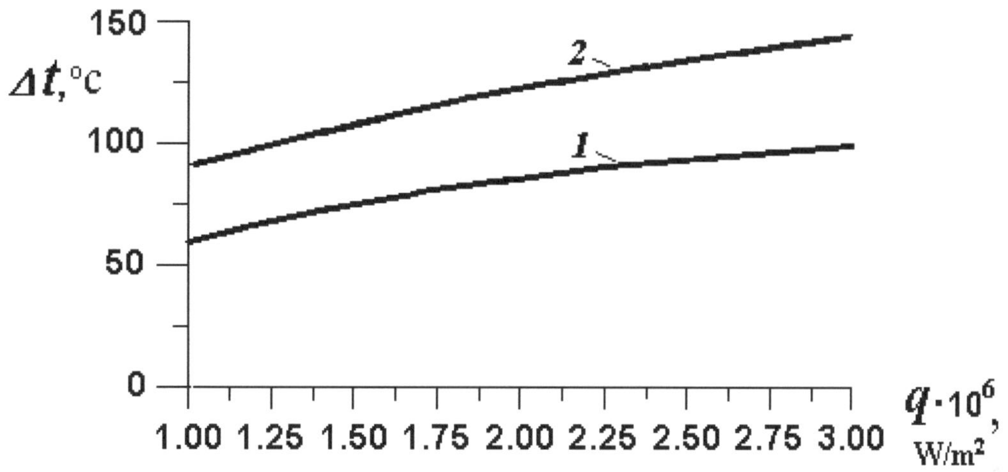

Fig. 1. Values of decrease of metal temperature *Δt* on a surface of cooled GTE blades depending on thermal flows *q* at thickness *h* = 0,14 mm of ceramic coatings ZrO_2 and different heat conductivity: *1* - 1,5 W/(m·K); *2* - 0,8 W/(m·K)

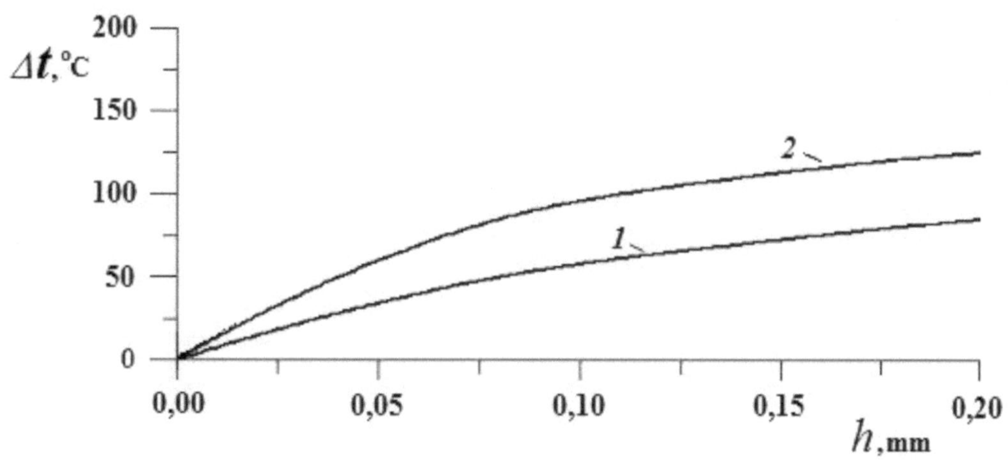

Fig. 2. Values of decrease of metal temperature *Δt* on a surface of cooled GTE blades depending on thickness *h* of ceramic coatings ZrO_2 at gas thermal flow $q = 1{,}8 \cdot 10^6$ W/m² and different heat conductivity of coatings: *1* - 1,5 W/(m·K);*2* - 0,8 W/(m·K)

steady-state and transient conditions, the application of ceramic TBC can diminish temperature gradients over the blade surfaces as well as reduce thermal stresses in them. A typical design of a TBC is presented in Fig. 3. The ceramic coating deposited directly on the superalloy surface does not show the required service life. Penetration of oxygen through the ceramic layer to the superalloy surface results in its quick oxidation and in spallation of the ceramic layer. That is why, as a rule, a TBC consists of at least two layers. An inner aluminide heat-resistant bond coat may be formed by different techniques. It may be either a diffusion or an overlay coating, depending on the requirements of its physical-mechanical properties and protection targets. The requirements of bond coat properties and protective coatings properties are much the same, yet the bond coat should meet some special requirements. First of all, it must be highly heat resistant; the oxides formed on its surface should have high adhesion to both the bond coat and the outer ceramic layer. When choosing a bond coat composition, one should pay special attention to its yttrium content as well as to the contents of the other elements, which guarantee high oxide adhesion to the surface and reactive element effect (Stringer, 1989). It is of special importance for bond coats deposited by the electron beam technique, because their yttrium contents depend on the yttrium content of the liquid bath and vary within wide limits (Malashenko et al., 1997). In this case, the required yttrium content of 0.2 to 0.3% is guaranteed by different technological procedures, such as direct yttrium addition to the liquid bath (Tamarin & Kachanov, 2008). Under these conditions, it is noteworthy that high yttrium contents of the liquid bath cause slag formation on its surface, thus resulting in occurrence of microdrops. These microdrops on the bond coat surface may provoke defects in the ceramic coating (layer).

It should be taken into consideration that TBCs are usually applied to the blades of high-temperature turbines. The blades of such turbines feature directionally solidified or single-crystal structures, thin walls, and high cooling efficiency. Under service conditions, high thermal stresses and strains arise in these blades, especially in their surface layers. That is why thermomechanical fatigue characteristics are as important in choosing a bond coat composition as its heat resistance. During thermal cycling, the bond coat should not experience considerable plastic strain. For example, the effect of a "rippled" blade surface (Fig. 4) always entails spallation of the ceramic layer. The outer zirconium oxide/yttrium oxide (ZrO_2-Y_2O_3) system base ceramic layer can-be applied by two techniques (Fig 4 and Fig 5): air plasma spraying of powders (APS-technique) or vapor condensation at electron beam evaporation of ceramic pellets (EB-technique). For this system, ceramic coating service life depends on Y_2O_3 content. The ZrO_2-(6 to 9%) Y_2O_3 compositions are usually applied, because they have demonstrated maximum service lives in the tests carried out (Miller, 1983 &Stecura, 1986). However, one should bear in mind the fact that the coating service life depends not only on its chemical composition but also on its structure and adhesive strength at the ceramic layer/bond coat interface, which depends on deposition technique. For coatings deposited by different techniques, the optimal chemical compositions may be other than that stated previously. The ceramic layer deposition technique determines such characteristics as ceramic layer structure and adhesive strength, its corresponding service life, thermal stresses in the ceramic layer, and its surface roughness. The main difficulty in designing TBCs for turbine blades lies in the combination of the ceramics on the blade surface and the superalloy that they are made of. At heating-up/cooling-down cycling, considerable difference between the ceramics and superalloy expansion coefficients~ $5.0 \cdot 10^{-6}$ 1/°C causes the generation of high thermal stresses in ceramics, which in turn results in ceramic layer spalling from the surface. To reduce thermal stresses, various technological procedures are used. In the ceramic layer deposited by the APS technique, special heat treatment is used to form a network of microcracks that break the ceramics into isolated fragments (Ruckle & Duvall, 1984). In the ceramic layer deposited by the EB technique, some specific columnar structure is formed that is readily fragmentizing when tensile stresses arise in (Strangman, 1982). The point crucial to success in the development of TBCs lies in obtaining the required adhesive strength of the ceramic coating/heat-resistant bond coat, providing for holding of the ceramics on the blade surface during all the blade service life. As a rule, in aircraft engine manufacturing, the technique of plasma deposition is used for nozzle vanes; in aircraft engine turbine blades, the EB technique is considered to be preferable. This is due to the fact that the following properties can be rendered ramie layer. The specific columnar structure, with the crystallites oriented perpendicular to the surface, forms in the ceramic vapor-deposited coating. In the case of tensile stresses, the ceramic layer is readily fragmentizing, thus reducing ceramic tearing stress during thermal cycling. In the temperature range of 850 to 950 °C, which is below the blade heating temperature at ceramic layer deposition, compressive stresses arise in it.

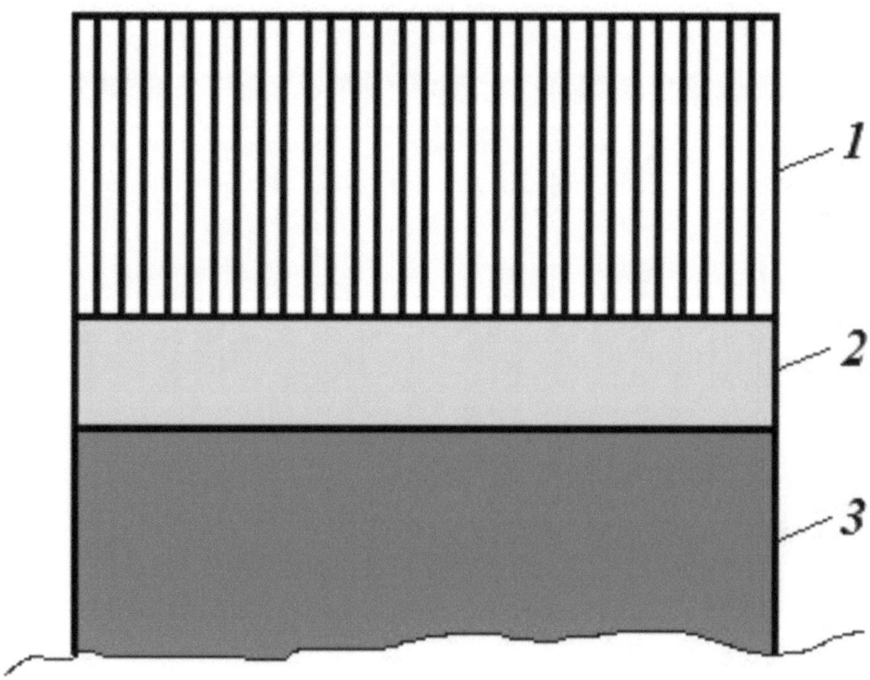

Fig. 3. Thermal barrier coating system: *1* – ceramic coating ZrO_2-8%Y_2O_3, 2 – bond coating MCrAlY, *3* - superalloy

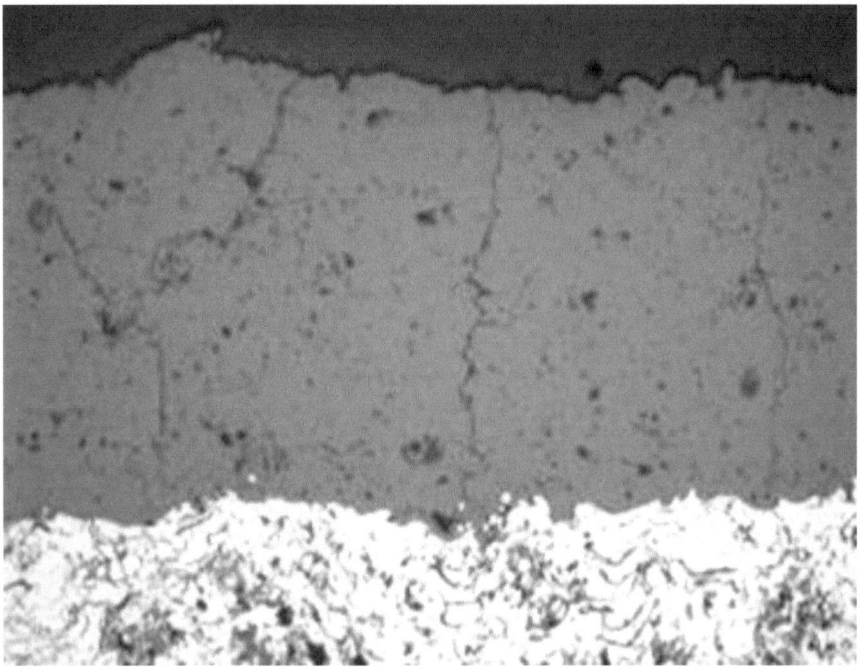

Fig. 4. Thermal barrier ceramic coating ZrO_2-Y_2O_3 (APS-technique)

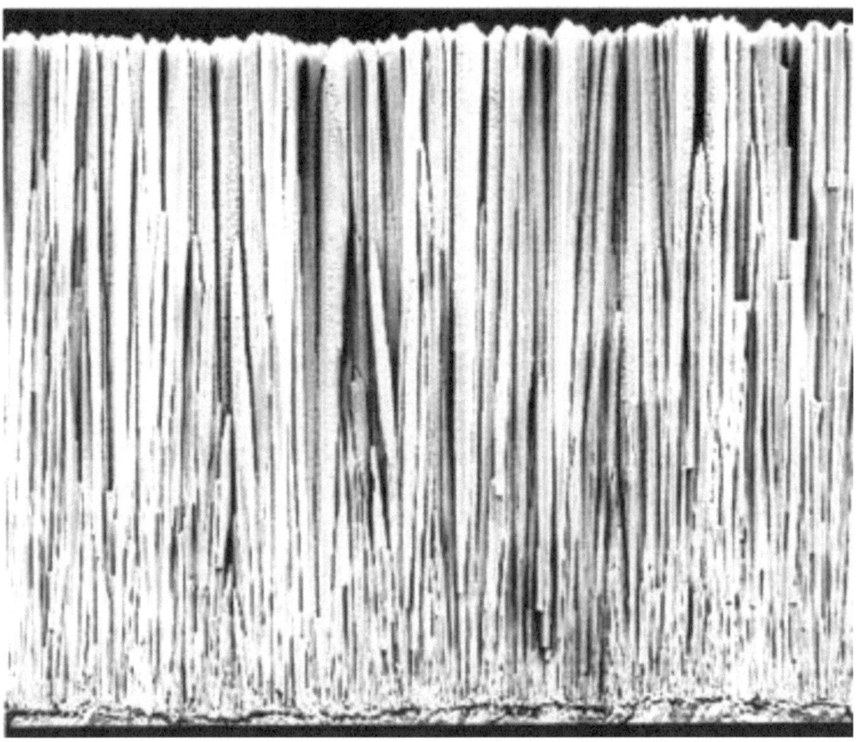

Fig. 5. Thermal barrier ceramic coating ZrO_2-Y_2O_3 (EB-technique)

Their generation is due to the different values of the ceramic and superalloy thermal expansion coefficients. These stresses do not relax on subsequent process annealing and under service conditions. The adhesive strength of the ceramic coating is controlled by physical-chemical reactions occurring between the ceramics and the metallic bond coat. As-deposited ceramic coating adhesive strength is above 50-100 MPa. The surface roughness of the ceramic coatings does not exceed 1.5 µm after their deposition. The need for maintaining the parameters of condensation and ceramic coating crystallite growth at a steady level requires a certain layout of relative positions of the blades, the vapor generator, and the EB guns for blade heating. As is shown in (Schulz, 1997), substrate rotation speeds have the same effect as temperatures. This behavior is caused by the effect of rotation on the time of growing crystal presence in the zones with different vapor density. The higher the temperature and rotation rate, the larger the diameter of an individual crystallite of the condensing ceramics (Fig. 6). A deposited ceramic coatings at the temperatures t_3 = 850-950 °C, t_2 = 0,85· t_3, t_1 = 0,7· t_3 and rotational speeds are shows on the Fig. 6.

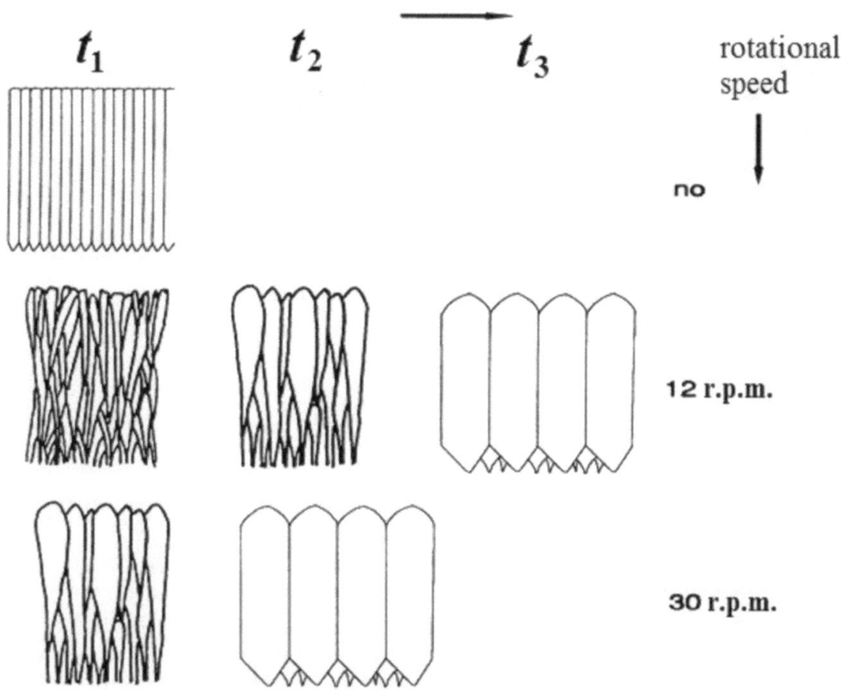

Fig. 6. Influence substrate and rotational speed on columnar microstructure of the deposited thermal barrier ceramic coating(EB technique)

Using different rotation speeds, structural characteristics of the ceramic coating can be governed. From the experience of ceramic coating deposition and taking into consideration intricate blade profiles and a need for simultaneous coating deposition on several blades, the best results can be achieved by combining blade revolution around the evaporator and rotation about then-axes. An illustration of blade arrangement and their revolution/rotation is given in Fig. 7.

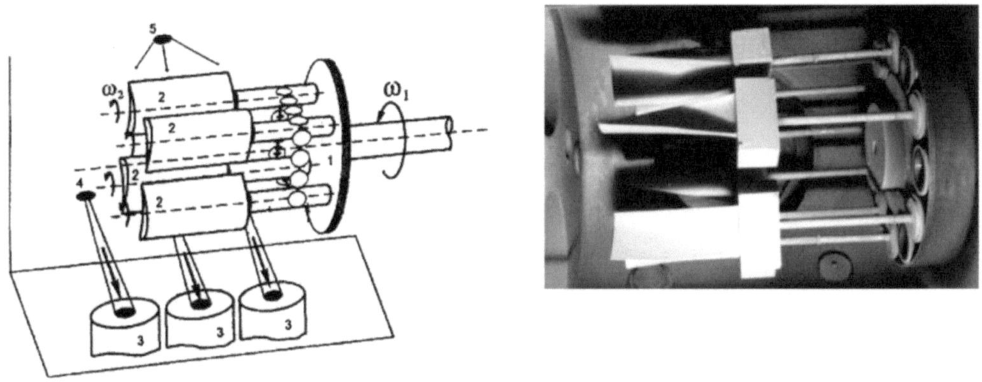

Fig. 7. Scheme for depositing of ceramic thermal barrier coatings (EB-technique) : 1 - rotor, 2 - blades, 3 - ceramic, 4 - electron beam gun of evaporator, 5 - electron beam gun for blade heating

The fixture in use revolves in the vapor flow with the speed of ~ 12 rpm. At each fixture revolution the blades additionally revolve once around the fixture axis. The choice of blade rotation conditions depends also on the requirements to the ceramic coating thickness and its spread over the blade surface. The ceramic structure features the pronounced texture of growth perpendicular to the surface (Fig. 5). Some individual ceramics crystals are preferably oriented in a [100] direction. Their diameters are in the range of 0.6 to 1.2 μm. They do not vary much along the full crystal lengths. Ceramic coating crystallites should have high cohesive strength and withstand an attack of a high-temperature gas flow. That is why the ceramic evaporation process feature is a requirement to its continuity. Unlike metallic bond coat deposition, in which no process interruption is harmful for the coating quality, any interruption of ceramic coating deposition forms an additional boundary, in the ceramics. The strength of this boundary is much lower than the crystallite strength. Thus, under these conditions, the ceramic coating will never meet the requirements of its properties. In the case of any pause in ceramics evaporation, all the lot of blades being coated are rejected and sent to ceramic layer removal procedure, followed by its redeposition. When ceramic layer deposition is carried to its completion, the blades are removed from the unit and passed to heat treatment. After treatment its color changes from dark gray to white. Two-step annealing does not change ceramic layer structure and phase composition. Check operations in the TBC quality control include a visual inspection to guarantee that its surface is free from ceramics droplets; measurements of ceramic layer thickness in the specified blade zones; and a bend test of a flat check sample on the radius of 3 and 10 mm to assess its adhesive strength. On its bending to the angle of 90°, ceramic coating spallation is prohibitive. Some cracking of the ceramic coating is allowed. The results of thermal conductivity studies for different ceramic coatings formed by the EB technique are presented in (Nichols et al., 2001). On the basis of the studies (Lawson et al., 1996), a two-zone model of a ceramic coating is suggested. Thermal conductivity of a dense, inner ceramic zone that forms at the starting moment of condensation is much lower than thermal conductivity of an outer zone (Fig. 8). This effect is attributed to the presence of numerous boundaries in the dense zone. Therefore, for reducing thermal conductivity of EB ceramic layers, it is advantageous to form thin layers 0.2 to 2.0 μm thick in the crystallites. Their boundaries ensure effective phonon scattering. Multilayer structure may be formed by plasma discharge to vary the density of the ceramics during deposition. According to the research, the efficiency of thermal conductivity reduction by means of multilayer structure may be high.

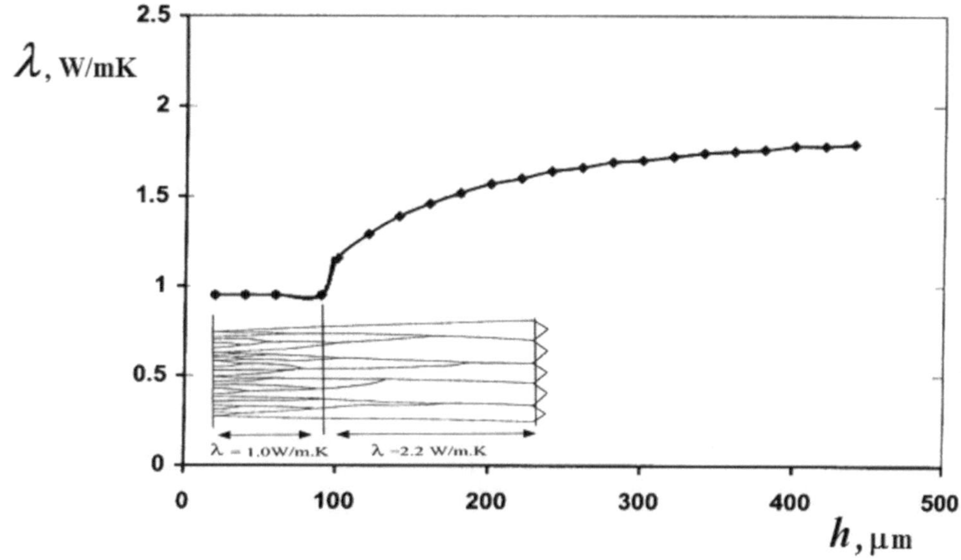

Fig. 8. Thermal barrier ceramic coating ZrO_2-Y_2O_3 (EB-technique): λ - thermal conductivity of a two-layer model, h – thickness of a coating

2.3 Methods of heating for investigations of ceramic thermal barrier coatings of parts

For providing the above-indicated heating conditions, there are various ways of heating such as gasdynamic heating and radiant heating, for example, in a reflective furnace electrical current (AC or DC) or induction heating with the use of high-frequency currents. Gasdynamic (flowing hot gas) heating has been used for more than 50 years. When using this method, a more accurate simulation of the heat exchange conditions from gas flow to the part is realized relevant to the gas-turbine engine. The rigs with gasdynamic heating enable a high heating rate to be provided to the part, to investigate the influence of oxidation in gas flow, but at the same time it is difficult to provide mechanical loading of parts. The cost of tests using such rigs is very high and the bench equipment needs to be frequently repaired or replaced. Alternating current (AC) or direct current (DC) resistance heating is effective for testing solid and tubular specimens. In accordance with this method, there is no need to use expensive and complex equipment. It enables tests to be conducted both at in-phase and at out-of-phase change of temperatures and mechanical loads. This method provides ease for inspection of the specimen surface. At the same time, this method cannot be used for tests of gas-turbine engine parts. The direct passing of electrical current can influence on the mechanical properties of the specimen material. In addition, this method does not enable the actual conditions for part heating in gas flow to be simulated. When a specimen with a thermal barrier coating is heated by direct passing of electrical current, the coating temperature is lower than the base material temperature. Radiant heating of parts is of certain use when conducting the thermocyclic tests for specimens with a TBC. In so doing the surface is heated at a high rate, however, because of radiation focusing during test of a part (or a part model) it is difficult to simulate the required temperature field. Additionally, the heaters have a low cyclic lifetime. Evidently, induction heating with the use of high-frequency currents in the surface of a part is of greatest use to heat parts and models of parts when conducting tests for thermomechanical fatigue. Such a method may be used to test both standard specimens and engine parts. When it is used, the surface part heating realized under service conditions is well simulated. In so doing, heat releases directly in the part. There is no need to use expensive heating equipment, and the equipment used features of high durability. The mechanical loading device can be used in the rig with inductor heating. It provides the possibility of conducting thermomechanical fatigue tests of turbine blades. In so doing, with the use of a special inductor the temperature field is simulated for the blade section under the service conditions of which the strength margin is minimum and with the use of a suitable loading device, the centrifugal load is simulated in this section. It is worth noting that induction heating is only effective for testing of metallic alloys. For tests of parts made of ceramic materials, it is recommended in a number of papers to use dielectric heating (in Mega Hertz frequency range) or heating with the use of a susceptor. In the latter case, it is not possible to provide suitable heat-up rates for the temperature of the part. As conducted investigations showed that when using currents of more than 400 kHz to heat a metallic part with a TBC, both heating of metal located under the external layer coating and the effective heating of the dielectric (TBC) take place. Correlation of heat shared depends on the thermophysical properties of the base and coating materials and the frequency at which heating is performed, and a number of other factors. The experiments showed that the ceramic ZrO_2 -based thermal barrier coatings on specimens and parts made of high-temperature nickel-based alloys are effectively heated at frequencies between 0.4 and 2.0 MHz. Use of a higher frequency requires a complicated rig design. Consequently, it seems that in spite of a lack of data concerning the absence of a knowledge of the influence of induction heating on mechanical properties of the materials under investigation, this method of heating can be successfully used for tests of specimens and engine parts (primarily for comparative tests for selection of coatings and materials, design solution, manufacture and repair of engine parts with a TBC by production processes). The cost of the tests conducted with the use of high-frequency heating is by an order lower than the cost of the tests conducted on a gasdynamic rig.

3. Investigations of ceramic thermal barrier coatings of parts with the use of HF induction heating

3.1 HF induction heating of ceramic thermal barrier coatings of parts

At present, the cyclic fatigue life of thermal barrier coatings in the course of their development has been studied using radiant heating with a low rate (less than 20 K/s), which does not correspond to actual operating conditions. At such low heating rates, thermal stresses are almost completely absent and the main damage factor is the oxidation of a sublayer, which leads to spalling of the coating. Actually, these processes are heat resistance tests at variable temperatures. Under real conditions, the rate of change in the temperature of parts lies in the range 100-200 K/s. In this case, there arise cyclic thermal stresses and deformations of the base material and coating, which are accompanied by the appearance of alternating stresses. The results of tests for thermal fatigue of parts with thermal barrier coatings can differ significantly from the results of tests for cyclic heat resistance, which have been obtained by developers at a low rate of change in temperature. Therefore, in the design of thermal barrier coatings, it is necessary to investigate their heat resistance together with a protected material under the conditions providing high rates of heating and cooling. The tests performed in a gas-dynamic flow are expansive and require a long time. The high-frequency induction heating is significantly lower in cost and requires a shorter time. The process of high-frequency heating involves not only induction heating of conductive materials but also heating of dielectrics, including ceramic materials. The dynamics of heating of the coating and the base material depends on the electrophysical and thermophysical properties of the material, its volume, the cooling conditions, the rate of heating of the object, the dielectric properties of the ceramic coating, and the frequency of the electric current used for heating. The calculated simulation of the heating conditions for parts with thermal barrier ceramic coatings has not been adequately developed as compared to thermal calculations of the parts operating in a gas dynamic flow. More reliable data on the temperature state of parts with thermal barrier ceramic coatings during their heating in a high-frequency electromagnetic field and on their heat resistance can be obtained from experimental investigations. In order to create prerequisites that are necessary for the development of computational methods used for determining the thermal and thermostressed states of parts with thermal barrier coatings in the course of their heating in a high-frequency electromagnetic field and for the experimental evaluation of the thermal cyclic fatigue life of these parts, in this work we set the problem of the development of a technique for high-frequency heating and thermophysical measurements in tests of blades and models of other parts with thermal barrier coatings based on zirconia. The develop of a design-experiment method is necessary for modeling of high-frequency induction heating and determination of fatigue and thermophysical measurements in thermal cycling tests of blades of gas turbine engines, to perform experimental investigations on the determination of the temperature state of blades and models with zirconia thermal barrier coatings with the use of a thermal vision system during high-frequency heating of parts with ceramic coatings, to determine the ratio between the processes of high-frequency and dielectric heatings, to obtain a generalized dependence of the temperature gradient across the ceramic coating thickness on the frequency of the electric current from multivariant calculations, and to compare the thermal cyclic fatigue lives of parts with a thermal barrier coating and without it.

3.2 Technique and results of investigations

The design-experiment method involves complex interrelated physical processes (such as heating of metal and ceramic materials in a high-frequency electromagnetic field, dielectric heating of the ceramic material, and interactions of nonstationary fields of temperatures and thermal stresses in a metal-ceramic part with cooling holes) and takes into account the electrophysical and thermophysical properties of the materials in thermal cyclic tests [Kuvaldin &Lepeshkin, 2006). New tasks on the determination of the ratio between the processes of high-frequency and dielectric heatings and on the identification of the dielectric heating effect and its influence on the distributions of heat fluxes and temperatures in a metal-ceramic part are set in computational and experimental studies. By using multivariant calculations, it is necessary to obtain a generalized dependence of the temperature gradient across the ceramic coating thickness on the frequency of the electric current.

3.2.1 Numerical simulation

The computational part of the method consists in sequentially solving the following problems: the electromagnetic problem based on the Maxwell equations, the transient heat problem based on the solution of the heat conduction equation, and the problem of determination of the thermostressed state. The first problem was solved with due regard for the recommendations proposed in (Kuvaldin & Lepeshkin, 2006). By solving this problem (taking into account the gap between the inductor and the part and the electric current frequency of 440 kHz), we determined the distribution of internal heat sources (specific heat power) over the thickness of the base metal (the refractory nickel alloy of the flame tube) with an intermediate refractory metal coating NiCoCrAlY, as well as in the ceramic coating due to the induction heating (as a result of the change in the electrical resistivity ofzirconium oxide with an increase in temperature) and dielectric heating (as a result of the change in the permittivity and the dielectric loss tangent with an increase in temperature). The obtained distributions of internal heat sources are nonstationary; i.e., they depend on the heating time. During the solution of the coupled electromagnetic and heat problems at each computational step, the value of the current temperature was transferred from the module of the solution of the heat problem to the module of the solution of the electromagnetic problem in order to correct the electrophysical properties of the materials. The computational investigations allow one to refine the thermal and thermostressed states of thermal barrier ceramic coatings on cooled blades and models during high-frequency induction heating with the inclusion of dielectric heating. The initial data used in the performed calculations were electrophysical, thermophysical, and strength properties of the ceramic coatings and the material of cooled parts, the characteristics of bench conditions for heating and cooling, and the parameters of the test thermal cycle. The electrophysical and dielectric properties of the ceramic (zirconia) coating were taken from (Rubashev at al., 1980), and the electrophysical, thermophysical, and strength properties of the ceramic coatings were taken from (Tamarin & Kachanov, 2008). The parameters of the permittivity ε and dielectric loss tangent $tg\delta$ of a zirconium oxide depending on temperature are shown in Fig. 9. Electrical resistivity of a zirconium oxide makes at the temperatures: 100 °C - 10^{11} Ohm·cm,1000 °C - 10 Ohm·cm.

The calculations performed by author using the finite element method implemented in the ANSYS program and the distribution of the heat flux from the inductor between the zirconia coating and the metal of the cooled part at an induction current frequency of 440 kHz made it possible to investigate the nonstationary thermal state of the coating and the cooled part with the inclusion of the parameters of the test thermal cycle. The boundary conditions for the solution to the heat problem were as follows: the temperature of the ambient air was 20 °C, the heat-transfer coefficients of the ambient air were equal to 20-30W/(m^2 K), the heat-transfer coefficients of the cooled air inside the model of a flame tube were equal to 1800-2000 W/(m^2 K) (according to the experimental data), the specific heating power on the surface of the refractory metal coating was 9 x 10^5 W/cm^2, and the specific heating power in the ceramic coating was 1.8 x 10^5 W/cm^2 (these specific heating powers were obtained from the solution of the electromagnetic problem). The minimum and maximum heating temperatures of the metal surface of the part in the thermal cycle were equal to 350 and 900 °C, respectively. The mathematical simulation of the thermal state of the ceramic coatings takes into account the specific features of the electrophysical properties of zirconia. In particular, an increase in the temperature results in an increase in the permittivity, the dielectric loss tangent, and the electrical conductivity (Kuvaldin & Lepeshkin, 2006). On the whole, the ceramic coating in the course of the test thermal cycle was heated by means of both the heat transfer from the metal of the part and the dielectric heating. The computational scheme for a fragment of the cooled part with the thermal barrier coating is shown in Fig. 10.

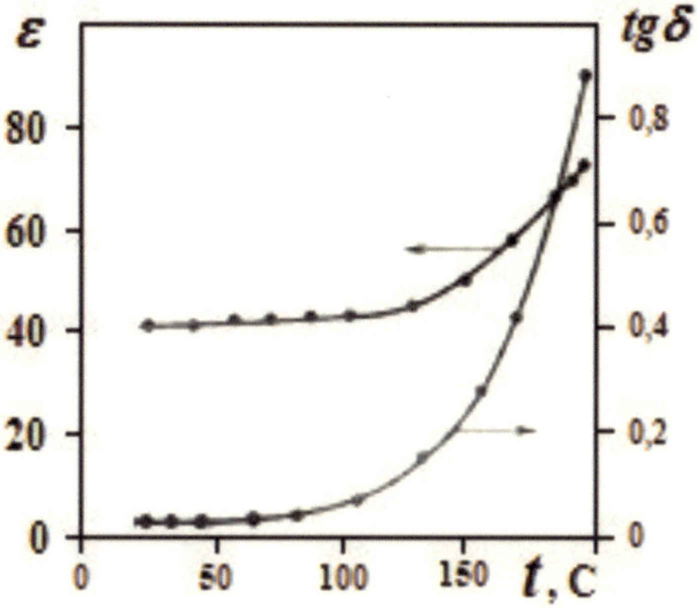

Fig. 9. Parameters ε and $tg\delta$ depending on temperature

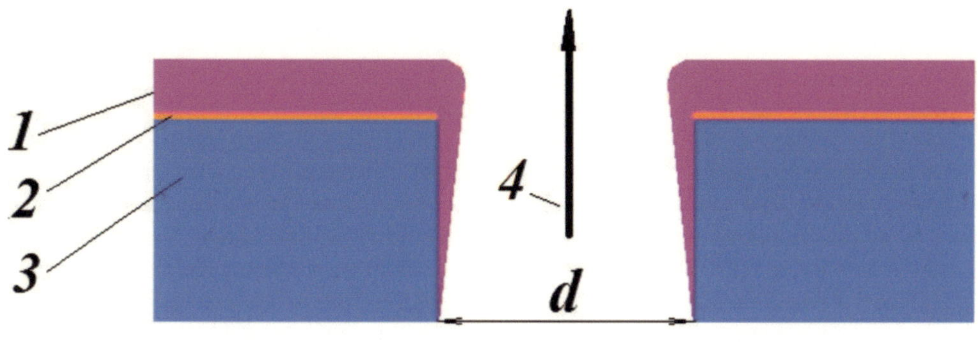

Fig. 10. Schematic diagram of a fragment of the cooled part with the thermal barrier ceramic coating: *(1)* ceramic coating, *(2)* metal of the workpiece, *(3)* refractory metal layer, and *(4)* direction of the flow of cooling air in the hole. Designation: *d* is the hole diameter

The performed calculations of the nonstationary thermal and thermostressed states of the models of cooled parts with thermal barrier ceramic coatings (Fig. 11 and Fig. 12) have demonstrated that, at the maximum temperature of the thermal cycle, the temperature of the outer surface of the ceramic coating at the end of heating is higher than the temperature of the metal and, consequently, there arise temperature gradients across the ceramic coating thickness.

Fig. 11. Calculated temperature distribution of the fragment of the cooled part with the thermal barrier ceramic coating in the region of the cooling hole

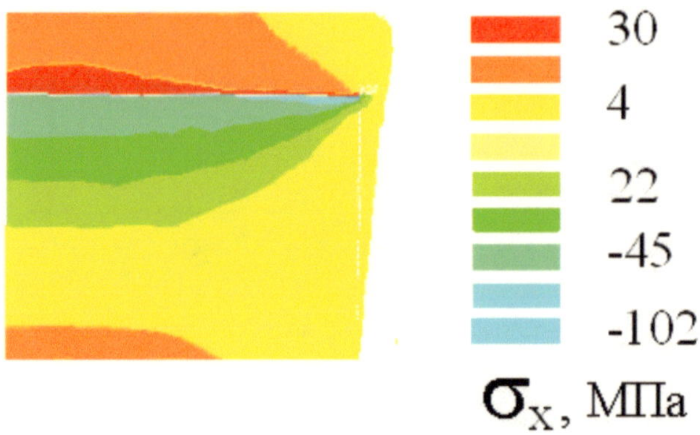

Fig. 12. Calculated thermostress distribution of the fragment of the cooled part with the thermal barrier ceramic coating in the region of the cooling hole

The temperature gradients depend on the thermal conductivity coefficient, the coating thickness, and the heat loss due to the environment on the surface of the coating. The heat losses were calculated taking into account the convective heat exchange, the radiative heat exchange, and the maximum experimental temperature of uncooled plates of the inductor (300 °C) at the end of heating in the first stage of the thermal cycle. At an induction current frequency of 440 kHz and taking into account the ratios between the heated masses of the base material from the refractory alloy and the coating, as well as their electrophysical and thermophysical properties and cooling conditions, the calculated distribution of the high-frequency electromagnetic energy over the sample from the nickel-based alloy with the zirconia thermal barrier coating approximately corresponds to 80 % for the metal (the high-frequency energy is released in the metal of the sample) and 20 % for the coating (the high-frequency energy is released in the zirconia ceramic coating due to the induction heating (10 %) and the dielectric losses (10 %)). According to the results of the numerical calculations under the aforementioned conditions at a heating rate of 100 K/s, the temperature on the outer surface of the model of the part with the thermal barrier coating in contact with the environment is approximately 60-80 °C higher than that at the "metal-thermal barrier coating" interface; i.e., the temperature state of the part is simulated in operation. In this case, compressive thermal stresses of 100 MPa on the metal surface and tensile thermal stresses of 30-35 MPa on the side of the thermal barrier ceramic coating are observed. A generalized dependence of the temperature gradient $\Delta t(f)$ across the ceramic coating thickness on the frequency of the electric current was obtained using multivariant calculations. This dependence $\Delta t(f)$ with variations in the frequency of the electric current from 200 to 2000 kHz is plotted in Fig. 13.

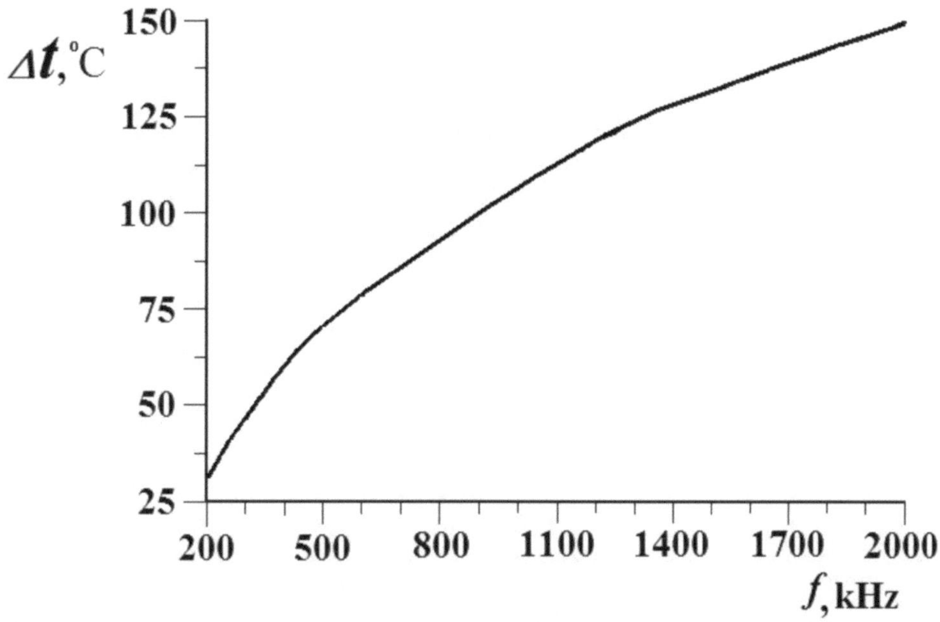

Fig. 13. Dependence of the temperature gradient across the ceramic coating thickness on the frequency of the electric current

3.2.2 Experimental investigations

The experimental part of the method provides simulation of high-frequency induction heating and the performance of thermophysical measurements in the course of thermal cyclic tests of blades and other cooled parts and takes into account the electrophysical and thermophysical properties of their materials. In the developed method, the contactless thermophysical measurements are carried out using a thermal vision system for the performance of investigations and for the confirmation of the calculated results for the nonstationary thermal state of the part with the thermal barrier ceramic coating (with the inclusion of the temperature gradient across the ceramic coating thickness) in thermal cyclic tests of rotating blades and models of flame tubes with thermal barrier ceramic coatings. The temperatures of the surfaces of the ceramic coating and the metal under the coating were measured simultaneously with a thermal imager lens through a hole in the inductor. In this method, we also proposed the design of a split uncooled plate inductor (Bychkov, 2008) with a hole for the examination of thermal and thermostressed states of the cooled and uncooled blades, including parts of gas turbine engines with coatings. The specific features of this method are as follows: the possibility of performing thermal cyclic tests of parts of gas turbine engines with retaining a guaranteed minimum constant gap between the inductor and the blade surface, which decreases the probability of distortion of the temperature field after the replacement of the parts and favors an increase in the efficiency of high-frequency induction heating; the fulfillment of the relationship $\Delta < 0.1 \cdot h$, where Δ is the depth of penetration of the electric current (the electromagnetic wave) and h is the minimum thickness of the metal wall of the part (the cooled blade) for the choice of the frequency of the electric current; etc. Two cooling circuits are used to create the required nonstationary thermal state of the cooled part and to provide the optimum parameters of the thermal cycle. The first cooling circuit ensures air supply into the inner cavity of the part, and the second circuit is responsible for supply of air passing between the inductor plates and the part for blowing the surface of the part and its cooling at the end of each thermal cycle. Preliminary, while heating the model in electric furnace there were obtained empirical data about degree of blackness for specimen with and without the thermal-protective coating which were used for thermal-imaging measurements. The values of blackness degree ε_b for specimen with the coating under temperatures, approximately, of 850÷900 °C close to peak ones in the cycle were equal to about 0.55 and for blade specimen in high-temperature-resistant alloy without the coating their value was about 0.80 (Fig. 14). The thermal cyclic tests of the blades with a thermal barrier coating and the models of cooled parts were carried out in the course of high-frequency induction heating of the object at a frequency of 440 kHz according to the developed technique on a setup (Lepeshkin, 2005) equipped with a VChG-10/0.44 high-frequency valve generator. In order to perform comparative thermal cyclic tests, the working surface of the models of flame tubes produced from a refractory alloy sheet 1.0 mm thick with preliminarily perforated holes (Fig. 15) was subjected to sand blasting with synthetic corundum, followed by the deposition of two variants of the thermal barrier ceramic coating with the intermediate refractory joining NiCoCrAlY layer and without it. Figure 10 shows a fragment of the model of a cooled part (the model of a flame tube with a thermal barrier coating) mounted inside the inductor connected to electric buses of the VChG-10/0.44 generator. Air with a controlled flow rate and a controlled pressure was fed inside the sample. This scheme provided the possibility of reproducing the operating fields of temperatures and thermal stresses for the model and the possibility of experimentally determining the thermal cyclic fatigue life of the models of a section of the flame tube with different variants of thermal barrier coatings and without them.

The temperature was controlled by a chromel-alumel thermocouple. The temperature state of the surface of the thermal barrier coating in the working region was controlled using an Agema thermal imager. For the experimental verification of this thermal state, we carried out contactless measurements of the temperature of the surface of the model with the thermal barrier coating based on zirconia with the use of an Agema 782 SW thermal imager operating in the spectral range from 3.0 to 5.6 μm. The specific features of the technique used for measuring temperatures with thermal vision systems are described in (Bychkov, 2008). The model of the cooled part (the flame tube with holes) is shown in Fig. 15. The diameter of two rows of holes is equal to 1 mm. The diameters of holes of the other rows are smaller than 1 mm. The refractory joining layer (coating) has a thickness of 0.06 mm.

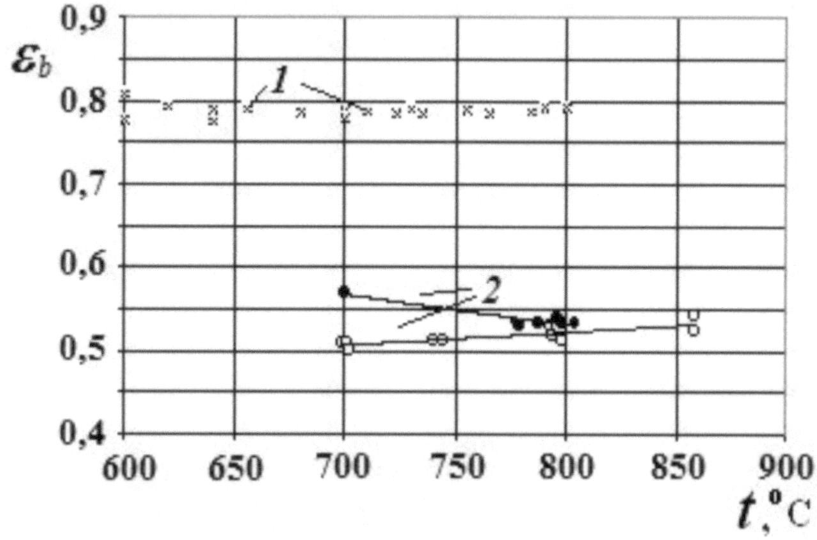

Fig. 14. Blackness degree ε_b depending on temperature

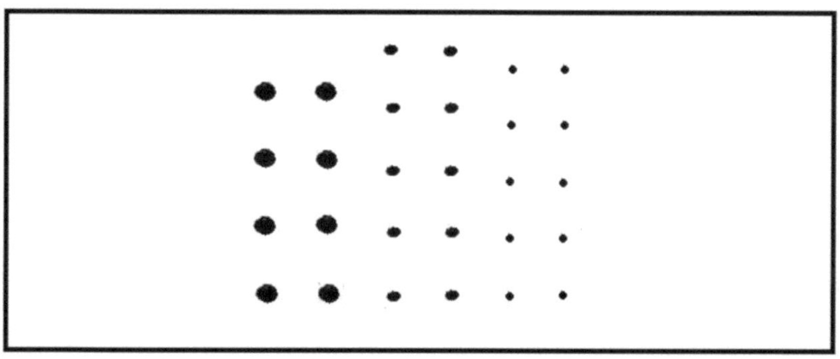

Fig. 15. Model of the cooled part with holes

The cooling air was fed into a box-like hollow model (with a rectangular cross section) of the flame tube. The model had a wall thickness of 1 mm and a cross section of 10 x 25 mm. In the examination of the temperature state of the part with the ceramic coating, the optical accessibility of the object during the thermal cycle was provided by a small hole (5 mm in diameter), which was drilled in the inductor and through which the surface region was scanned (Fig. 16). Thermal images in the course of thermal cyclic tests were recorded in a personal computer with a frequency of three to five frames per second. The analog signal of the thermal imager was digitized using an L-783 analog-to-digital converter board fabricated by the L-CARD Corporation. The complete cycle (from the beginning of heating of the sample to cooling) was recorded; however, only the frames in the vicinity of the peak value of the temperature were used in the processing. The parameters of the thermal cycle are presented in Fig. 17.

Fig. 16. Tests of the cooled part with the heat resistant ceramic coating: (*1*) inductor and (*2*) cooled part.

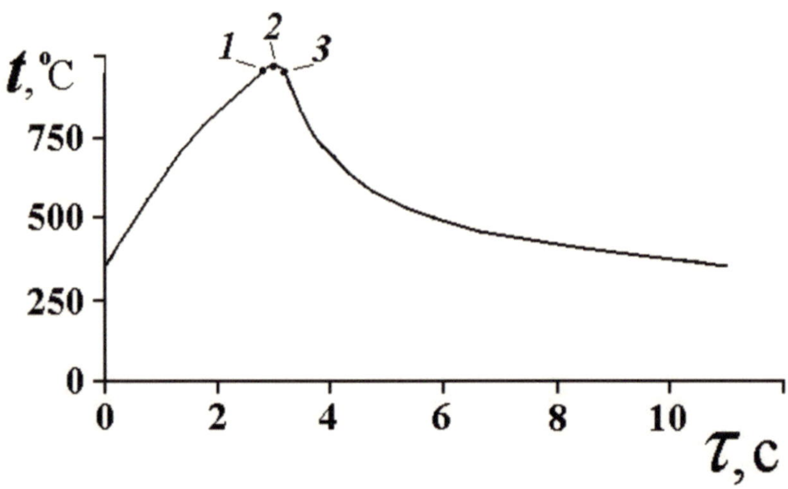

Fig. 17. Heating thermogram.

As an example, Fig. 18-20 displays three thermal images of the model with the thermal barrier coating during induction heating. The temperatures at the surfaces of the ceramic coatings shown in the thermal images in Figs. 18-20 are equal to 946.6 °C (Fig. 17, point *1),* 969.4 °C (Fig. 17, point *2),* and 953.9 °C (Fig. 17, point *3),* respectively. At the peak temperature, the indication of the control thermocouple on the sample (which corresponds to the lower edge of the thermal image) is approximately 60–70°C below the temperature of the outer layer of the coating (in the vicinity of the thermocouple), which is recorded by the thermal imager (Fig. 18-20).

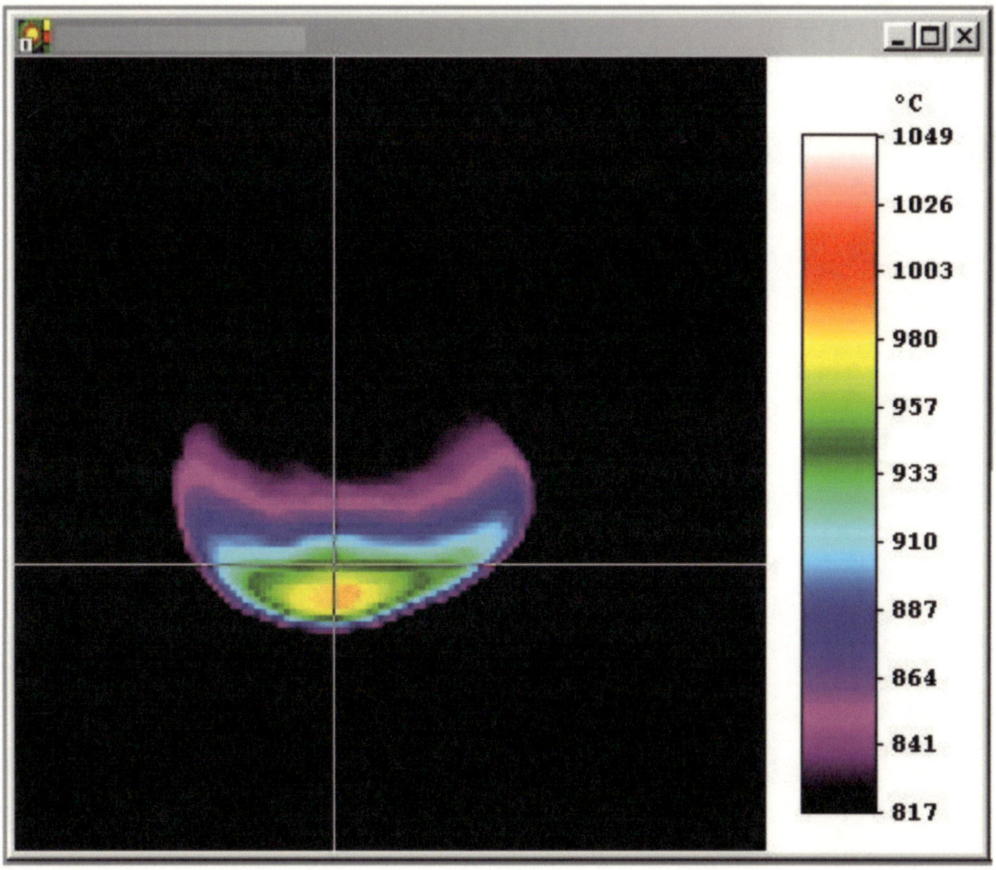

Fig. 18. Thermal image of the sample with the thermal ceramic barrier coating (0.2 s before the heating is switched off)

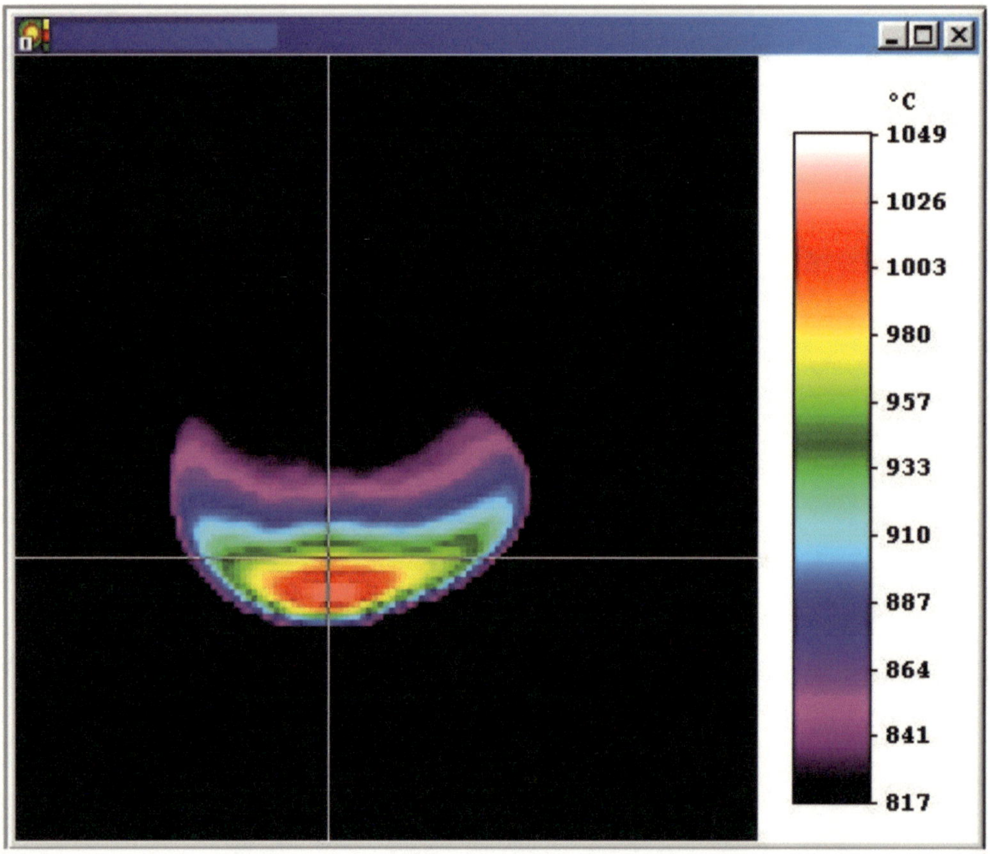

Fig. 19. Thermal image of the sample with the thermal ceramic barrier coating at the instant of switching off

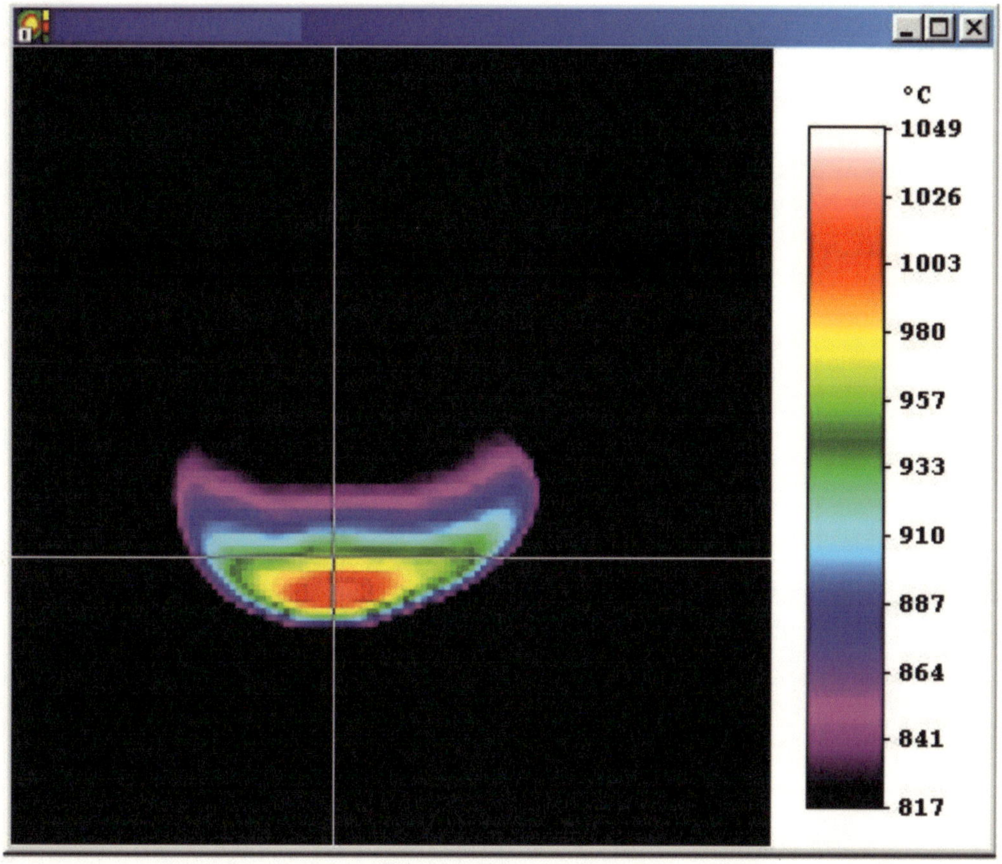

Fig. 20. Thermal image of the sample with the thermal ceramic barrier coating (0.2 s after the heating is switched off)

The control thermocouple is fixed on the metal surface (locally protected against the ceramic coating) of the refractory layer. The temperatures of the thermal barrier coating and the refractory layer were recorded on the thermal imager simultaneously. The performed experimental investigations and measurements of the temperature of the part with the thermal barrier coating with the use of the thermal imager in the course of thermal cycling confirmed the calculated value of the temperature gradient across the ceramic coating thickness. Thus, the analysis of the results obtained has demonstrated that the thermostressed state observed for parts of the hot gas section of gas turbine engines (combustion chambers, turbine blades, etc.) during blowing them by a high-temperature gas flow under operating conditions can be simulated under laboratory conditions on a setup with high-frequency heating. The temperature gradient across the ceramic coating thickness can be varied over a wide range by varying the flow rate of air supplied for cooling, the power of the high-frequency generator, and the wall thickness. Thus, the analysis of the results of thermal fatigue tests of blades (Fig. 21, Fig. 22) of gas turbine engines during thermal cycling according to the regime $t_{min} \leftrightarrow t_{max}$ (350 °C ↔ 900–1000 °C have shown that the thermal cyclic fatigue life of blades with a thermal-barrier ceramic coating deposited by the electron-beam method increases, on average, by a factor of 3.4 compared to the blades produced from the refractory nickel alloy without a coating. The photographs of turbine blades with a ceramic TBC and inductor under thermocyclic tests are presented in Fig. 22.

The results of thermal fatigue tests of the models of flame tubes during thermal cycling according to the regime $t_{min} \leftrightarrow t_{max}$ (350 °C ↔ 900 °C) have revealed that the service life of the models with a three-layer coating of the thickness h = 320-520 μm increases by a factor of approximately 2,7 compared to the models without a coating.

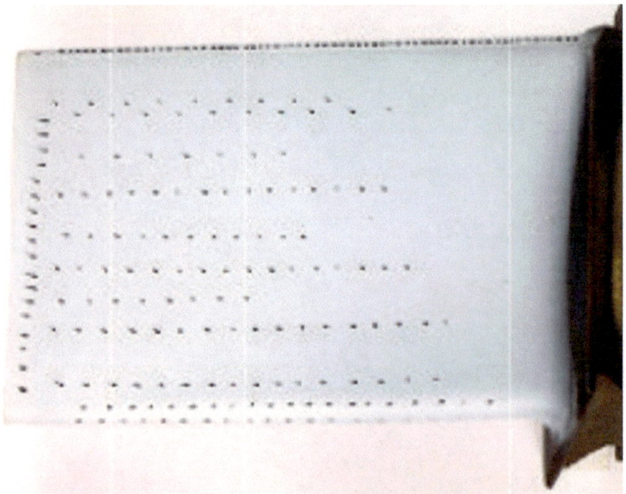

Fig. 21. Blade feather with a thermal ceramic barrier coating before thermocyclic tests

Fig. 22. Thermocyclic tests of the turbine blades with a thermal-barrier ceramic coating

4. Investigations of thermal barrier properties of ceramic coatings with the use of heating

4.1 Investigations of thermal barrier properties of ceramic coatings with the use pulse of heating

Thermal barrier coating application efficiency depends on ceramic layer thermal conductivity, which determines the cooled blade temperature drop and corresponding increase in its service life. To measure thermal conductivity of a TBC ceramic layer, a laser flash method is used (Siegwart et al., 2006) (Parker et al., 1961). The method is based on irradiating the surface of a flat sample surface with an energy pulse, followed by recording a temperature rise on its backside. Thermal diffusivity and heat capacity are determined experimentally using the pulse method of heating. Then, thermal conductivity *(λ)* is calculated from these characteristics:

$$\lambda = a \cdot \rho \cdot C_p \tag{1}$$

in W/m·K, where a is thermal diffusivity (cm^2/s); ρ is density (g/cm^3); C_p is heat capacity (J/g·K).

Measuring each thermophysical characteristic is an independent task. The most developed method is that of thermal diffusivity calculation, because the main formula for thermal diffusivity includes only one experimentally measured parameter. It is a period for the temperature to reach half of its maximum level:

$$a = 0.1388\ (\delta^2/\tau_{1/2}) \tag{2}$$

in cm^2/s, where δ is sample thickness, and $\tau_{1/2}$is the time required for the temperature of the sample backside to reach the level equal to one-half of the maximum temperature. Coefficient 0.1388 corresponds to an ideal case when the following conditions are met: instantaneous and uniform heat pulse, heat pulse absorption in a thin surface layer, and no heat losses. For experimental thermal diffusivity determination, one should know neither absolute temperatures nor parameters of a heat flow affecting a sample. Measuring heat capacity by the flash technique, especially for coated samples, is a much more complicated task. Analysis of thermal diffusivity - and thermal conductivitjrof cerarrricxoaflngs are discussed elsewhere (Pawlowski et al., 1984). For thermophysical studies of ceramics condensates, the TC-3000H unit manufactured by the Sinku-Riko Company was used. A ruby laser with a wave length of 6.943 μm was used as an energy source, and as a temperature pickup on the backside of the sample, either a thermocouple (Pt-PtRo) or an infrared sensor was used (Maesono, 1983). The tested sample is essentially a flat disc 10 mm in diameter and 0.8 to 2 mm thick. When thermal diffusivity is studied in this unit, two types of experimental errors are possible. The first type of errors results from some lack of information on the parameter values used in the design formulas. They are due to the available accuracy of sample thickness and time of $\tau_{1/2}$ measurements, exactness of the temperature rise assessment, and of catching the moment of the sample irradiation start. These errors are covered in detail in (Cape & Lehman, 1963). On the basis of the results reported in the literature, one can deduce that, with the modern data collection systems used, the contribution of this type error does not exceed 0.5%. The second type of errors is due to the difference between the experimental conditions and assumptions in the mathematical model used for calculating thermal diffusivity and heat capacity. These errors are related to the finite pulse duration and its spatial inhomogeneity, to heat losses (due to irradiation, mainly), and to violation of pulse absorption conditions in the thin surface layer. These errors may be avoided by using certain corrections (Clark & Taylor, 1975). For the TC-3000H unit, pulse duration and spatial inhomogeneity errors determined according to the Sinku-Riko Company recommendations are unessential (less than 1%). Heat losses in the experiment result in a quick temperature rise to its maximum and then a sharply defined smooth temperature decrease. The main cause that gives rise to measurement errors is radiation heat exchange, whose effect rises simultaneously with a temperature rise. The errors caused by radiation may account for 30%. To meet the requirements of pulse absorption in the thin surface layer, the ceramic samples, which are partially transparent, were coated with a thin layer (10 to 12 μm) of the NiAl intermetallic compound (20% Al). This layer ensured steady surface optical parameters of the samples as well.

4.2 Investigations of thermal barrier properties of ceramic coatings with the use of gas-flame heating

For maintenance of competitiveness of aircraft engines it is necessary to raise a gas temperature over 1700 K in front of the turbine. Thus serviceability of details of a high-temperature gas can keep only at perfection of their heat-protection. Thus the serviceability of details of a high-temperature gas probably keep only at perfection of their heat-protection.It is known, that in world practice the ceramic heat-protective coatings on basis ZrO_2 are widely used. At the same time the data on heat conductivity and thermal conductivity and efficiency of a heat-protection of details with help thermal barrier ceramic coatings at their heating in a gas flow are rather limited. The characteristics of heat conductivity thermal barrier ceramic coatings have received at use of various known laboratory methods are inconsistent. Basically the preference is given the thermal barrier ceramic coatings have deposited on a plasma technology. At use of laser pulse heating it has been received that heat conductivity of plasma coatings approximately in 3 times is lower than at the coatings have deposited on the electron beam technology. The laser pulse method is inexpedient to use for determination of the temperature in part transparentceramic coatings as the part of a beam flow warms up directly a metal on which it is deposited coating. The protective thin metal screen with thickness 10-15 μm deposited by researchers on surface of coating on the side of the laser at heating, itself starts to let out a beam flow. In real conditions the turbine blades and walls of combustion chambers are heated up by a gas flow. In the given work the developed technique by an objective estimation of efficiency of a heat-protective of metal with the help of coatings of plasma andelectron beam technology is resulted at gas-flame heating of object on the developed rig (Bychkov, 2008). The essence of the given original technique protected by the patent RU will be that through the demountable specimen (collected from two halfs) the high-temperature gas flow (Fig. 23) is passed.

The investigations for evaluating the efficiency of thermal protection of materials of the turbine blades and parts with use TBC (received on electron beam technology and plasma technology) against the convective and radiant components of the high-temperature gas flow were conducted. In this case, it is recommended to conduct the tests at the small-size

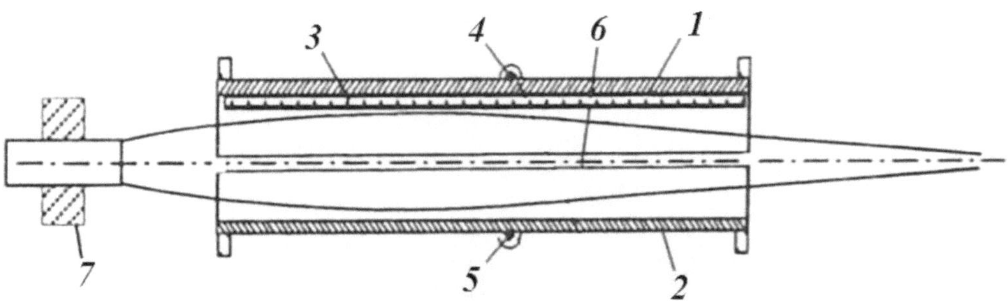

Fig. 23. Sketch of specimen: *1* - specimen half with coating, *2* - specimen half without coating, *3* - coating, *4, 5* - thermocouples, *6* - axis of specimen (flame), *7* - burner

rig and use the small-size specimens whose surfaces during tests are accessible for inspecting the thermal state both by the contact and contactless methods. This rig in particular is usable effectively for conducting the comparison thermal barrier propertiesand thermocycles tests of various coatings. The rate of change of the temperature in a thermocycle reaches 100 °C/s. For performing these investigations, a test rig has been developed with gas-flame heating of model specimens. The gas generator is a water electrolysis device equipped with a control system; it provides the variable flammable gas flow. Hydrogen has a high combustion temperature and this fact ensures high-speed heating of the specimens. This test rig has a system for providing enrichment of the flammable gas with different fuels. This makes it possible to attain the required gas composition. While testing, the burner is installed fixed, however the attachment allows its position to be adjusted. The hollow specimen is of an axisymmetrical form. Before the test, the burner is installed in a way ensuring coincidence of the specimen axis with the flame torch axis in the process of heating. While investigating the efficiency of influence of ceramic coating on specimen temperature state, the unit with specimens was fixed. A special specimen construction was developed for these tests. The hollow specimen was cut longitudinally in two equal portions. The ceramic coating under investigation was applied on one half of the specimen, the other half remained uncoated. The thermocouples XA by diameter of 0,2 mm weld on an external surface of halfs of a compound specimen (Fig. 23) and are connected to recording computer system. The half of a specimen is protected from products of combustion by a coating it is warmed up less, than unprotected half. The difference of temperatures Δt of protected wall with coating and unprotected wall characterizes the efficiency and thermal conducting of the thermal barrier coatings. Heat insulating material was placed between them to exclude the influence of heat transfer through the contacting edges of the specimen halves. While heating, temperature was measured on the outer (opposite to flame torch) specimen side. Conditions for heating the inner surfaces of both specimen halves by flame were the same, but with a difference in heat protection efficiency the outer surface of the specimen with a TBC had a lower temperature than the surface without a TBC. The after of lighting of a combustible gas the heating of an internal walls of both halfs of model begins. The difference of temperatures on lateral side grows until the heat transfer from a hot surface of a wall to cold surface is less, than a heat-conducting from an external surface in an environment. At absence of the organized cooling lateral side of a wall the maximal difference of temperatures Δt_{max} outside of both of halfs corresponds to a gradient of temperatures on TBC under these conditions. In experiment Δt_{max} it is reached at temperature of a cold wall 600 °C. The results of investigations are presented on Fig. 24 and 25.

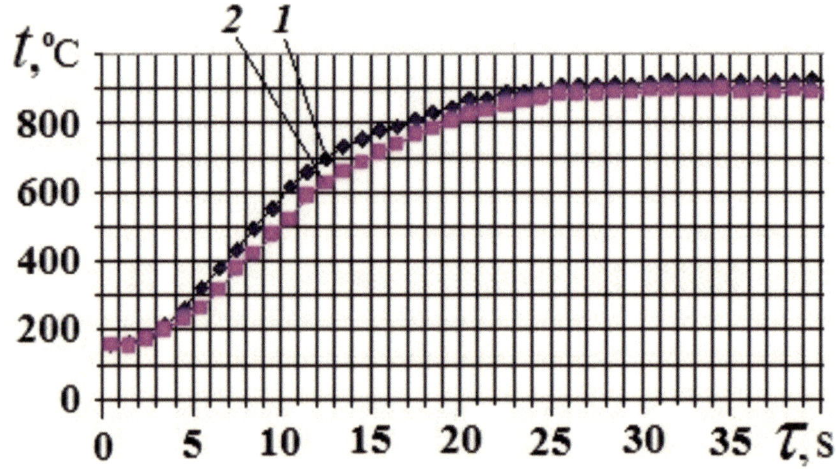

Fig. 24. The temperature on an external (cold) surface specimen: *1* - without coating, *2* - with ceramic coating (APS technique)

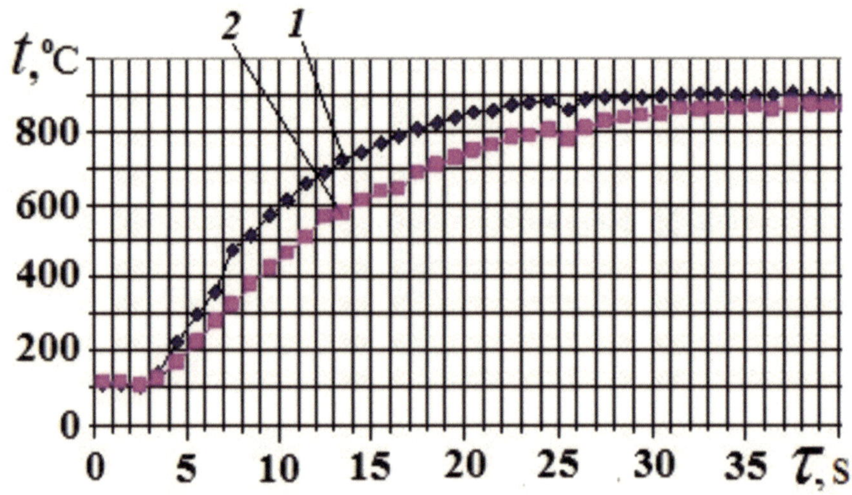

Fig. 25. The temperature on an external (cold) surface specimen: *1* - without coating, *2* - with ceramic coating (EB technique)

The models with ceramic TBC ZrO_2 + 8%Y_2O_3 deposited on plasma technology were made of Ni-alloy. Other models with TBC of columnar structure deposited on electron beam technology were made of other Ni-alloy. Unprotected halfs of each model were made of the same material as halfs with TBC. The tests on each model were repeated some times for maintenance of reliability of determination of heat-protective efficiency. At retesting the model was unwrapped about the axis on 180 °. The difference of temperatures at repeated measurements did not exceed 10 °C. The maximal temperature on the "cold" side of a wall made 900 °C. The temperature of a gas stream made 1773 K. The experimental investigations have shown that the efficiency of decrease of metal temperature at gas-flame heating after deposited TBC by thickness δ = 120 μm on plasma and electron beam technologies make correspondingly Δt_{Max} =

60-70 °C and Δt_{Max} = 100-110 °C. By he received results it is possible to estimate the thermal conductivity EB ceramic coatings which on the average in 1,6 times is lower than at APS coatings. Thus the received results of experimental estimation of the thermal conductivity and decrease of wall temperature of heat-resistant materials after deposited TBC of ZrO_2 + 8%Y_2O_3 by thickness about 120 μm show that at gas-flame heating of models the investigated EB coating of columnar structure protects metal is better than the tested APS coating. The developed original method of the experimental determination of thermal conductivity and estimation of efficiency of the thermal protection of details with thermal barrier ceramic coatings at gas-flame heating provides the reception of more exact data about thermophysic properties of ceramics under operating conditions of turbine details of aviation engines.

5. Calculated investigations of stress state of columnar structure of thermal barrier ceramic coatings with view of influence of centrifugal forces

The most effective protection of a detail material against a thermal flow occurs in case of use ofelectron beam method for depositing of ceramic coatings ZrO_2 (Tamarin & Kachanov, 2008). With the help of the specified method the ceramic coating of column structure on a surface of a metal sublayer (heat resisting coating) of working turbine blade is formed. The specified ceramic barrier coating is generated as columns (Fig. 26), are directed perpendicularly surface on which it is deposited. The columns of the ceramic coating possess low heat conductivity and provides the required durability at thermal cycles. The strength characteristics of ceramics are very low.

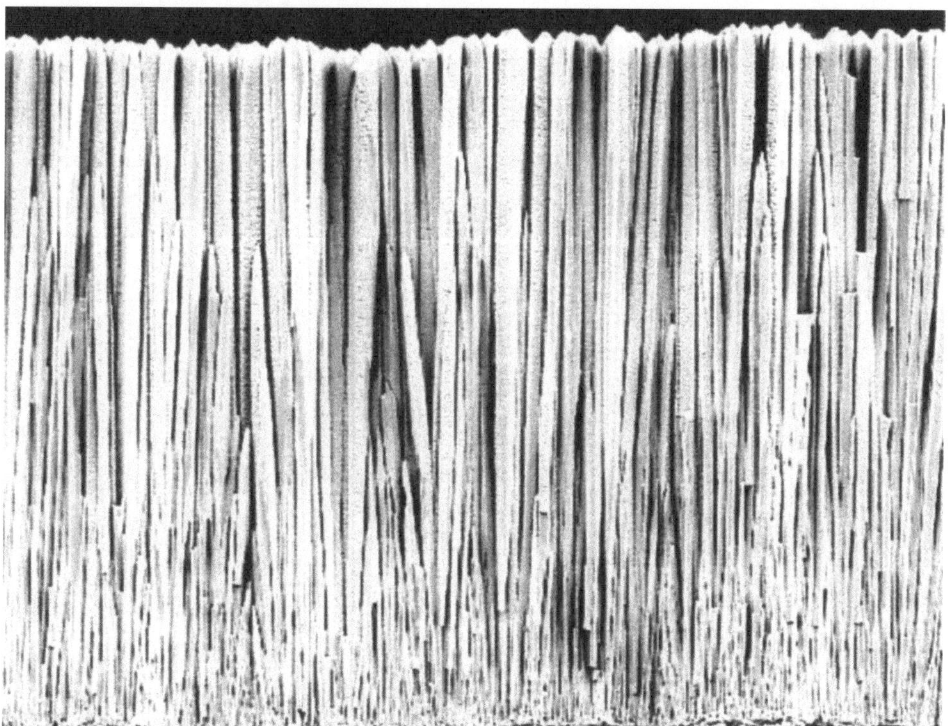

Fig. 26. Columnar thermal barrier ceramic coating (EB technique)

The calculated investigations of stress state of columnar ceramic coating of GTE blade on view of operated conditions were carried out when the materials of blade and ceramic coating are loaded by centrifugal forces (lepeshkin, 2010). The deformation of a sublayer under action of centrifugal forces together with temperature deformation is accompanied by increase of the distance between legs of columnar coating. Thus the jointed coating surface to crack on blocks and single columns (Fig. 26 and Fig 27).

Fig. 27. Columnar thermal barrier ceramic coating (surface)

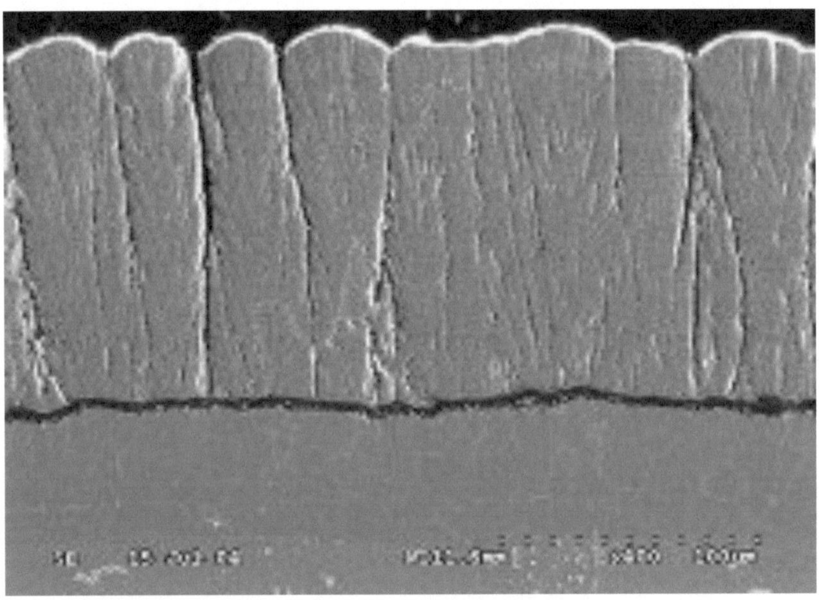

Fig. 28. The growth of cracks in a columnar coating leads to occurrence of blocks

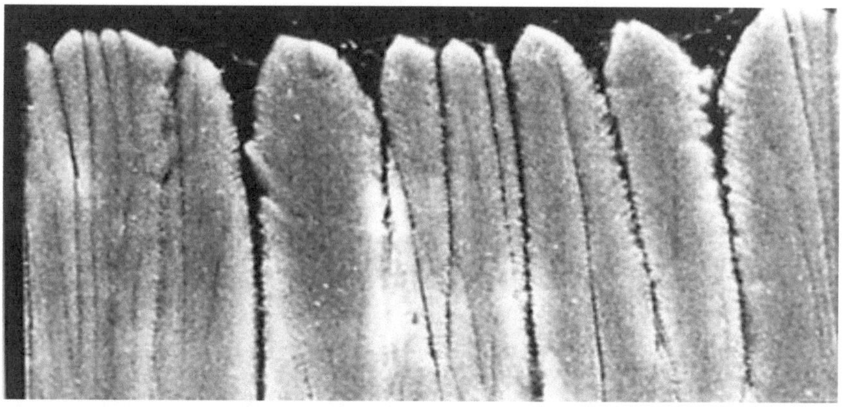

Fig. 29. The top part of a columnar coating

Under action of centrifugal forces the coating columns are exposed to a bend. The low strength of ceramics at a stretching ($\sigma_B \leq$ 50-200 МПа) leads to breaking off columns during a bend. Therefore the calculation of as much as possible allowable thickness "column" barrier ceramic coating should be carried out in view of operational loadings and also a conFiguration of coating columns in view of a metallized bottom ceramic layer of a coating by thickness 10-15 μm. The destruction of the columnar coating at height 10-20 μm under influence of operational loadings is shown in Fig. 30 and the spalling of the thermal barrier ceramic coating on the turbine blade is presented in Fig. 31.

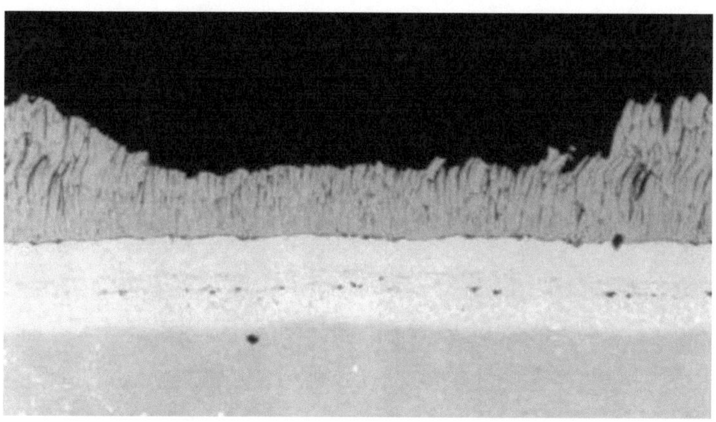

Fig. 30. Destruction of the columnar coating

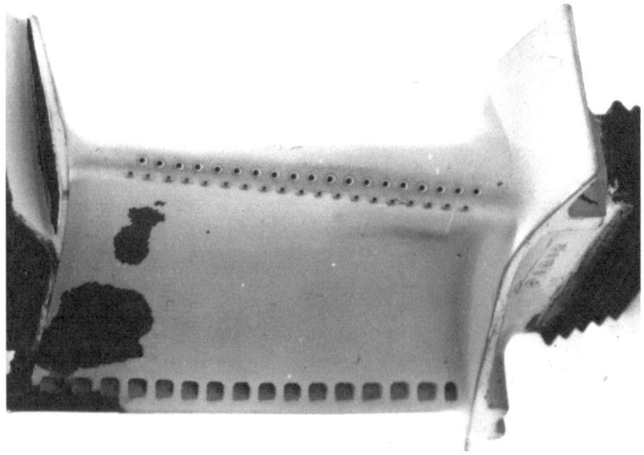

Fig. 31. Spalling of the thermal barrier ceramic coating on the turbine blade

The analytical calculations of stress state of columns of a ceramic coating of turbine blade were carried out under following conditions: frequency of rotation - 10000 r.p.m, radius - 400 mm from an axis of rotation, density of a coating - 4450 kg/m^3, parameters of columns: d_1 = 0,5 μm - diameter of the basis of a column, d_2 = 0,5÷5,0 μm - diameter of the top part of a column, l - height of a column (thickness of a coating). Two calculated cases were considered (Lepeshkin & Vaganov, 2010). In the first case the stress state of a single column was considered with fastening his leg in cantilewer. In the second case the calculation of stress state of a column in the block was carried out at fastening his leg in the basis of the block and his top part in a continuous surface of the block in view of a hypothesis of plane-parallel movement. The positions of the given hypothesis consist of the following. The top part of the block of a coating is formed by connection of the top parts of columns and represents a continuous surface. The roof of coating block starts to move in parallel the basis of the block under influence of centrifugal forces on the block. In view of the specified conditions the stress state of a column in the block in a field of action of centrifugal forces is calculated. The calculated circuits for determination of the stress state of columns of a ceramic coating are resulted in Fig. 32. From Fig. 26 follows that columns have the cone form that also is shown in circuits in Fig. 32.

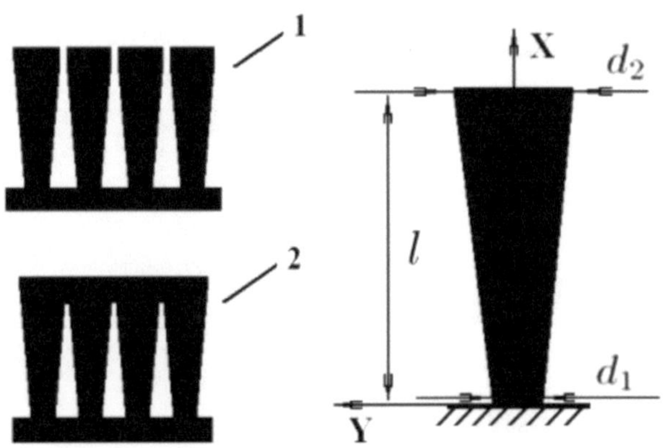

Fig. 32. Calculated schemes of determination of a stress state: 1 - single columns; 2 - columns in blocks; d_1 - diameter of the basis of a column, d_2 - diameter of the top part of a column,

The features of calculation of a stress state of coating columns in different cases of fastening at influence of centrifugal forces. The bending moment under action of centrifugal forces in section of a column at length x:

$$M(x) = M_B + \int_x^l \rho\omega^2 r(v - x)S(v)dv\,, \tag{3}$$

Here ρ - density of a material (ceramic) of a column,

ω- angular speed of rotation,

r – distance from a column up to an axis of rotation,

M_B - moment of action of external forces in top section,

v – current coordinate.

The bending moment from action of centrifugal forces in column cantilewer

$$M(0) = M_B + \int_0^l \rho\omega^2 rvS(v)dv\,. \tag{4}$$

Some auxiliary parameters:

the area of section of a column at length x

$$S(x) = \pi\frac{d(x)^2}{4} \tag{5}$$

moment of inertia of section at length x

$$J(x) = \pi\frac{d(x)^4}{64} \tag{6}$$

diameter of circular section at length x

$$d(x) = d_0 + \frac{d_2 - d_0}{l} \tag{7}$$

Then

$$y'' = \frac{M(x)}{EJ(x)}\,. \tag{8}$$

If the top part is free:

$$M(0) = \int_0^l \rho\omega^2 rvS(v)dv\,. \tag{9}$$

The condition of a fastening of columns in the top part (plane-parallel movement)

$$d(x) = d_0 + \frac{d_2 - d_0}{l}. \tag{10}$$

$$y'(l) = 0 \tag{11}$$

As y′(0) = 0 (rigid fastening in the basis), then

$$\int_0^l y''(v)dv = y'(l) - y'(0) = 0 \tag{12}$$

Hence, the condition for a moment

$$\int_0^l \frac{M(v)dv}{EJ(v)} = 0. \tag{13}$$

Substituting in (10) the formula (1) we receive the moment of action of external forces in the top section of a column:

$$M_B = -\frac{\rho\omega^2 r \int_0^l \frac{\int_v^l (u-v)S^2(u)du}{d(v)^4}}{\int_0^l \frac{dv}{d(v)^4}}, \tag{14}$$

Thus, it is known M(x):

$$M(x) = -\frac{\rho\omega^2 r \int_0^l \frac{\int_v^l (u-v)S^2(u)du}{d(v)^4} dv}{\int_0^l \frac{dv}{d(v)^4}} + \int_x^l \rho\omega^2 r(v-x)S(v)dv. \tag{15}$$

Knowing M(x) is possible to find the maximal stretching stress in section of a column:

$$\sigma_{max}(x) = \frac{M(x)d(x)}{2J(x)} \tag{16}$$

The results of calculated investigations have been conducted. The stress distributions on length of single columns and the columns which are taking place in blocks of a ceramic coating in height 100 microns at influence of centrifugal forces (as shown in Fig. 33-36) are received. The stresses (curves) in the basis of a column depending on length are submitted on Fig. 37. The taper of a columns is determined by a ratio d_2/d_1. From Fig. 35 follows that at increase of a ratio d_2/d_1 from 1 up to 10 the stresses in the basis of the column which is taking place in the block are reduced twice. The analysis of stress distribution on length of a column in 100 μm with the cross-section sizes d_1 = 0,5 μm, d_2 = 2,0 μm in Fig. 36 shows that the stress in the basis of a column in the block is less than in the basis of a single column in 7 times. For comparison of stresses in the basis of a column in different cases the curves 1 and 2 with ratios d_2/d_1 from 1 up to 2 for a single column and curves 3, 4 and 5 with ratios d_2/d_1 from 1 up to 10 columns in the block at increase in length of a column are shown on Fig. 37.

From the analysis of Fig. 37 follows that there are following restrictions on length of a column in view of stresses in the basis: more than length 120 μm at d_2/d_1 = 10 and 100 μm at d_2/d_1 = 4 and 80 μm at d_2/d_1 = 1 for length of a column in the block and 80 μm at d_2/d_1 = 10 and 40 μm at d_2/d_1 = 4 for a single columns.

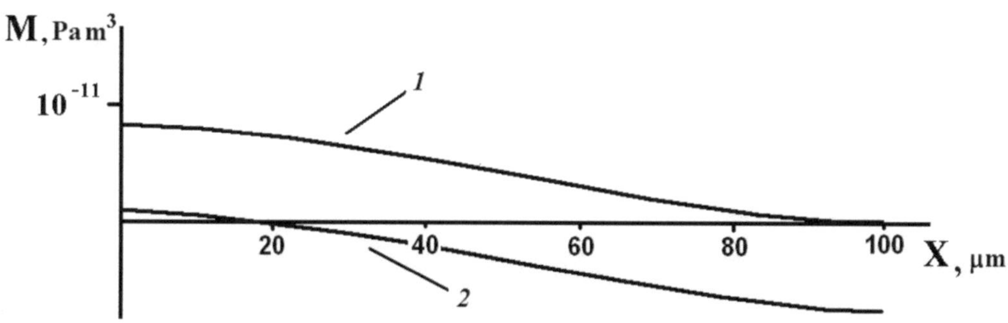

Fig. 33. Distribution of the bending moment in a column by length 100 μm (d_1 = 0,5 μm, d_2 = 2,0 μm): 1 - single column, 2 - column in the block

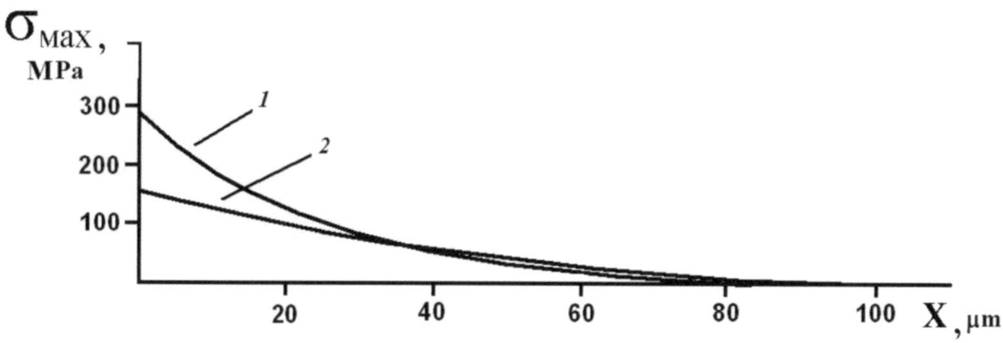

Fig. 34. Distribution of stresses in a single column by length 100 μm, the column of the different cross-section sizes: 1 - d_1 = 0,5 μm, d_2 = 1,0 μm; 2 - d_1 = 0,5 μm, d_2 = 0,5 μm

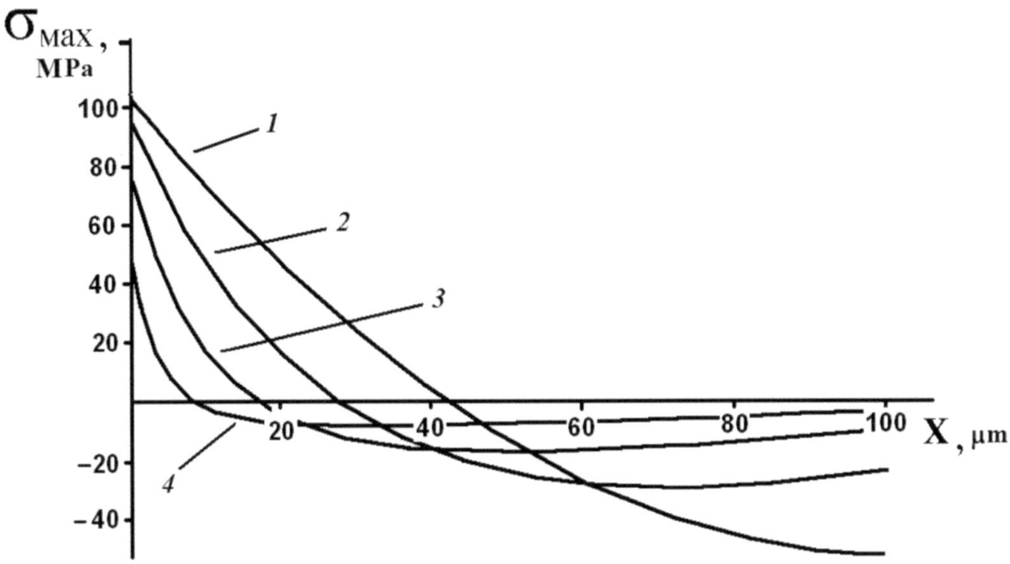

Fig. 35. Distribution of stresses in a column by length 100 μm (in the block), the column of the different cross-section sizes: 1 - d_1 = 0,5 μm, d_2 = 0,5 μm; 2 - d_1 = 0,5 μm, d_2 = 1,0 μm; 3 - d_1 = 0,5 μm, d_2 = 2,0 μm; 4 - d_1 = 0,5 μm, d_2 = 5,0 μm

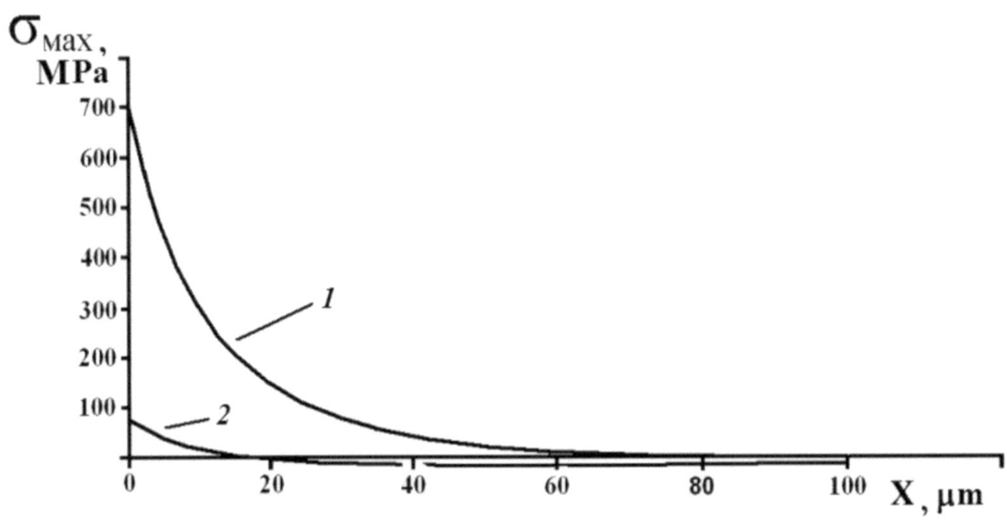

Fig. 36. Distribution of stresses in a column by length 100 μm (d_1 = 0,5 μm, d_2 = 2,0 μm): 1 - single column, 2 - column in the block

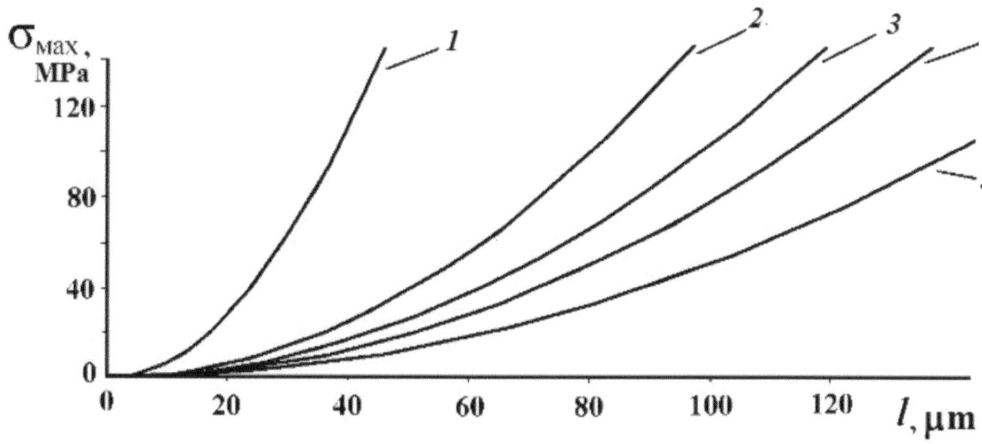

Fig. 37. Stresses in the basis of a column depending on length: Single column: 1 - d_1 = 0,5 μm, d_2 = 2,0 μm; 2 - d_1 = 0,5 μm, d_2 = 0,5 μm; Column in the block: 3 - d_1 = 0,5 μm, d_2 = 0,5 μm; 4 - d_1 = 0,5 μm, d_2 = 2,0 μm; 5 - d_1 = 0,5 μm, d_2 = 5,0 μm

The formed single columns are exposed to a bend under influence of centrifugal forces under operating conditions at cracks in coatings and at their length more than 40-100 μm can break. In the educated blocks the columns are loaded in a field of action of centrifugal forces and at their length no more than 100-140 μm can be kept without destruction. Thus the probability of destruction of columns in the block decreases at increase taper - ratio d_2/d_1. It is shown that at designing columnar ceramic coverings their allowable thickness should make no more than 100-140 microns in conditions of influence of the centrifugal forces on the basis of carried out investigations.

6. Thermal barrier ceramic coatings of variable thickness

One of the directions of future investigations are improvement of the designs and rise of durability of ceramic coatings and the development of advanced technologies for their depositing. The blades and turbine components of gas turbine engines have a ceramic coating of uniform thickness. The disadvantages of the part design are the increased thickness of the ceramic coating that maintains or increases the unevenness of temperature distribution and thermal stresses in the metal blade. In addition, in the increased ceramic coating mass of the turbine blades will arise the increased stresses under the influence of centrifugal forces. The factors can lead to lower life of a coating and blade. On a turbine parts is possible to put ceramic coatings of variable thickness. The coatings of variable thickness allow to raise the durability of a turbine part due to increase of strength of a covering and reduction thermal stresses in part metal. The durability raises due to variable thickness of a coating and due to uniformity of temperature distribution in a junction of a coating with metal of a part and due to performance of a coating of the maximal thickness in zones of the maximal temperatures and the minimal thickness in zones of the minimal temperatures on a surface of a coating. In result the temperature drops and thermal stresseson a profile and height of a part (blade) are reduced. Also the probably of occurrence of defects, cracks and chipping of ceramics in the areas of stress concentration decreases and the durability of a coating and blade increases in view of influence of centrifugal forces and variables thermal stresses. The examples of designs of turbine parts of with thermal barrier ceramic coatings of variable thickness are considered below. The turbine blade and flame tube of a combustion chamber of gas turbine engine with coating variable thicknessare presented in Fig. 38 and Fig. 39 (Lepeshkin, 2005).

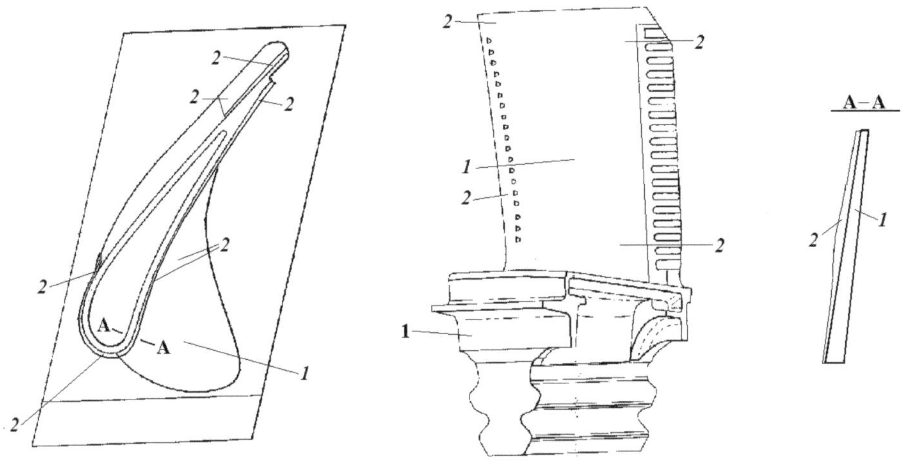

Fig. 38. Turbine blade with thermal barrier ceramic coating of variable thickness:*1*- blade,*2* - coating

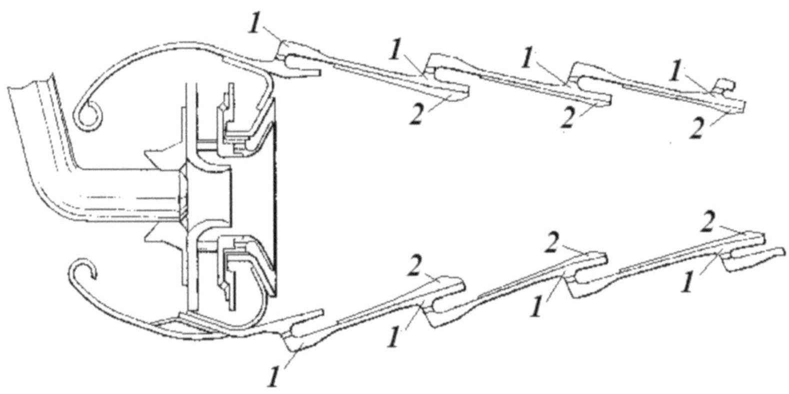

Fig. 39. Flame tube of a combustion chamber with thermal barrier ceramic coating of variable thickness:*1*- flame tube, 2 - coating

7. Conclusion

The analysis of methods of heating for investigations of ceramic thermal barrier coatings of parts has been conducted. The induction heating was demonstrated to be effectively used during investigations of a thermal ceramic barrier coatings and tests for thermofatigue and thermo-mechanical fatigue of such gas-turbine engine parts with a TBC as the cooled turbine blades and vanes and combustion liner components. When using induction heating the surface heat-up of the part in the high-temperature gas flow is well modeled. It was shown that induction heating at a frequency 440 kHz may be successfully used for tests of the parts with a

TBC. It was experimentally shown that when using such a heating method, the temperature of the outer ceramic coating surface exceeds the temperature of the metallic bond coat by 50-70°C. Thus, a design-experiment method of high-frequency induction heating and thermophysical measurements in thermal cycling tests of blades and other cooled parts with a TBC has been developed taking into account the electrophysical and thermophysical properties of their materials. The results of thermophysical measurements, design-experiment studies of the nonstationary thermal state of the parts with coatings with the use of a thermal vision system, and thermal cycling tests of blades and models of flame tubes with thermal barrier ceramic coatings are presented. The generalized dependence of the temperature gradient across the ceramic coating thickness on the frequency of the electric current was obtained using multivariant calculations. The test rig with induction heating at a frequency 440 kHz was used also for making the comparison evaluation of the influence of various thermal ceramic barrier coatings (APS and EB techniques) on thermocyclic strength of the parts with a TBC (to select material and coating for the part and mature the repair technology). Gas-flame heating is considered to be preferable when investigating the gas-turbine engine parts with a TBC for thermofatigue in the special cases when both the convective and radiant components of thermal flow are of great importance. The small-size rig with gas-flame flow made it possible to conduct the comparison investigations with the purpose of evaluating the efficiency of thermal protection of the ceramic deposited thermal barrier coatings on APS and EB techniques. The developed design-experiment method was introduced in bench tests of turbine blades of gas turbine engines. The use of the developed method for high-frequency induction heating in thermal cycling tests of blades and models with thermal barrier ceramic coatings made it possible to reduce the duration of tests and their cost and to obtain the experimental evaluation of the service life of ceramic coatings with due regard for their nonstationary thermal and thermostressed states. The results of successful tests were used in operation of aircraft engines. At designing columnar ceramic coverings it is necessary to consider that their allowable thickness depending on conditions of influence of the centrifugal forces on the basis of carried out investigations. The further investigations will be carried out in the following directions: comparison of the results of the evaluations of the thermal cyclic fatigue life of the cooled parts during gas and high-frequency induction heating, improvement of the technique developed for determining the thermal conductivity of ceramic coatings deposited by the electron-beam technique and design of new types of coatings.

8. Acknowledgment

This work has been performed at CIAM. I is very thankful to Dr. N.G. Bychkov for his help in carrying out of the investigations of a thermal ceramic barrier coatings.

9. References

Stringer, J.; Streiff, R.; Krutenat, R. & Gaillet, M. (1989) The Reactive Element Effect in High-Temperature Corrosion, *High Temperature Corrosion.* No.2, pp. 129-137. Elsevier Science Publishers

Malashenko, I.S.; Vashchilo, N.P.; Belotserkovsky, V.A. & Yakovchuk, K.Y. (1997) Effect of Yttrium on Functional Characteristics of Vacuum Condensates and Protective Coatings MCrAlY and MCrAlY/ZrO_2-8%Y_2O_3 at Thermocyclic Loading, *Adv. Spec. Electrometali,* Vol.1, pp. 24-33

Miller, R, A.; Garlick, R.G. & Smialek, J.L. (1983). Phase Distribution in Plasma-Sprayed Zirconia-Yttria. *Am. Ceram. Soc. Bull.,* Vol.63, No.12, pp. 1355-1358

Ingel, R.P.; Lewis, D.; Bender, B.A. & Rice, R.W. (1984). Physical, Microstructural and Thermomechanical Properties of ZrO_2 Single Crystals, *Advances in Ceramics,* Vol.12, pp. 408-414, American Ceramics Society, Inc.

Stecura, S. (1986). Optimization of the Ni-Cr-Al-Y/ZrO_2-Y_2O_3 Thermal Barrier System, *Adv.Ceram. Mater.,* Vol.1, pp. 68-76

Ruckle, D.L. & Duvall, D.S. (1984). Quench Cracked Ceramic Thermal Barrier Coatings, U.S. Patent 4457948, August 3, 1984

Strangman, T.E. (1982) Columnar Grain Thermal Barrier Coatings, U.S. Patent 4321311, March 23, 1982

Schulz, U.; Fritscher, K.; Ratzer-Scheibe, H.J., et al. (1997). Thermocyclic Behavior of Microstructurally Modified EB-PVD Thermal Barrier Coatings, *High Temperature Corrosion,* pp. 957-964, No.4

Parker, W.J.; Jenkins, R.J.;Bulter, C.P. & Abbott, G.L. (1961). Flash Method of Determining Thermal Diffusivity, Heat Capacity and Thermal Conductivity, *J. Appl. Phys.,* pp. 1679-1684, Vol.32, No.9

Pawlowski, L.; Lombard, D.; Mahlia, A. et al. (1984). Thermal Diffusivity of Arc Plasma Sprayed irconia Coatings, *High Temp.* High Press., pp. 347-359, Vol.16

Maesono, A. (1983). Measurement of Thermal Constants by Laser Flash Method, SinkuRiko Co. LTD, April, 1983

Cape, A. & Lehman, G.W. (1963). Temperatureand Finite-Pulse Time Effect in the Flash Method for Measuring Thermal Diffusivity, *J. Appl. Phys.,* p 1909-1913, Vol.34, No.7

Clark, L.M. & Taylor, R.E. (1975). Radiation Loss in the Flash Method for Thermal Diffusivity, *Appl. Phys.,* pp. 714-719, Vol.46, No.2

Nicholls, J.R.; Lawson, KJ.; Johnston, A. & Rickerby, D.S. (2001). Low Thermal Conductivity EB-PVD Thermal Barrier Coatings, *High Temperature Corrosion,* Ed., Trans Tech Publication, pp. 595-606, No.5

Lawson, K.J.; Nicholls, J.R. & Rickerby, D.S. (1996). The Effect of Coating Thickness on the Thermal Conductivity of CVD and PVD Coatings, *Fourth International Conf. on Advances in Surface Engineering,* Newcastle, U.K.

Kuvaldin, A.B. & Lepeshkin, A.R. (2006). *High speed Regimes of Heating and Thermal stresses in Articles,* NGTU, 286 p., ISBN 5-7782-0626-7, Novosibirsk

Lepeshkin, A.R.; Bychkov, N.G. & Perchin, A.V. (2005). Method of Test Parts with a Thermal barrier coating for durability, R.U. Patent 2259548, July 27, 2005

Bychkov, N.G.; Nozhnitsky, Y.A.; Perchin, A.V. et al. (2008). Investigations of Thermomechanical Fatigue for Optimization of Design and Production Process Solutions for Gas-Turbine Engine Parts, *International Journal Fatique,* No.30

Tamarin, Yu.A. & Kachanov, E.B. (2008). Properties of the Thermal Protective Coatings Rendered by Electron Beam Technology (New Technological Processes and Reliability of Gas Turbines Engines), CIAM, pp. 125-143, issue 7,Moscow

Rubashev, M.A.; Presnov, V.A. & Rotner, Yu.M. (1980) Thermal Dielectrics and Their Junction in New Technologies, Atomizdat, Moscow

Lepeshkin, A.R. & Vaganov, P.A. (2010). The Calculation of Stressed State of the Cone-Columnar Structure of the Thermal Barrier Ceramic Coatings of GTE Blades under Service Loads, *Russian Conference "New materials and technology",* MATI, p. 57, Moscow

Lepeshkin, A.R.; Bychkov, N.G. & Perchin, A.V. (2005). Turbine Blade, R.U. Patent 2259481, July 27, 2005

Lepeshkin, A.R.; Bychkov, N.G. & Perchin, A.V. (2005).Flame Tube of a Combustion Chamber, R.U. Patent 2260156, September 10, 2005

6

Thermal Spraying of Oxide Ceramic and Ceramic Metallic Coatings

Martin Erne and Daniel Kolar
Institute of Materials Science, Leibniz University of Hannover
Germany

1. Introduction

Thermal Spraying is called a group of processes by means of that thin ceramic and ceramic metallic (cermet) coatings can be applied on a vast variety of materials, so called substrates. The goal is to reach considerably different characteristics on the surface of the component part regarding the resistance against abrasion and corrosion, the electrical conductivity and many more. This chapter intends to give an overview of the different processes, the processable feedstock materials, the different areas of application and new developments in the field of Thermal Spraying.

2. Thermal spray processes and coatings´ microstructure

All together thermal spray processes make use of heat and kinetic energy to warm-up and propel feedstock material to build up a coating on the substrate. Often the goal is to melt the feedstock thoroughly due to reach a dense microstructure, but in some cases the feedstock impinges in solid state and is deformed by the kinetic energy as the particles reach supersonic velocity before impact. Dependent on the source of energy distinctly different process characteristics and therefore visibly diverse microstructure and properties of the coatings can be obtained.

In the norm DIN EN 657 "Thermal Spraying" the different processes are distinguished by the means of the energy source. The processes being widely in operation are based on the energy sources flames and electric or gas discharges. Although the lasers assisted spraying techniques are coming more and more into operation, they cover only a small segment compared to the conventional techniques. By means of molten bath and "cold" or, in other terms, kinetic spraying only metallic feedstock can be used. Therefore both processes are not covered in this chapter. In fact the focus of this chapter lies on workings done in the fields of thermal spraying by means of Atmospheric plasma spraying (APS) as well as High velocity oxyfuel spraying (HVOF, see markings in **Figure 1** on the following page).

2.1 Achieving near net shape coatings

Besides metallic feedstock for their electrical and tribological properties as well as repair purposes many ceramic materials can be sprayed. The commonly used feedstock can be divided into oxide ceramics and the embedding of covalent bound materials like carbides and borides in metallic binder phases (so called cermets derived from "ceramic metals"). Besides the different hardness of the hard phase and the two-phase nature of cermet coatings, the feedstock itself is manufactured by totally different production routes. For both the molten and crushed oxide ceramics and the usually agglomerated and sintered cermet powders there is the trend to use finer grain sizes to reach denser and better microstructures of the coatings on the one hand (Gell, M., et al., 2001; Tilmann et al., 2008a). On the other hand with fine feedstock powders near net shape coatings can be sprayed, showing a comparable low surface roughness, both allowing to reduce the costs of finishing workings (Matthäus, G., Wolf, J. & Ackermann, D., 2010; Tilmann et al., 2008b).

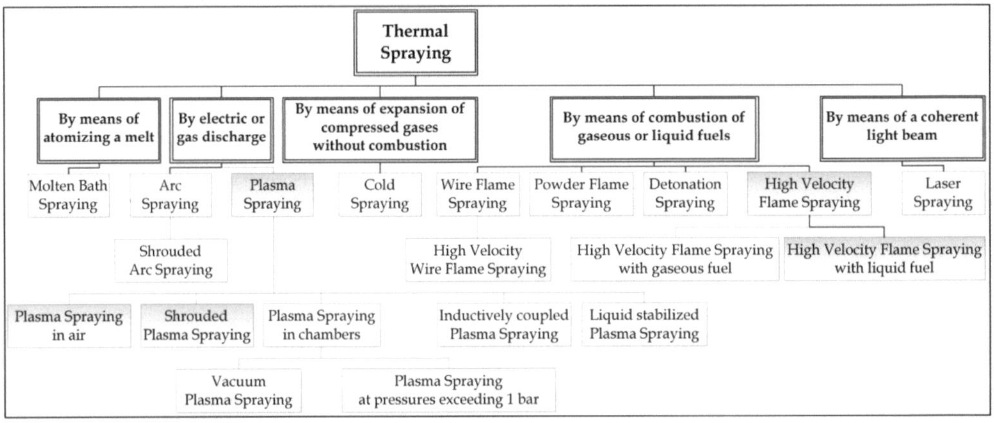

Fig. 1. Classification of Thermal Spray processes with regard to the source of energy (after DIN EN 657 "Thermal Spraying")

In the following the differences in the performance of abrasion and corrosion resistant coatings regarding the deposition efficiency, surface roughness, hardness, porosity, wear behavior and corrosion resistance will be discussed regarding the inset feedstock grain size and the resulting coatings´ microstructure. Several feedstock materials typically used for named fields of operation (WC-CoCr 86/10/4, Cr_3C_2-Ni20Cr 75/25 and Cr_2O_3) were considered for developing near net shape coatings. In contrast to grain sizes commonly used in thermal spray processes of up to appr. 50 µm, the grain sizes of all examined powders were specified with a maximum of 25 µm (-15+5 µm, -20+5 µm and -25+5 µm). Different types of conventional and one specialized powder feeder were investigated regarding their abilities of continuous feeding. For the coating experiments the kerosene fuelled HVOF-gun K2 (GTV GmbH, Luckenbach, Germany) was used to apply the carbide based feedstock materials (WC-CoCr and Cr_3C_2-NiCr), whereas the conventional APS-gun F4 (Sulzer Metco AG, Wohlen, Switzerland) was used to apply Cr_2O_3 coatings. Compared to coatings being sprayed using conventional fractionated feedstock, the coatings based on fine feedstock showed better results concerning their key characteristics.

2.2 Comparison of microstructures, phase contents and deposition rates

The micrographs on the following page show the microstructures obtained by spraying fine feedstock with particle sizes < 25 µm (left side) and conventional fractionated feedstock (-45+5 µm in case of chromia and -45+25 µm for the cermets, right hand side).

The parameter settings for the spraying experiments were investigated using methods of designed experiments (for a detailed discussion see chapter 3). After conducting tests regarding a continuous feeding of the different feedstock powders, preliminary test series were conducted to evaluate the effects of the main process parameters regarding the feedstock grain size, the amperage in case of APS and the air-fuel-ratio in case of HVOF, spraying distance and powder feed rate. The results of these experiments were investigated regarding the coatings criteria named at the beginning. For finding optimal parameter sets the economic relevant criteria deposition efficiency and surface roughness were given the highest priority as well as reaching sufficiently high indention hardness at the same time.

The microstructures of the optimum parameter sets for the fine grained feedstock compared to coatings sprayed with conventional fractionated feedstock are shown in **Figure 2**. The metallographic cross sections of the coatings showed, that the porosity of the coatings can be decreased by processing fine powders. Measurements by means of image analysis revealed, that the ratio of porosity in case of the near net shape coatings is approximately only on quarter to one third compared to the conventional coating systems reaching values of 0.1 % in case of the WC-CoCr coating. At the same time the roughness of the top layers described by the profile parameters roughness average (R_a) and height (R_Z) is also considerably lower. For all fine feedstock powders R_a values in the range of 2.5 to 2.7 ± 0.1 µm of the as sprayed coatings could be reached, whereas for the usually applied powders the values were significantly higher with 4.5 ± 0.3 µm in case of chromia and 6.7 ± 0.4 µm for the cermet coatings. Furthermore the uniformity of the coatings is significantly better when spraying the fine feedstock permitting the goal of applying near net shape coatings. But these efforts are accompanied by considerably lower deposition rates caused by lower mass throughputs and the difficult heat transfer to the relative high melting NiCr-matrix in case of HVOF spraying of the Cr_3C_2-NiCr feedstock. On the other hand this disadvantage can be equalized by the aim of achieving coatings of lower thickness resulting in comparable times for the spraying process for both the fine and the coarse fractionated feedstock.

Then again when spraying the finer powders in the spray process, there is also a higher risk of overheating the small spray particles. In particular the composition of the carbide based coatings can be changed because of decarburization and oxidation effects. The examination of the metallographic cross sections under this aspect showed, that especially the coatings based on fine Cr_3C_2-NiCr powder showed strong oxidation (see the dark-gray phases in **Figure 2b** left hand side). In order to achieve more information about these phase changes the carbide based samples were analyzed by X-ray diffraction. The obtained X-ray diffraction patterns are shown in **Figure 3**. The pattern of the Cr_3C_2-NiCr sample sprayed with feedstock -15+5 µm (see lower pattern in Fig. 3 a) shows noticeable Cr_2O_3 peaks indicating that a strong oxidation of the spray particles took place during the spray process. Furthermore decarburization effects were also stronger when using the fine powder. In the sample sprayed with the standard feedstock, the dominating carbide phase was Cr_3C_2, whereas the other coating was dominated by the lower carbide phase $Cr_{23}C_6$. For the WC-

CoCr samples the effect of decarburization was examined by determining the intensity ratio of the strongest WC in relation to the W_2C peak ($I_{W2C(100)}/I_{WC(100)}$, see Fig. 3 b). Similar values of 0.22 in case of the fine and 0.17 for the coarser fractionated feedstock were obtained indicating a stronger decarburization when spraying the fine feedstock. Compared to the Cr_3C_2-NiCr samples these effects of phase chances were quite low.

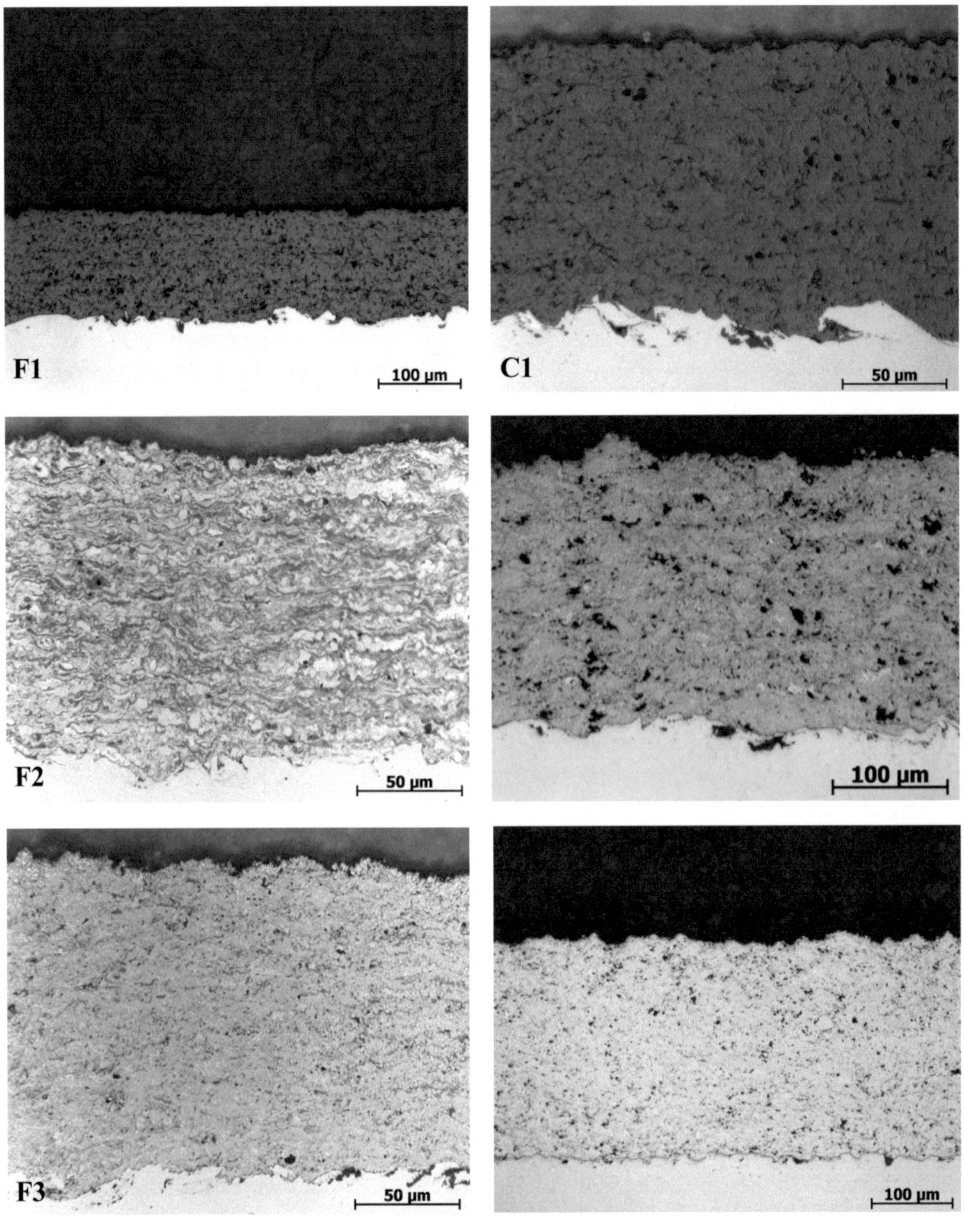

Fig. 2. Comparison of achievable microstructures using fine powder feedstock < 25 µm grain size (left) and conventional more coarsely fractionated feedstock -45+5/20 (right) for Cr_2O_3 (a), Cr_3C_2-NiCr (b) and WC-CoCr (c)

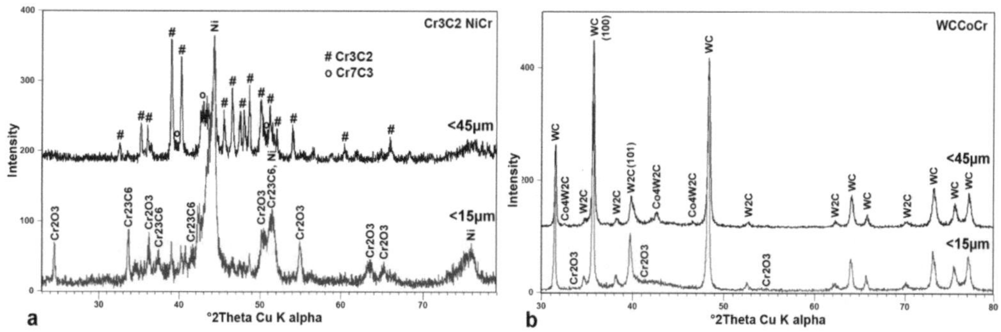

Fig. 3. X-Ray diffraction patterns of Cr_3C_2-NiCr (a) and WC-CoCr (b) coatings

2.3 Indentation hardness

One characteristic criterion determining the wear resistance of thermal sprayed coatings is the hardness, which is usually measured by indentation techniques. The Vickers hardness indentation test is well-established both in the course of the quality management of job shops as well as in the characterization of coatings reported in literature. Another technique is the superficial Rockwell hardness testing, by means of that the coatings can be analysed without metallographic preparation. To investigate the suitability of both methods and the influences on the measurement results, a cause-and-effect diagram was established for the indentation testing of thermal spray coatings (see **Figure 4**). The goal of the workings was the reduction of the variability of the measuring results to enhance the comparability.

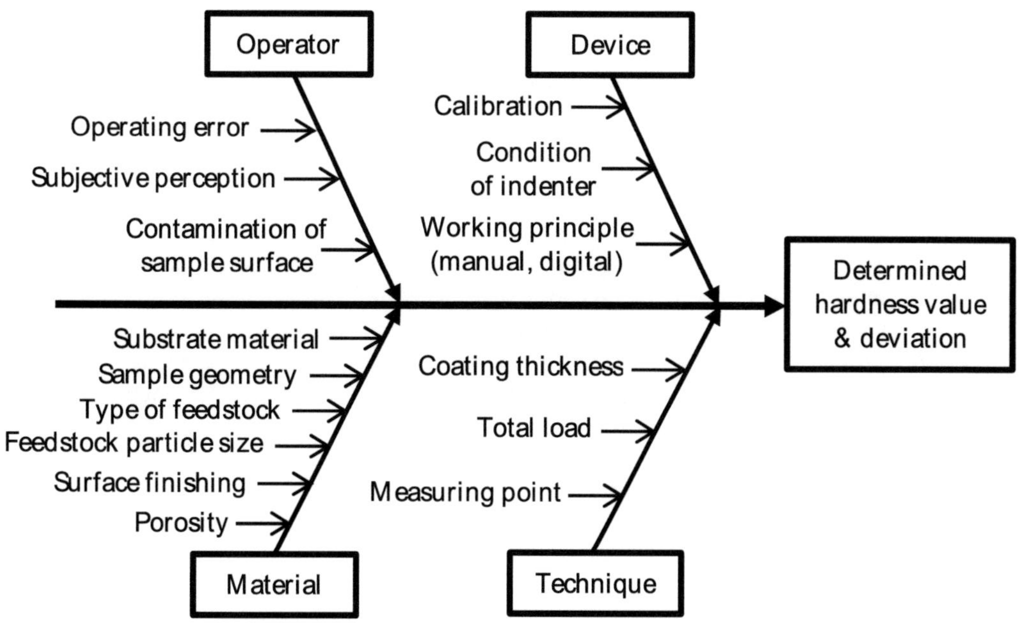

Fig. 4. Cause-and-effect diagram of the indentation hardness measurement of coatings

A large number of predominantly oxide ceramic and cermet coating systems were investigated concerning the different sources of variation depicted in Figure 4. In the following especially the influence of the microstructure, the loading force, the inset type of hardness tester and the necessary quantity of measurement repetitions on one sample regarding the increase of the repeatability are discussed. In the following **Table 1** the indentation hardness values for the coatings shown in **Figure 2** are listed.

Feedstock	Cr_2O_3						Cr_3C_2-NiCr						WC-CoCr					
Fraction/ Sample	F1			C1			F2			C2			F3			C3		
Technique[1)]	1	2	3	1	2	3	1	2	3	1	2	3	1	2	3	1	2	3
Mean	89,7	85,9	1222	87,4	90,2	1424	87,9	87,2	1032	85,6	85,8	867	87,5	86,8	1172	88,2	85,3	1245
SD 5 values	1,1	1,4	45	2,6	1,9	74	1,8	2,3	31,8	2,0	3,9	76	2,3	2,3	26,5	4,6	1,9	148,5
SD 10 values	1,0	1,7	37	2,3	2,0	81	2,0	2,5	57,3	1,8	3,5	73	1,9	2,9	84,3	3,4	2,9	142

[1]Techniques:
1 = Superficial Rockwell HR15N manual
2 = Superficial Rockwell HR15N digital
3 = Vickers HV0.3

Table 1. Comparison of indentation hardness values derived by superficial Rockwell and Vickers testing

The mean values were derived from 10 measurements for each sample and measurement technique. The results of the measurements of different experimental series were investigated regarding their distribution and the appearance of outliers using the span of standard deviation and Grubb´s test. In most cases the values are not normal distributed, but also hardly any outlier can be detected. Therefore the goal was chosen to reduce the standard deviation of the measurements as the repeatability between different operators, hardness testing devices etc. is expected to increase with decreasing standard deviation. For first evidence the standard deviations of the first 5 measurements and of 10 measurements were compared to get information about the necessary number of measurements to receive robust results.

The results of the two different types of Rockwell hardness testers (one manual Wilson device and a digital type STRUERS DuraJet with closed loop control of the applied force) do not differ very much. The standard deviation is lower than approximately 4 % of the mean measured value and is often higher when it was calculated from ten values instead of the first five ones. This might be due to influences of the microstructure on the results like unmelted particles in the case of chromia and the bimodal hardness distribution of the cermet type coatings. Furthermore the derived mean indentation hardness value is comparable for both the fine and the coarse fractioned feedstock. In case of the Vickers testing the same effect was established. As the Vickers measurements were performed by a minor experienced operator, the tests were repeated by another more experienced person. For the sample C1 a significant lower value of 1124 HV0.3 with comparable standard deviation values of 68 and 79, respectively, were derived. In a further series on the same sample the standard deviation could be reduced significantly to 27 for both 5 and 10 measurements by excluding nonuniformly shaped indentation pits showing different lengths of the two diagonals. The mean value of 1152 HV0.3 seems to be the most reliable one. When increasing the loading force to 0.5 kp, the standard deviation is comparable low with 24 to 27. But the calculated mean value of 1327 HV0.5 is considerably higher than all values derived with 0.3 kp loading force. Nevertheless the comparison with other testing series showing less dense microstructures resulted in the conclusion, that the standard deviation of Vickers measurements is lowest with 0.3 kp loading force. When applying 0.1 kp, the indentation pits are too small to be analysed correctly, and the standard deviation rises againg. With the higher force of 0.5 increased cracking occurs due to not optimal cohesion of the coatings. Therefore it is the best solution to choose 0.3 kp loading force to obtain results of high reproducibility.

To investigate the necessary number of measurement repetitions in correlation to the porosity as weakening effect of the coatings cohesion, samples with extraordinary high and relative low porosity were measured 50 times with all techniques. The relative uncertainty of the derived mean value is plotted over the number of repetitions (see **Figure** 5). It is calculated as follows:

Calculation of standard deviation: $$s=\sqrt{\frac{\sum_{i=1}^{n}\left(x_{mean}-x_i\right)}{n-1}} \quad (1)$$

Calculation of mean standard deviation: $$s_{mean}=\frac{s}{\sqrt{n}} \quad (2)$$

Calculation of relative uncertainty: $$\varepsilon=\frac{s_{mean}}{x_{mean}} \quad (3)$$

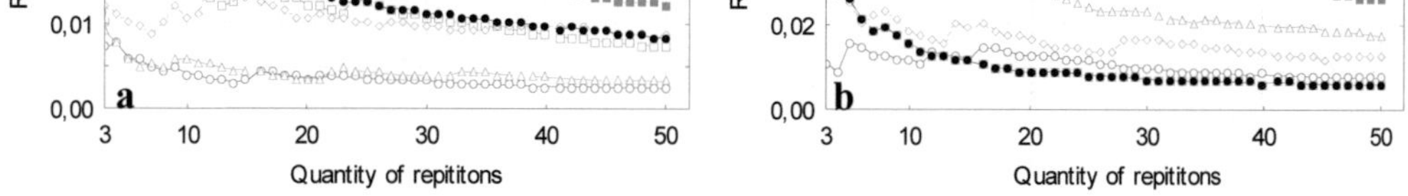

Fig. 5. Relative uncertainty of indentation hardness values for chromia (a) and Cr_3C_2-NiCr (b) coatings in relation to the measurement technique, the quantity of measuring repetitions and the coatings porosity given in brackets

As expected the relative uncertainty of the derived mean hardness value is significantly higher for the samples with high porosity compared to the more dense coatings. The values tend to remain static when more than approximately 20 repetitions are made, whereas this trend is reached after circa the half quantity of measurements when testing the denser coatings. Furthermore not the same degree of certainty can be reached when testing coatings with the high porosity. In further workings this tool will be investigated to work out a measuring concept to classify the reliability of indentation hardness testing of thermal sprayed coatings.

2.4 Corrosion and wear behaviour

The corrosion resistance of the coatings was determined with salt spray tests according to DIN EN ISO standard 9227. For this purpose mild and stainless steel substrates were coated and were exposed for 240 hours to a corroding atmosphere produced by spraying a sodium chloride solution. The appearance of corrosion products was evaluated every 24 hours. In addition the samples were weighed before and after the test period to determine mass increasing effects caused by formation of corrosion products. During the testing period the mass of the samples increased because of the formation of corrosion products, **Table 2**. In the case of the carbide based coatings the use of the fine fractionated feedstock lead to a considerable improvement in terms of corrosion resistance, the samples sprayed with fine powders showed significant less mass increases than the standard fractionated samples. The Cr_2O_3 coatings showed a quite contrary behaviour. This is due to the fact, that the chromia coatings received no sealing treatment leaving, so that the salt media could reach the substrate through the thin coating more easily compared to the thicker conventional sample. The coatings on stainless steel substrates showed the same behaviour like the coatings on mild steel substrates. But of course the actual values were lower due to the higher corrosion resistance of stainless steel.

Coating system	Feedstock grain size / Sample	Mass increase on substrate 1.0037 / 1.4301 [mg]
Cr_2O_3	-25+5 µm (F1)	90 / 8
	-45+5 µm (C1)	38 / 5
Cr_3C_2-NiCr	-15+5 µm (F2)	19 / 16
	-20+45 µm (C2)	111 / 23
WC-CoCr	-15+5 µm (F3)	36 / 8
	-20+45 µm (C3)	233 / 22

Table 2. Results of the corrosion tests: mass increase of coated samples exposed 240 h in salt spray fog.

The wear resistance of the coatings was evaluated by ball-on-disk wear tests according to ASTM standard G 99. The ball-on-disk test is a model test for determining friction and wear of two solid surfaces being in sliding contact (ball against coated disk). A sintered WC6Co ball (10 mm in diameter) fixed into a steady ball holder was pressed against the coated and polished sample disk (105 mm in diameter) with a normal load of 40 N. The disk rotated 2500 cycles with a linear speed of 0.1 m/s. After the experiments the wear track was examined by microscopic analysis in order to determine the wear volume loss.

The results of the wear tests after 2500 cycles showed different results for each spray feedstock material. The Cr_2O_3 coatings regardless whether based on fine or standard powder fractions showed almost no volume loss. According to optical micrographs of the wear scars a tribofilm was formed consisting of plastically deformed debris and splats. This tribofilm was smoother than the original surface and was placed slightly above the mean line of the unworn surface protecting the surface from further wear, see **Figure 6a**. During the experiments of Cr_3C_2-NiCr samples measurable wear scars were formed, **Figure 6b**. The volume loss was higher on the coating based on the fine powder (0.33 mm^3) whereas the wear rate of the standard sample was a bit lower (0.24 mm^3). Apparently the above-mentioned phase changes which occurred while processing the fine Cr_3C_2-NiCr powder influenced the wear behaviour negatively. On the contrary the WC-CoCr coatings regardless whether based on fine or standard powder fractions did not suffer a measurable volume loss. In fact the sintered WC6Co ball was abraded instead of the coatings. This is probably due to the surface finish of the coatings as fine and carbide rich hard grooves (see **Figure 3c)** abraded the Co-matrix of the ball.

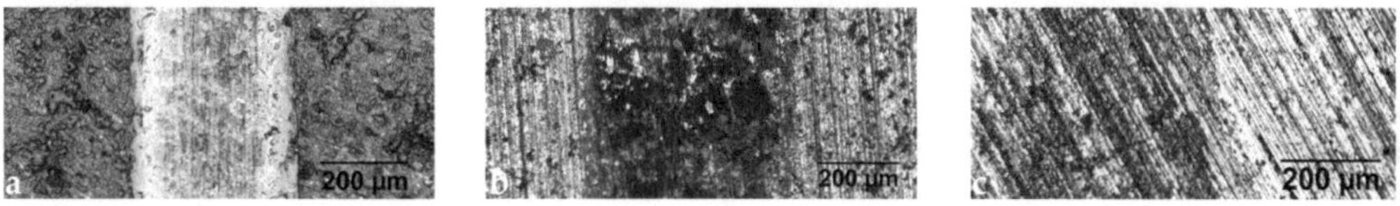

a) Cr_2O_3 validation sample (E1)
b) Cr_3C_2-NiCr validation sample (E2)
c) WCCoCr validation sample (E3)

Fig. 6. Optical micrographs of wear scars after 2500 cycles in ball on disk-tests

2.5 Conclusion

Fine Cr_2O_3, Cr_3C_2-NiCr, and WC-CoCr feedstock with grain sizes below 25 µm were processed in order to investigate the spraying of near net shape coatings. The characteristics of the coatings based on fine powders were analysed and compared to standard coatings based on -45+5/20 µm powder fractions. Compared to standard coatings it was possible to improve the key coating characteristics porosity, surface roughness and corrosion resistance significantly. Other coating properties like hardness or wear resistance showed comparable behaviour as that of standard samples. In case of spraying cermet feedstock, especially Cr_3C_2-NiCr, optimized parameter sets are necessary to control decarburization and oxidation.

3. Design and optimization

Thermal Spraying is an indirect process, where only the basic conditions can be controlled by altering the process parameters. A deterministic control of the transfer of heat and kinetic energy to the feedstock particle is not possible. Due to the vast variety of process parameters - some time said to be more than one hundred (Lugscheider & Bach, 2002) - sophisticated approaches of designed experiments are a good tool to both understand the complex interdependencies between the parameters and to optimize coatings properties due to the demands. In the following the basic considerations and the proof of suitability of statistical design of experiments are given for controlling and optimization of thermal spraying processes.

3.1 Basic considerations

The goal of conducting experiments is to get information about the functional relation between the process conditions and the resulting coatings properties determining both the economical effectiveness of the coating process as well as the coatings behaviour under operational conditions. For example the deposition efficiency (DE) of the feedstock material in case of plasma spraying of oxide ceramics, i.e. the percentage of the inset feedstock contributing to the coating buildup, is dependent of the chosen federate as well as the achievable heat transfer from the plasma to the feedstock particles. Therefore it can be assumed, that there is a functional correlation between the powder feedrate, the applied amperage to the plasma and the chosen plasma and secondary gas mixture (species, total flow and ratio) controlling the specific heat and therefore the capacity of heat transfer of the plasma. Two further parameters defined by the inset feedstock are its heat of fusion and median grain size as the heat is transferred from its surface into its volume. The spraying distance is parameter controlling the time of flight of the particles in the plasma and therefore the time of exposure to heat, but there is a strong interdependency with the applied amperage. The higher the amperage, the higher is the temperature and heat capacity of the plasma, but also its velocity and therefore the time of flight for the particles decreases with raising the amperage. All together the functional dependency of the DE can be stated as follows:

$$DE = f_{\text{(amperage, plasma gases, spraying distance, particle size, ...)}}$$

or in other terms:

$$y\ DE = f\,(x1, x2, \ldots, xn) \tag{4}$$

One approach to derive information about the correlation of the coating´s criteria DE with the parameters is to vary the process parameters one by one in every single spraying experiment holding two complications: The number of experiments is large and the interdependency between distinct parameters cannot be estimated. Therefore the use of statistically designed experiments is a good alternative, as both goals can be realized utilizing this tools (for an example see Heimann, 2008). The experiments are arranged in matrices with a deterministic alteration of the factors (i.e. parameters to be investigated) on distinct levels. Afterwards the coatings criteria are measured and the results are analysed regarding the factorial effects (i.e. the correlation with the parameters). The ways to obtain the correlation can be divided into factorial analysis by means of multiple regressions on the one hand and by analysing the variance of the measured results according to the variation of process parameters (ANOVA) methods. The usability of the second approach in the field of thermal spraying is shown in the following using the examples of the experimental series described in chapter 2. For a comprehensive overview of the methods including model testing etc. see (Dean & Voss, 1999; Mason, 2003; National Institute of Standards and Technology [NIST], 2011).

3.2 Robust quality control basing on ANOVA techniques

The variability of thermal spray processes regarding coatings characterics and quality is a well-known problem in application. In the field of the designing and development of feedstock and coating systems designed experiments are sophisticated tools to achieve sufficient coating qualities in specified tolerance regions. Besides the gathering of the relevant know-how regarding the spraying of certain feedstock etc., the processes must be insensitive against deviations over longer periods of time to reach this goal. For example, in **Figure 7** a quadratic functional correlation between the deposition efficiency of feedstock and the applied amperage in the APS process is assumed.

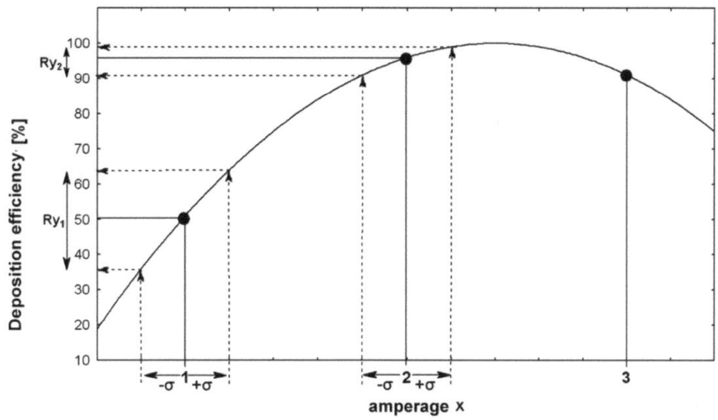

Fig. 7. Assumed correlation between applied amperage and relative deposition efficiency

The sketch shows, that the same magnitude of deviation of the applied amperage from the chosen control value results in two different deviation spans of the resulting DE (R_{y1} and R_{y2}). Another point is the influence of noise factors, which also can disturb the known relation between process parameters and the expected result. Following the approach after G. Taguchi, the effects of the control factors (i.e. process parameters) are extended by the effects of noise and signal factors (see **Figure 8**).

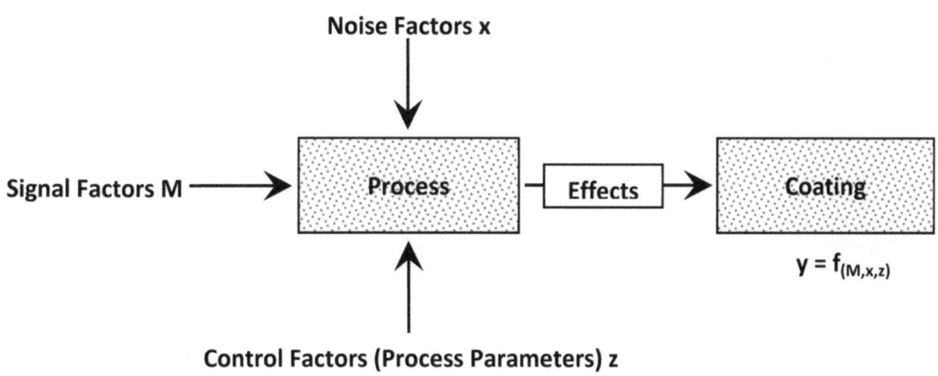

Fig. 8. Scheme of the effects of control, noise and signal factors on the coating process (after Phadke, 1989)

The basic tools of the method are the so called orthogonal arrays. Like the conventional matrices of designing experiments for factorial designs, the levels of the parameters to be investigated are arranged by given plans. But unlike the methods of DoE, the functional correlation between factors and the measured results are expressed in terms of a signal-to-noise ratio. The goal of the method is not to optimize one response regardless of other coatings criteria, but to achieve results being robust against the effect of noise factors like e.g. the wear of parts like the electrodes of the plasma gun etc. The signal factors also show effects on the results, but are kept normally constant, like e.g. the traverse speed of the gun relative to the substrate. In the following the results of applying the method are discussed.

3.3 Applying orthogonal arrays for optimizing coatings

Taguchi techniques were utilized in order to reduce the number of experiments and to evaluate and to adjust main process variables. The effectiveness of these techniques could be verified by spraying validation samples successfully. A Taguchi experimental design was used to reduce the number of coating experiments. Four main process variables, or factors, were identified and varied on three levels in an L_9 orthogonal array. This matrix dictated the combination of levels, at which the factors should be set for each experiment, **Table 3**. Furthermore the process output variables, or responses, which should be optimized, were defined. For near net shape coatings a surface roughness being as low as possible is requested. So the aim of the experiments was to obtain a set of spray parameters for each material, which allows the spraying of coatings with low surface roughness under consideration of cost-effective deposition rates. It was also tried to improve coating properties like hardness and porosity. The results were analysed by means of ANOVA to determine the relative contributions of the various main factors and interactions among them. This allowed a prediction for an optimal parameter set for each investigated spraying feedstock.

Factor ⇒ Experiment No.⇓	Particle size	Current (APS) / Air-Fuel-Ratio (HVOF)	Spray distance	Powder carrier gas
1	1	1	1	1
2	1	2	2	2
3	1	3	3	3
4	2	1	2	3
5	2	2	3	1
6	2	3	1	2
7	3	1	3	2
8	3	2	1	3
9	3	3	2	1

Table 3. Matrix used for the Taguchi experimental designs

For validation samples were coated using the predicted optimum spray parameters shown in **Table 4**, the results are shown in **Table 5**. The values predicted and actually measured proved to be quite consistent. It can be reasoned that the validation experiments were able to confirm and to reproduce the predicted values.

While comparing validation and standard samples the latter showed higher deposition rates. Of course this has to be ascribed mainly to the fact that coarser spraying feedstock were used to spray the standard samples. The validation samples showed significant lower surface roughness. Especially the carbide based coatings showed low R_a values (about 2.7 μm) compared to the R_a values of the standard samples (near 7 μm). The hardness of the coatings did not vary much regardless of which powder fraction was processed.

Feedstock	Particle size μm	Current, A (APS) or lambda (HVOF)	Spray distance, mm	Powder carrier gas, slpm
Cr_2O_3	+25-5 (3)	650 (3)	100 (2)	12 (2)
Cr_3C_2-NiCr	+15-5 (1)	1.4 (1)	270 (1)	10 (1)
WC-CoCr	+15-5 (1)	1.4 (2)	310 (2)	11 (2)

Table 4. Predicted optimum spray parameters (the numbers in brackets show the corresponding parameter level)

Feedstock	Sample	Deposition rate, μm per pass	Surface roughness Ra, μm	Hardness HR 15 N
Cr_2O_3	Predicted	18.4	3.2	87
	Measured	18.7	2.5	89
	Standard	16.9	4.5	87
Cr_3C_2-NiCr	Predicted	3.6	2.8	73
	Measured	3.3	2.7	77
	Standard	23.9	6.7	76
WC-CoCr	Predicted	11.8	3.1	89
	Measured	11.3	2.7	87
	Standard	26.9	6.8	88

Table 5. Predicted and measured results obtained from validation and standard samples

It can be summarized, that by applying the method of signal-to-noise ratios derived from the evaluation of orthogonal arrays, the workings could be reduced to nine experiments while investigating the effects of four quantitative parameters on three levels. The results show, that by applying this technique, reproducible forecasts regarding the optimisation of thermal spray coatings can be derived.

4. Developments of new applications

As stated at the beginning there is the goal to make use of finer grain sizes of feedstock powders to reach denser coatings showing higher cohesion and adherence to the substrate. The lower limit of feeding powders into the process is in the one-digit micrometer range. By dispersing of the feedstock in a liquid outer phase or the formulation of feedstock direct in a suspension by chemical methods, use can be made of nanometer sized feedstock. In the following the efforts are shown in new results regarding the achievement of coatings, which could not be realized by means of thermal spraying before.

4.1 Suspension plasma spraying of triboactive coatings

Up to now no coating systems are marketable in the field of metal forming like the direct hot extrusion process, which provide both surface protection of the parts being in contact to the billet (i.e. container and die), and a significant reduction of the frictional losses being induced by the billet passing along the container walls. To dispense the use of lubricants and to enhance the usable forming capacity of the process, different oxide ceramics were given in one suspension and plasma sprayed. The aim is to reach a mixing of the feedstock to obtain deterministic solid solutions of the oxide phases which show a reduction of their coefficient of friction under dry sliding conditions. To reach this goal the high surface-to-volume ratio of feedstock with primary particle sizes below 100 nm was used. By means of x-ray diffraction it could be proven, that the desired phases could be synthesized. The coatings showed a considerable lowering of their frictional coefficient in tribological testing against steel 100Cr6 in the region of the operation temperatures for the hot extrusion of aluminium alloys. Besides the experimental work the fundamentals of the mixing process of different oxides regarding crystallographic aspects are discussed.

Thermal sprayed coatings are not commonly used in the field of massive forming due to the high demands concerning the cohesion and adhesion of tool coatings. The cause is adhesive wear being induced by the elevated temperatures of operation and high relative velocities between the work piece and toolings resulting in high tensile and shear stresses. Nevertheless, there is the challenge to establish coatings to reduce both the wear of tools and frictional losses in the processes. For example in the case of direct hot extrusion, up to 60%

of the forming force have to be applied to counterbalance frictional losses. To come up against that losses different lubricants and material separating agents are used, but with the disadvantages of a higher degree of reworking of the semifinished extruded product and a limited thermal stability of the substances. To overcome these disadvantages, the usability of specific oxide ceramic phases basing on titania was tested, which show a reduction of their frictional coefficient under tribological operation and elevated temperatures. The desired phases should be synthesized in the suspension plasma spraying process by mixing different oxide feedstock with titania in one suspension.

4.1.1 Crystallographic aspects

In the system Titanium-Oxygen different non stoichiometric phases are known, which show the ability for deformation under mechanical stress due to a shearing of crystal lattice planes. These phases show a reduction of the frictional coefficient in dry sliding conditions under elevated temperatures of some 100 degrees Celsius. The beneficial effect was linked to the shearing processes being temperature induced (Gardos, 1988), the fundamental mechanism of the shearing processes are discussed elsewhere (Anderson, S. and Tilley, R. J. D., 1972). As the phases are expected to be not thermodynamically stable (for a discussion of redistribution effects of titanium and oxygen see Wood, G. J. et al., 1982), another approach was intended for these workings. By addition of a second cation besides Ti^{4+}, phases can be obtained which are homologues to the nonstoichiometric titaniumoxides. Those so called Andersson-phases were first described for the system Ti-Cr-O (Andersson, S., Sundholm, A. & Magnéli, A., 1959) showing a composition of $Ti_{n-2}Cr_2O_{2n-1}$.

As chromium exhibits a high steam pressure with rising temperature and therefore may tend to evaporate out of the lattice, the homovalent substitution of the Ti^{4+}-cation in the rutile base lattice was aspired. Several cations where chosen based on the rules for substitution processes stated by V. M. Goldschmidt (Goldschmidt, V. M., 1926), besides Cr^{3+} primary Ni^{3+}, Co^{3+} and Zr^{4+} by considering the ionic radii and coordination given in (Shannon, R.D., 1976). The goal is to reach phases with a similar composition compared to the Andersson-type phases on the one hand and a sufficient stability in temperature ranges up to 800° C on the other, which are commonly used for hot extrusion of aluminum and copper based alloys.

The assumption that the applicability of substitution processes may lead to the formation of solid solutions of the desired stoichiometry can be proven by means of the Inorganic Crystal Structure Database. In **Figure 9** for example the structures of the cubic Co(II)-oxide and of tetragonal rutile (i.e. Ti(IV)-oxide) are shown on top, where the oxygen is represented by the larger balls.

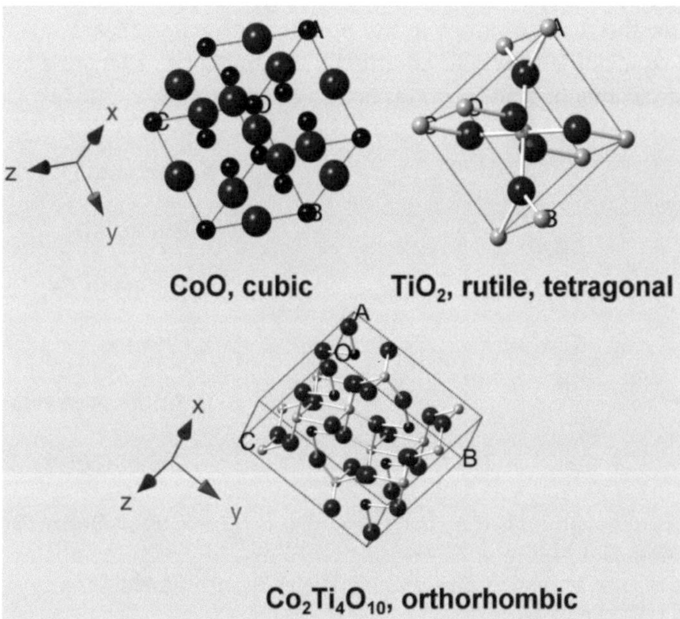

Fig. 9. Structures of Co- and Ti-oxide (top) and of the "mixed" solid solution oxide (bottom)

From the structure it can be inferred, that both cations have similar radii, which is - besides the valence and the coordination by the surrounding ions - the key requirement for the dissolution of the oxides. When both oxides are mixed, a structure of lower symmetry (orthorombic) is formed with a composition of $Co_2Ti_4O_{10}$. The difference compared to the aspired composition of $Co_2Ti_4O_{11}$ for n = 6 is due to the fact, that the divalent cobalt is incorporated in the structure instead of the trivalent ion. Like the most structures being crystallographic possible solid solution of rutile with the named oxides, the cobalt-titanium-oxide with trivalent Co-ions is not refined yet. Without the feasibility to refine the structures, the full quantitative Rietveld analysis by means of X-ray diffraction of the sprayed coatings is not possible.

4.1.2 Results

4.1.2.1 Phase analysis

Three different mixtures of titania (rutile) with the named oxides of trivalent cobalt, nickel and chrome where sprayed on structural steel SJ235R. X-ray diffraction analysis were performed on the coating systems using copper radiation, the diffraction patterns are plotted in **Figure 10** with an offset of 500 counts between the samples. The patterns where checked regarding the presence of unmelted or recrystallized feedstock, the possible solutions as well as reduced oxides and reaction products of the feedstock with the flux melting agent. Because of the marginal coating thickness of some tens of micrometres, the influence of the substrate is recorded in the patterns. As the relative intensity of reflection (RIR) of ferrite is considerably higher than that of the other phases present in the coatings, its peaks are of highest intensity (see the peaks at approximately 45 und 75° 2θ). Since no

structure data is available for the aspired solid solutions, the reference intensity ratio stated in the ICDD PDF4 database entries where used to perform semi quantitative analysis. As no RIRs are given for the solid solutions in the powder diffraction files, values of phases with nearly identical stoichiometry where assumed. The fractions of ferrite were deducted and the adjusted phase contents of the coatings are given in **Table 6**.

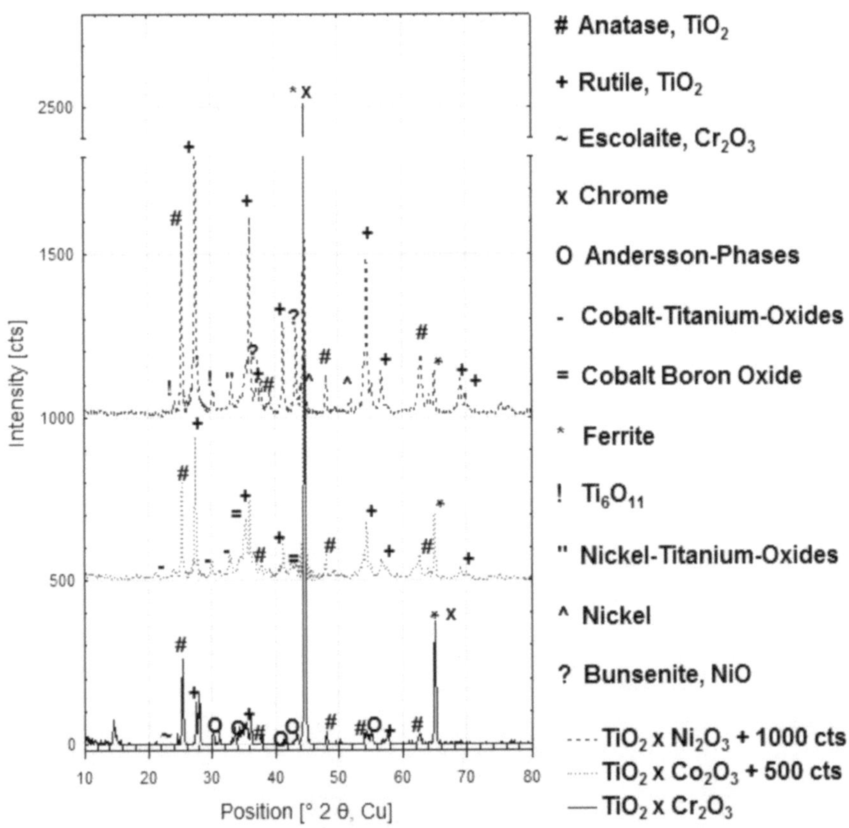

Fig. 10. Diffraction patterns of three suspension plasma sprayed coating systems

	Phase Contents (atomic percent)				
Sample	Ti-Oxides	Ni, Co-Oxides, Cr	Ti-(Ni,Co,Cr)-Oxides	Borates	Traces
$TiO_2xNi_2O_3$	55	7	26	10	2 % Ni
$TiO_2xCo_2O_3$	39	0	29	33	-
$TiO_2xCr_2O_3$	13	10	75	0	2 % Cr_2O_3

Table 6. Phase contents of the three coating systems

In case of the Ni- and Co-containing coatings, significant amounts of Ti(IV)-oxides were measured, of which approximately one third is anatase. As stated in (Bolelli, G., et al., 2009), in case of rutile feedstock the phase content of anatase especially in suspension sprayed coatings can be explained by slow cooling due to re-solidification of molten droplets in the process, compared to formation of rutile in rapid quenching on the substrate. Considering this explanation another assumption might be the influence of elevated substrate temperatures in the SPS process leading to a more slowly cooling of molten titania particles after impinging on the substrate. To distinguish both possible mechanisms further investigations will be conducted considering the thermodynamics of the phase changes of both titania species. In the case that the anatase content correlates well with the content of re-solidificated particles in the coating, the anatase-to-rutile ratio can be used to optimize the injection and spraying parameters.

For the coatings containing nickel, about 7% percent of Ni(II)-oxide were found, whereas in titania-cobalt-oxide systems no remains of the Co-feedstock was detected. The employed trivalent oxides of both cations decompose towards the divalent oxide at temperatures above approximately 600° in case of the Ni-oxide and 1910° C for the Co_2O_3. Otherwise the contents of borates formed by reactions of the boron oxide with the feedstock oxides is three times higher for the Co-based system compared to the titania-nickel-oxide coating. As the absolute value of the enthalpy of formation of the cobalt-borate is higher than that of the Ni-borate (Hawk, D. and Müller, F.; 1980; Paul, A., 1975), the Co-oxide feedstock is diluted in the boron oxide to a much higher extent compared to the Ni-containing system, and no remaining Co_2O_3 is embedded in the coating. In contrary to that the contents of Ni-borates are small in the titania-Ni-oxide coating, and remains of the Ni(II)-oxide are recorded. The phase contents of the aspired solid solutions are below 30 % for both coatings systems.

Compared to the Ni- and Co-containing coatings the mixing of titania with chromia leeds to different phase compositions. Due to the marginal miscibility of chromia with boron oxide (Tombs, N. C.; Croft, W. J. & Mattraw, H. C.; 1963), no borates and also just small amounts of the feedstock powders are found. The Andersson-phases with the mentioned stoichiometry of $Ti_{n-2}Cr_2O_{2n-1}$ amount to three quarters of the total coatings composition. Therefore it can be concluded, that the degree of mixing of the feedstock is significantly higher for the titania-chromia system. If the melted phase of the boron oxide supports the mixture process of the both oxide ceramics without further reaction cannot be clarified. Possibly the heat of the process is better transferred to the coarser feedstock of approximatly 100 nm median crystallite size compared to 30 to 60 nm of the feedstock of the Ni- and Co-containing coatings. As the heat transfer degreases drastically when the agglomerate size of the feedstock particles falls below a critical limit (so called Knudsen effect, Fauchais, P. et al., 2008), this might be a supposable explanation of the higher degree of feedstock mixing in the case of the titania-chromia system.

In addition, with approximately 10 % significant amounts of chromium are present in the coatings, being formed by reduction of the chromia feedstock. This effect is only detected when spraying the suspension with the Triplex-II and not when using the DELTA-Gun, and further on when besides chromia titania is present in the suspension. This result is probably due to the large gap between the absolute values of the Gibbs free energy of the two oxides. Hence the chromia is reduced in the presence of titania. By means of visible spectroscopy protons where found supposedly originating from the vaporization of the water of the suspension, but no ions of oxygen where detected. Together with the lamellar flow of the plasma jet of the Triplex resulting in marginal entrainment of surrounding air, apparently the conditions are given for the reduction of the chromia towards chrome.

4.1.2.2 Tribological testing of Andersson type coatings

Since the contents of the solid solutions of titania with another oxide were the highest in case for the titania-chromia system, tribological testing for recording the coefficient of friction dependent on the temperature of operation were conducted with coatings of this Andersson type phases using a ball on disk configuration. The coatings rotated against a ball of 100Cr6 (1.3505, diameter = 5 mm) with 0.1 m/s, the loading force was 5 N. The coefficient of friction was recorded in three runs on different samples at room temperature, 600° and 800° C (see **Figure 11**).

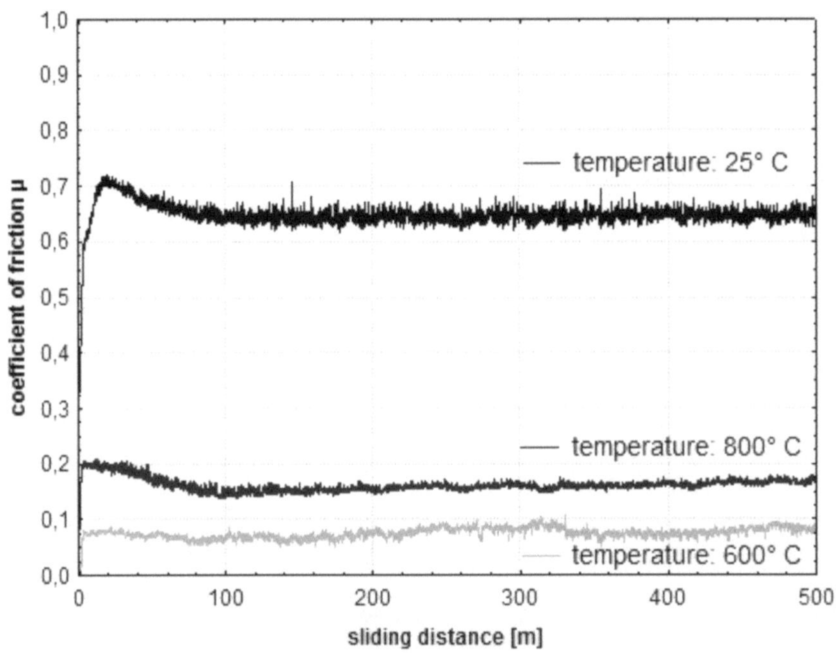

Fig. 11. COF of Andersson type coating systems measured against 100Cr6 at RT, 600° and 800° C

The friction pairing shows a COF of more than 0.6 when running at room temperature. When rising the temperatures up to 600° C, the ratio of the frictional force to loading force drops considerably to below 0.1. On the one hand this effect is surely due to the softening of the ball (see the debris of the ball on the coating in the second picture from top on the left hand in **Figure 12**), but this effect is desired as the billet in the extrusion process shows a comparable behavior. For this reason, the testing of the coatings in tribometer experiments is not directly comparable to the hot extrusion process, as the soft consistency of the flowing billet above yield stress cannot be tested because the ball would be abraded promptly and its holder would scratch the coating. But compared to the given values of operating unlubricated containers of more than 0.2 (Bauser, M.; Sauer, G. and Siegert, K., 2006), a significant lower frictional force was measured. Besides the tribological activity of the coating it shows good material separating properties against 100Cr6. When rising the temperatures to 800° C, the COF rises again to values of nearly 0.2. An explanation is the formation of black ferrous oxide (supposable magnetite) instead of the red oxide (presumably haematite), showing higher hardness and unfavourable tribolological properties (Barbezat, G., 2006).

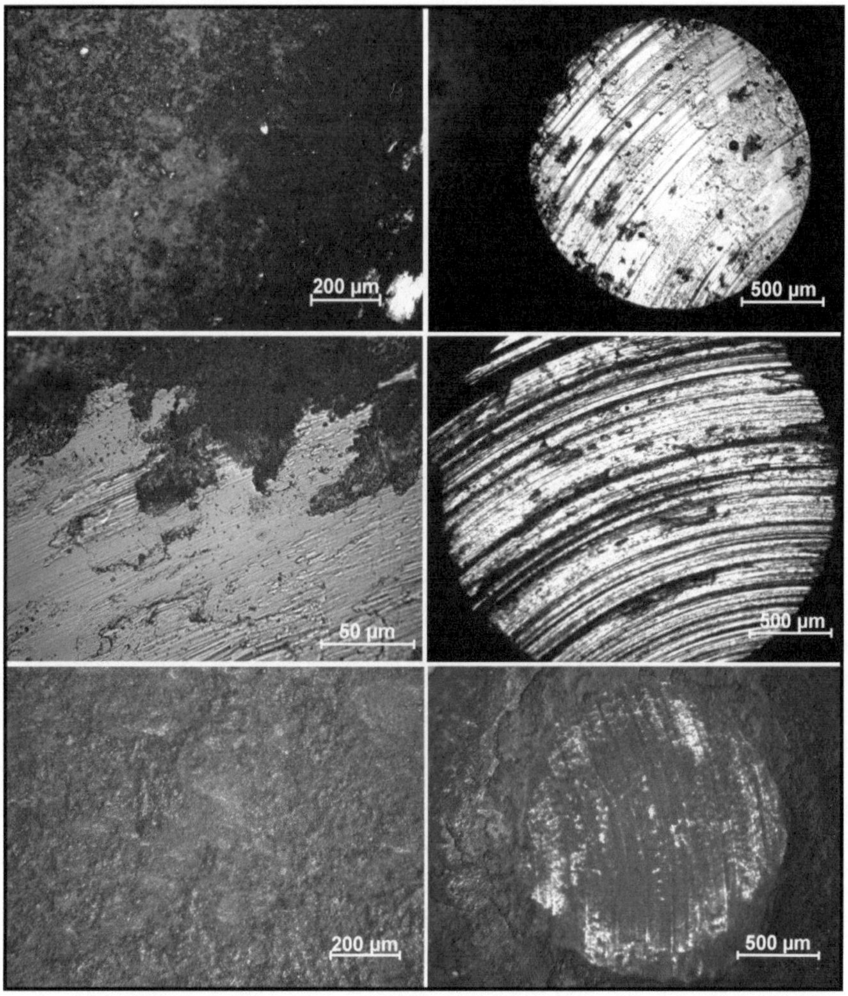

Fig. 12. Top views of the scare tracks left and corresponding friction surfaces of the counterparts from room temperature (top) to 800° C (bottom)

This finding is another example for the low comparability between tribometer experiment and hot extrusion, as the billet is pressed under air exclusion in the container. Another guess is a lack in thermal stability of the Andersson phases. Other SPS coatings sprayed with the same feedstock composition where tempered at different temperatures (300, 500 and 800° C) for several hours. The colouring of the coatings changed with temperature (see top views on the left hand side of **Figure 13**), as the phase composition changes (see the corresponding diffraction patterns on the right). The marked peaks in the diffraction pattern of the sample tempered at 500° C are caused by the Andersson phases. As clearly can be seen, this peaks are significantly smaller in the sample being not tempered and that one tempered at 800° C. So it can be stated, that the tribological active phases can be formed with rising temperature within the coatings during operation, but also may decompose with further increased temperatures.

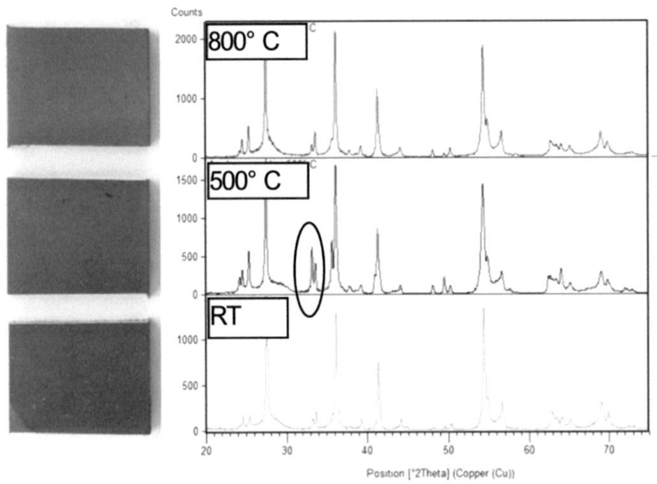

Fig. 13. Alteration of the phase composition of Andersson type coatings in relation to temperature (right side) and corresponding top views of the tempered samples

4.1.3 Conclusion

By means of x-ray diffraction analysis it could be proven, that the mixing of titania and other oxide feedstock in the SPS process could be realized. For example the achieved Andersson type coating system sprayed with titania and chromia containing suspensions showed a temperature induced lowering of their coefficient of friction when rotated against 100Cr6. Further experiments will be conducted to better understand the parameters controlling the mixing process of the feedstock on one hand and regarding tribological experiments using aluminium and copper based extrusion alloys.

4.2 Comparison of multielectrode plasma guns for development of new coatings

When high throughput is intended, three cathode guns are a supposable solution. Due to their stationary plasma jet and elevated power characteristics, higher feeding rates concurrent with sufficient deposition efficiencies can be realized compared to one-cathode plasma guns. On the contrary to those well-known equipments a newly marketable system makes use of three anodes to combine high power inputs into the plasma as well as stable process conditions. Besides a more narrow nozzle outlet diameter compared to multi-cathode designs hydrogen can be used as secondary plasma gas, both resulting in higher plasma velocities and net powers. The conceptional designs of two guns are discussed as well as their suitability for suspension and shrouded plasma spraying. The efforts in achieving new plasma sprayed coating systems are presented.

4.2.1 Design of marketable multielectrode plasma guns

To overcome the disadvantages of conventional plasma guns especialy regarding the discontinuity of the free jet due to plasma arc root rotation, mulitelectrode guns were developed. Since more than ten years guns basing on the three-cathode-design guarantee high plasma net powers combined with stable feedstock injection conditions. Until now the guns have two disadvantages concerning the use of expensive helium as secondary gas accompanied by low plasma arc voltages on the one hand and the restriction of the minimal nozzle outlet diameter on the other. For example three single plasma fingers originate from the single cathodes being passed through a cascaded neutrode in case of the second generation of the Triplex-design (Sulzer Metco AG, Wohlen/Switzerland). Hence a minimal nozzle outlet diameter of the anode of 9 mm can be realized because of the thermal design of the gun. Another approach is the inverted design of a plasmatron, where one arc originates from a single cathode and is divided on three anodes after passing the cascade. Therefore for the DELTA-Gun (GTV GmbH, Luckenbach/Germany) a minimal nozzle outlet diameter of 7 mm can be achieved resulting in higher plasma velocities at the nozzle outlet. Furthermore hydrogen can be used as secondary gas and high brut plasma powers of 80 kW can be applied to the torch.

4.2.2 Experimental

The workings concentrated on the investigations, to what extent both gun concepts are appropriate for inert and reactive shrouded plasma spraying as well as the processing of nanoscaled suspensions. Feedstock was used being not commonly applied in plasma spraying to identify the potential of plasma spraying for possibly new applications. For demonstration purposes coating systems of titanium and chromium as well as their nitrides and Indium-Tin-Oxide (ITO) showing electrical conductance were chosen. For the chromium coatings feedstock obtained from GTV GmbH with two different particle size distributions (-25+5 µm and -45+5 µm) were investigated. As titanium feedstock a powder of -45+10 µm came into operation, which is manufactured and distributed by TLS Technik Spezialpulver GmbH (Bitterfeld/Germany). For SPS suspensions containing 5 wt.-% ITO (ANM PH 15695, Evonik Degussa GmbH, Marl/Germany) and Al_2O_3 (Saint Gobain, Weilerswist/Germany) with primary crystallite sizes of some tens for the first and approximately 150 nm for the latter were used.

4.2.3 Results

4.2.3.1 Shrouded plasma spraying

For both guns modules have been designed and machined to apply shroud gases around the plasma free jet (for details see **Figure 14** on the following page). The attachments consist of water-cooled bodies, in which the shrouding gas is injected helically to ensure a sufficient shielding against the surrounding air after exiting the shroud. The feedstock injection is realized over middle sections between the exit of the gun nozzle and the shroud gas inlet to avoid interferences with the shroud gas flow. The body housings of the shrouds are integrated in the cooling circuit of the spraying equipment.

The spraying experiments were conducted applying plasma brut powers of approximately 25 to 50 kW (see **Table 7** for spraying parameters). When operating the Triplex high helium flows of 20 SLPM were used to guarantee a sufficient heat transfer to the feedstock, but for the DELTA-Gun no secondary gas was applied due to the formation of black depositions with a consistency of soot when hydrogen was applied. To guarantee an adequate shielding effect in the case of argon shroud gas on one hand and an effectual entrainment of nitrogen for reactive spraying of the feedstock on the other, high shroud gas flows of 90 SLPM were applied for spraying with both guns.

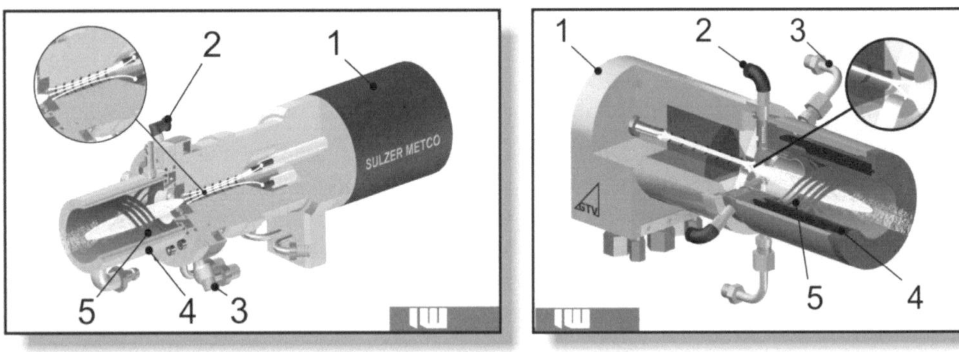

Fig. 14. 3D sectional view of the shroud modules of Triplex (top) and DELTA-Gun (bottom): 1 Gun, 2 shroud gas inlet, 3 powder injector, 4 water-cooled shroud, 5 helical injected shroud gas

Parameter	Triplex	DELTA-Gun
Argon flow	35 slpm	40 slpm
Helium flow	20 slpm	-
Current	250 – 350 A	210 – 330 A
Plasma brut powers	36 kW - 51 kW	24 – 48 kW
Spraying distance (shroud/substrate)	10 mm	10 mm
Traverse speed	48 m/min	48 m/min
Shroud gas flow	90 SLPM	90 SLPM

Table 7. Parameters for Shrouded Plasma Spraying

The obtained coatings microstructures are illustrated in Figure 15. The micrographs on the left hand side show coatings sprayed in an inert atmosphere of argon, the ones on the right hand side the results of reactive spraying using nitrogen. The coatings of a and b were sprayed using the Triplex gun, whereas for spraying of the coatings c to f the DELTA-Gun was used. When spraying in inert atmosphere, the coatings show a uniformly and homogenous microstructure with a level of porosity comparable to conventional plasma sprayed coatings. Otherwise when nitrogen is supplied, coatings with high levels of open cavities and microstructures comparable to metallic sponges are built. This is supposedly due to a turbulent entrainment of the nitrogen shielding gas when the feedstock reacts towards the nitride. To proof the existence of the aspired nitride phases, semi-quantitative measurements by means of energy dispersive x-ray analysis (EDX) were performed on coatings sprayed with the Triplex. The results revealed contents between approximately 13 and 17 at.-% nitrogen. Further on, the nitrogen contents of the coatings were investigated using a N/O/H-Analyser (LECO Instruments Corp, St. Joseph/USA). Unfortunately the contents of nitrogen in the titanium coatings could not be measured to the high melting point of the titanium nitride, but for the chromium coating (section f) the nitrogen content was determined to account for approximately 10 at.-%.

The existence of the nitride hard phases was also verified by indentation hardness measurements. When spraying the titanium with argon as shrouding gas, mean hardness values of 360 HV 0.1 in case of the Triplex and 270 HV 0.1 for the DELTA-Gun were measured. Otherwise with the employment of nitrogen significantly higher maximum values of more than 1200 and 1000 HV 0.1 were detected.

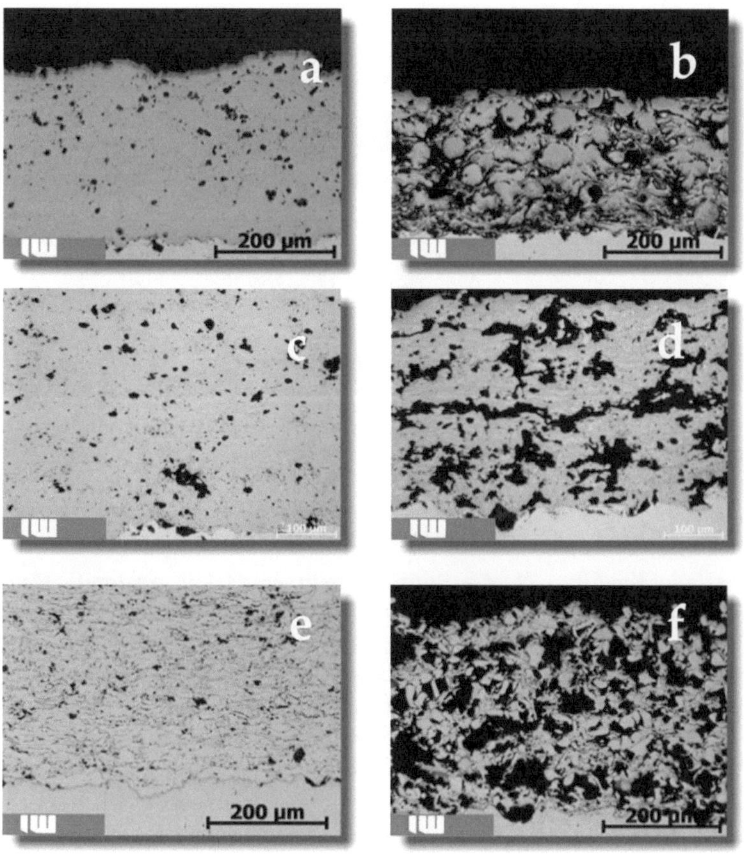

Fig. 15. Micrographs of shrouded plasma sprayed titanium (field a – d) and chromium (field e + f) feedstock using argon (left hand side) and nitrogen (right hand side) as shroud gas

To characterize the adhesion of the coating systems tensile adhesive tests according to DIN EN 582 on grid blasted 1.4301 substrates were conducted. Again slightly better values were achieved for the coatings sprayed with the Triplex-II gun, as the mean values of more than 50 MPa were measured for the titanium coatings compared to approximately 35 MPa for both the titanium and the coarsely grained chromium feedstock when spraying was performed with the DELTA-Gun. This might be due to the problems of injecting the feedstock in the case of the DELTA, as the gun uses a gas flow supporting the cooling of the anodes. Together with the plasma and the shrouding gas, the gas throughputs through the shroud module are high and a proper feedstock injection is not easily achieved. Therefore, further optimization potential is given for the shrouded spraying in case of the DELTA-Gun. Otherwise when using the fine fractionated chromium feedstock, the tensile adhesion of the coatings reach nearly 50 MPa, comparable to the coatings sprayed with the titanium feedstock with the Triplex gun.

When spraying the titanium on polished substrates instead of the grid blasted samples, even higher tensile adhesive strengths of nearly 60 MPa were measured. This result being not expected is probably due to diffusion phenomena of the titanium into the austenitic substrate. The effect was not recorded when using ferritic steels. In Figure 16 the backscattered electron micrograph (left hand side) and an EDX line scan analysis (right hand side) of the interface section of a titanium coating on 1.4301 steel substrate is shown. The EDX analysis confirms the findings of a zone of some micrometers depth, in which the titanium diffused. It can be stated as remarkable result, that with the limited heat transfer to the substrate enough potential is given for the diffusion process. This is due to the high diffusion coefficients of both titanium and chromium in 1.4301 austenitic steel (Kale, G.; 1998).

Fig. 16. BSE image of the interface of a titanium coating on 1.4301 austenitic steel (left) and corresponding EDX line scan (right hand side)

To investigate the alteration of the feedstock in the suspension plasma spraying process, Indium-Tin-Oxide (assumed composition of 9:1) was suspension plasma sprayed. ITO is used to coat glass for electrically conductive coating beeing transparent in the visible spectrum. The coatings are commonly deposited by sol-gel methods and are used in touch-screen purposes. The goal was to reach thin optical transparent ITO coatings showing electrical conductance. When overheating the feedstock it tends to build coatings with a yellowish color, whereas the coating system shows no conductance when it is not uniformly deposited. To find optimal conditions the relevant parameters (solid content of feedstock, species of the outer phase, injection conditions, applied amperage and spraying distance) were varied. The melting behavior of the feedstock was tested with wipe tests (see SEM images on top of **Figure 17**).

Fig. 17. SEM images of wipe tests of suspension plasma sprayed ITO feedstock and top views of ITO coatings (Triplex-II left side, DELTA-Gun on the right hand side)

With optimized parameter sets the coatings were sprayed on slides of borosilicate glass with both plasma guns. The coatings were uniformly deposited (see top view SEM images in **Figure 17**), showing homogenous structures. The coatings were measured by a project partner regarding their thickness and electrical conductance. It could be proven, that coatings with a thickness of approximately 400 nm and a sheet resistance of 850 Ω could be achieved.

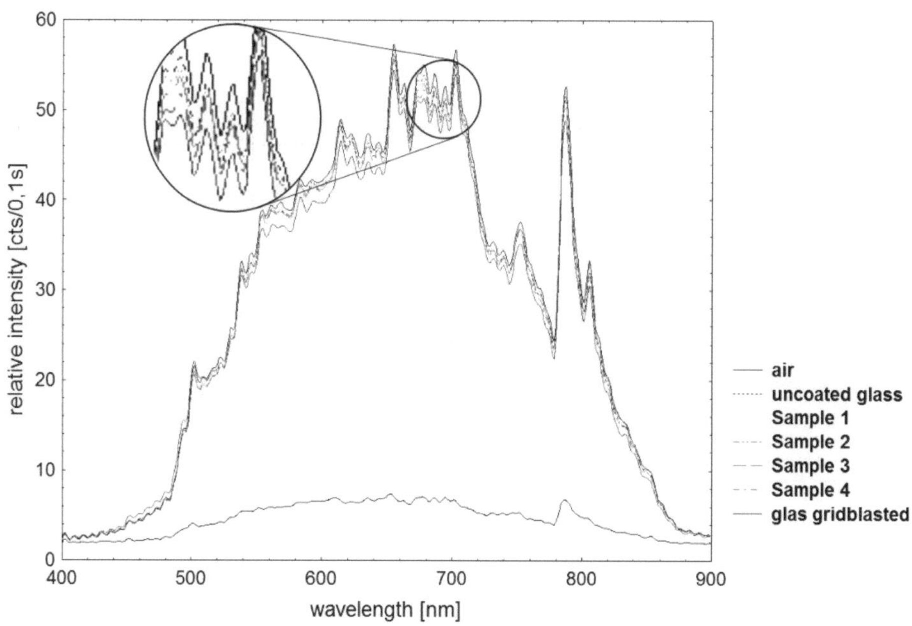

Fig. 18. Transmission spectra of four ITO coated glass slides compared to uncoated and grid blasted glass

To determine the optical transparency of the coatings, four samples were measured using a VIS-spectrometer and the results were compared to uncoated and grid blasted glass (see transmission spectra in Figure 7). As source the tungsten lamp of the calibration module of a Tecnar DPV-2000 was used delivering a stable spectrum covering the whole visible range. The coated samples show a high degree of transparency over the whole visible spectrum. For example in the red range below 700 nm (see marking), the relative intensity measured is maximal 1 to 3 counts lower than that of the uncoated glass. This equals to a grade of transparency of 95 to 98%. It can be stated, that both requirements regarding the electrical conductance as well as the optical transparency of the coatings systems were fulfilled. These findings show, that by suspension plasma spraying new coating systems can be realized in fields of operation, where up until now coating deposition processes like CVD and PVD are used.

4.2.4 Summary

With the adaption of shroud gas modules to the multieletrode plasma guns Triplex-II and DELTA-Gun it could be proven, that the spraying feedstock being susceptible to chemical reactions can both sprayed in inert atmospheres using argon and reactively sprayed by applying nitrogen as shrouding gas. The inert conditions led to the formation of coatings showing a homogenous microstructure comparable to conventionally APS sprayed metallic coatings. In the case of the use of nitrogen, no dense coatings could be achieved, but the presence of nitride phases in the coatings could be proven. Further on by means of suspension plasma spraying glass was coated with electrically conductive coatings reaching optical grades regarding their transparency. These efforts show that by means of plasma spraying new coating systems can be achieved.

5. Acknowledgements

The workings carried out for this contribution where funded by the German Research Foundation under the reference numbers BA 851/93-1 and SPP 1299 BA 851/94-1) and the AiF Arbeitsgemeinschaft industrieller Forschungsvereinigungen „Otto von Guericke" e.V (IGF Nos. 14.509 N and 16.411 N). This support is gratefully acknowledged by the authors.

6. References

Anderson, S. and Tilley, R. J. D. (1972). Crystallographic shear and non-stoicheiometry, in: Roberts, M. W. and Thomas, J. M. (eds.): Surface and Defect Properties of Solids, The Chemical Society, London, 1972, ISBN 978-1-84755-696-7

Andersson, S., Sundholm, A. & Magnéli, A. (1959). A Homologous Series of Mixed Titanium Chromium Oxides Ti(n-2)Cr2O(2n-1) Isomorphous with the Series Ti(n)O(2n-1) and V(n)O(2n-1), Acta Chemica Scandinavica, Vol. 13 (1959), pp. 989-997

Bauser, M.; Sauer, G. and Siegert, K. (eds., 2006). Extrusion, ASM International, Ohio, second edition, 2006, ISBN 978-0-87170-837-3

Barbezat, G. (2006). Thermal Spray Coatings for Tribological Applications in the Automotive Industry, Advanced Engineering Materials, Vol. 8 (2006), No. 7, pp. 678–681

Bolelli, G., et al. (2009). Deposition of TiO2 Coatings: Comparison between High Velocity Suspension Flame Spraying (HVSFS), Atmospheric Plasma Spraying and HVOF-spraying, in: *Proc. of the ITSC 2009*, 04.-07. May 2009, Las Vegas, ISBN 978-1-61503-004-0

Dean, A. M. & Voss, D (1999). *Design and Analysis of Experiments, Springer, ISBN 978-0-387-98561-9, Berlin, Germany, DOI: 10.1007/b97673*

Fauchais, P. et al. (2008). Parameters Controlling Liquid Plasma Spraying: Solutions, Sols, or Suspensions, *Journal of Thermal Spray Technology*, Vol.17 (2008), No. 31, pp. 31-59

Gardos, M. N. (1988). The Effect of Anion Vacancies on the Tribological Properties of Rutile (TiO2-x). *Tribol. Trans.* Vol. 31(4), 1988, pp. 427-455

Gell, M., et al. (2001). Development and implementation of plasma sprayed nanostructured ceramic coatings. *Surface and Coatings Technology*, Vol. 146-147 (2001), pp. 48-54

Goldschmidt, V. M. (1926). Die Gesetze der Krystallochemie, Naturwissenschaften, Vol. 14 (1926), No. 21, pp. 477-485, in German

Hawk, D. and Müller, F. (1980). Thermochemie des Systems CoO-B2O3, *Z. anorg. allg. Chem.*, Vol. 466 (1980), pp. 163-170, in German

Heimann, R.B. (2008). Plasma Spray Coating - Principles and Applications, pp. 389. WILEY-VCH Verlag, ISBN 978-3-527-32050-9, Weinheim, Germany

Kale, G. (1998). Interdiffusion studies in titanium 304 stainless steel system. *Journal of Nuclear Materials*, Vol.257, No.1, (1998), pp. 44-50, ISSN 0022-3115, DOI: http://dx.doi.org/10.1016/0022-3115(88)90072-4

Lugscheider, E. & Bach, Fr.-W. (eds., 2002). *Handbuch der thermischen Spritztechnik.: Technologien - Werkstoffe – Fertigung*, Verlag für Schweißen und Verwandte Verfahren, DVS-Verl., ISBN 3871551864, Düsseldorf, Germany

Mason, R. L. et al. (2003). *Statistical Design and Analysis of Experiments*, John Wiley & Sons, ISBN 9780471372165, Hoboken, NJ, USA, DOI: 10.1002/0471458503

Matthäus, G., Wolf, J. & Ackermann, D. (2010): *Near-net-shape HVOF coating and finishing techniques for highly stressed components in aircraft industry*, Proceedings of the International Thermal Spray Conference 2010, ISBN 978-3-87155-590-9, May 03 - 05 2010, Singapore

NIST (July 2011). *NIST/SEMATECH e-Handbook of Statistical Methods, 01. July 2011. Available from: <http://www.itl.nist.gov/div898/handbook/>*

Paul, A. (1975). Activity of nickel oxide in alkali borate melts, *Journal of Materials Science*, Vol. 10 (1975), pp. 422-426

Phadke, M. S. (1989). *Quality Engineering Using Robust Design*, Prentice Hall, ISBN 978-0137451678, New Jersey, USA

Shannon, R.D. (1976). Revised effective ionic radii and systematic studies of interatomic distances in halides and chalcogenides, *Acta Cryst. A*, Vol. 32 (1976), pp. 751-767

Tilmann, W. et al. (2008a). Influence of the HVOF gas composition on the thermal spraying of WC-Co submicron powders (-8 + 1 micron) to produce superfine structured cermet coatings. *Journal of Thermal Spray Technology*, Vol.17, No.5-6, (2008), pp. 924-932, ISSN 1059-9630

Tilmann, W. et al. (2008b). Near-Net-Shape and Dense Wear Resistant Thermally Sprayed Coatings. *Key Engineering Materials*, Vol.384-384 (2008), pp. 117-123, ISSN 1013-9826

Tombs, N. C.; Croft, W. J. & Mattraw, H. C. (1963). Preparation and Properties of Chromium Borate, *Inorg. Chem. 2*, Vol. 4 (1963), pp. 872–873

Wood, G. J. et al. (1982). Mechanism of oxidation of the crystallographic shear phase Ti4O7, *Philosophical Magazine A*, Vol. 46, (1982), No. 1, pp. 75-86

7

Ceramic Coating Applications and Research Fields for Internal Combustion Engines

Murat Ciniviz[1], Mustafa Sahir Salman[2],
Eyüb Canlı[1], Hüseyin Köse[1] and Özgür Solmaz[1]
[1]*Selcuk University Technical Education Faculty,*
[2]*Gazi University Technical Education Faculty*
Turkey

1. Introduction

Research for decreasing costs and consumed fuel in internal combustion engines and technological innovation studies have been continuing. Engine efficiency improvement efforts via constructional modifications are increased today; for instance, parallel to development of advanced technology ceramics, ceramic coating applications in internal combustion engines grow rapidly. To improve engine performance, fuel energy must be converted to mechanical energy at the most possible rate. Coating combustion chamber with low heat conducting ceramic materials leads to increasing temperature and pressure in internal combustion engine cylinders. Hence, an increase in engine efficiency should be observed.

Ceramic coatings applied to diesel engine combustion chambers are aimed to reduce heat which passes from in-cylinder to engine cooling system. Engine cooling systems are planned to be removed from internal combustion engines by the development of advanced technology ceramics. One can expect that engine power can be increased and engine weight and cost can be decreased by removing cooling system elements (coolant pump, ventilator, water jackets and radiators etc.) (Gataowski, 1990; Schwarz et. al. 1993).

Initiation of the engine can be easier like shortened ignition delay in ceramic coated diesel engines due to increased temperature after compression because of low heat rejection. More silent engine operation can be obtained considering less detonation and noise causing from uncontrolled combustion. Engine can be operated at lower compression ratios due to shortened ignition delay. Thus better mechanical efficiency can be obtained and fuel economy can be improved (Büyükkaya et. al., 1997).

Another important topic from the view point of internal combustion engines is exhaust emissions. Increased combustion chamber temperature of ceramic coated internal combustion engines causes a decrease in soot and carbon monoxide emissions. When increased exhaust gases temperature considered, it is obvious that turbocharging and consequently total thermal efficiency of the engine is increased.

Combustion characteristics is the most important factors which affect exhaust emissions, engine power output, fuel consumption, vibration and noise. In diesel engines, combustion characteristics depended on ignition delay at a high rate (Balcı, 1983). Ignition delay is determined mostly by temperature and pressure of compressed air in combustion chamber. Conventional diesel engines have lower temperature and pressure of compressed air just because engine cooling system soaks considerable heat energy during compression to protect conventional combustion chamber materials. When the lost heat energy, useful work are taken into account, the idea of coating combustion chambers with low heat conduction and high temperature resistant materials leads to thermal barrier coated engines (also known as low heat rejection engines). Thermal barrier coated engines can be thought as a step to adiabatic engines. To achieve this aim, ceramic is a preferred alternative. Thermal barrier coating is mostly done by ceramic coating of combustion chamber, cylinder heads and intake/exhaust valves. If cylinder walls are intended to be coated, a material should be selected which has proper thermal dilatation and wear resistance. Some ceramic materials have self lubrication properties up to 870 ⁰C (Hocking et. al., 1989).

Exhaust gas temperature changing between 400-600 ⁰C for conventional diesel engines while it is between 700-900 ⁰C for thermal barrier coated engine. This temperature value reaches to 1100 ⁰C in turbocharged engines. When exhaust gas temperatures reaches these high levels, residual hydrocarbons and carbon monoxides in the exhaust gases are oxidized and exhaust emission are become less pollutant regarding hydrocarbons and carbon monoxide. In Figure 1, energy balance diagrams for conventional diesel engine and ceramic coated engine are given (Büyükkaya, 1994). Beside these advantages of ceramic coated low heat rejection engines, mechanical improvements also gained by light weight ceramic materials. By their high temperature resistance and light weight, moving parts of the engine have more duration owing to low inertia and stable geometry of the parts. Bryzik and Kamo (1983) reported 35% reduction in engine dimensions and 17% reduction in fuel consumption with a thermal barrier coated engine design in a military tank.

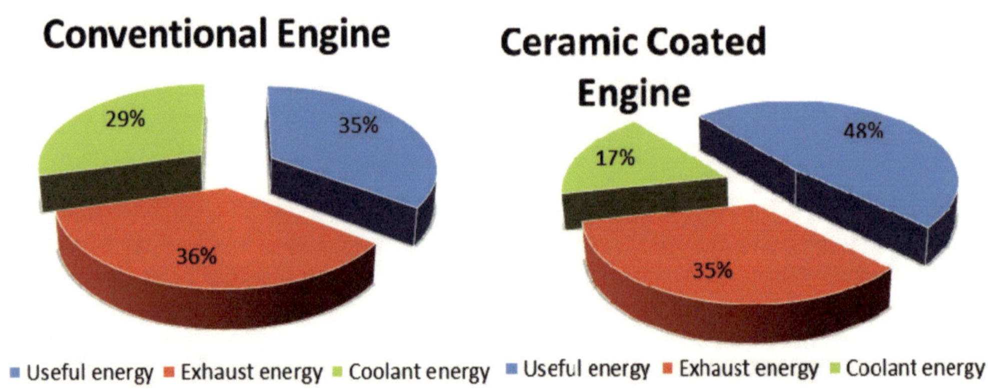

Fig. 1. Energy balance illustration for conventional engine and ceramic coated engine

1.1 Advanced technology ceramics

Ceramics have been used since nearly at the beginning of low heat rejection engines. These materials have lower weight and heat conduction coefficient comparing with materials in conventional engines (Gataowski, 1990). Nowadays, important developments have been achieved in quantity and quality of ceramic materials. Also new materials named as "advanced technology ceramics" have been produced in the last quarter of 20th century. Advantages of advanced technology ceramics can be listed as below;

- Resistant to high temperatures
- High chemical stability
- High hardness values
- Low densities
- Can be found as raw material form in environment
- Resistant to wear
- Low heat conduction coefficient
- High compression strength (Çevik, 1992)

Advanced technology ceramics consist of pure oxides such as alumina (Al_2O_3), Zirconia (ZrO_2), Magnesia (MgO), Berillya (BeO) and non oxide ones. Some advanced technology ceramic properties are given in Table 1.

Material	Melting Tempe-rature (^0C)	Density (g/cm^3)	Strength (MPa)	Elasticity Module (GPa)	Fracture Toughness ($MPa\ m^{1/2}$)	Hardness (kg/mm^2)
SiO_2	500	2,2	48	7,2	0,5	650
Al_2O_3	2050	3,96	250-300	36-40	4,5	1300
ZrO_2	2700	5,6	113-130	17-25	6-9	1200
SiC	3000	3,2	310	40-44	3,4	2800
Si_3N_4	1900	3,24	410	30-70	5	1300

Table 1. Some advanced technology ceramic properties

Zirconia has an important place among coating materials with its application areas and properties essential to itself. The most important property of zirconia is its high temperature resistance considering ceramic coating application in internal combustion engines. Ceramics containing zirconia have high melting points and they are durable against thermal shocks. They have also good corrosion and erosion resistances. They are used in diesel engines and turbine blades to reduce heat transfer.

1.1.1. Zirconia (ZrO2)

Zirconia can be found in three crystal structure as it can be seen in Fig. 2. These are monolithic (m), tetragonal (t) and cubic (c) structures. Monolithic structure is stable between room temperature and 1170 ^0C while it turns to tetragonal structure above 1170 ^0C. Tetragonal structure is stable up to 2379 ^0C and above this temperature, the structure turns to cubic structure.

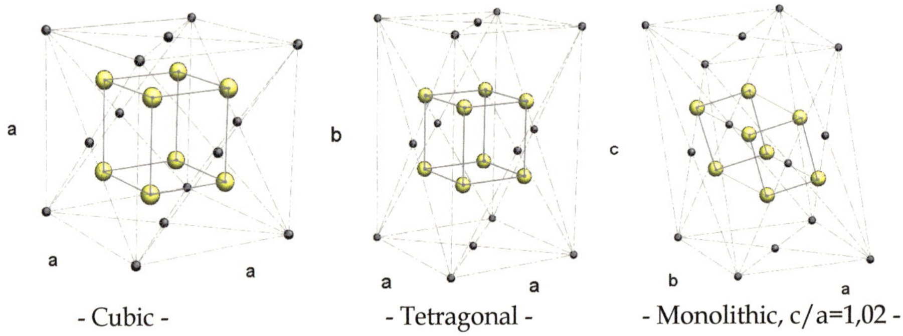

Fig. 2. Cubic, tetragonal and monolithic zirconia

Usually cracks and fractures are observed during changing phases because of 8% volume difference while transition to tetragonal structure from monolithic structure. To avoid this and make zirconia stable in cubic structure at room temperature, alkaline earth elements such as CaO (calcium oxide), MgO (magnesia), Y_2O_3 (yttria) and oxides of rare elements are added to zirconia. Zirconia based ceramic materials stabilized with yttria have better properties comparing with Zirconia based ceramic materials which are stabilized by magnesia and calcium oxide (Yaşar, 1997; Geçkinli, 1992).

Mechanical properties of cubic structure zirconia are weak. Transition from tetragonal zirconia to monolithic zirconia occurs at lower temperatures between 850-1000 ^{0}C and this transition has some characteristics similar to martensitic transition characteristics which are observed in tempered steels. In practice, partially stabilized cubic zirconia (PSZ) which contains monolithic and tetragonal phases as sediments, is preferred owing to its improved mechanical properties and importance of martensitic transition. Partially stabilized zirconia has been commercially categorized since early 70s. Table 2. contains partially stabilized zirconia types and their properties. Structural properties of these materials are;

- Zt35: Contains 20% (t) phase in cubic matrix. Particle dimensions are about 60-70 μm.
- ZN40: Contains 40-50% (t) phase.
- ZN50: Particle dimensions are about 60-70 μm and a thin film (m) phase lays on the borders of particles.
- ZN20: Is developed for thermal shocks. Contains (m) phase.

Material	**Code**	**Elasticity Module (GPa)**	**Fracture Toughness (MPa $m^{1/2}$)**	**Vickers Hardness (HVat 22 ^{0}C)**	**Expansion Coefficient (22-1000 ^{0}C)**
Ca/Mg-PZS	Zt35	200	4,8	1300	$9{,}8 \times 10^{-6}$
Mg-PZS	ZN40	200	8,1	1200	9,8
Mg-PZS	ZN50	200	9	900	7
Y-PZS	ZN100	190	9,7	-	9,3
Mg-PZS	ZN20	180	3,5	-	5,5

Table 2. Partially stabilized zirconia types and properties

1.1.2 Yttria (Y_2O_3)

Melting point of yttria is 2410 ^{0}C. It is very stable in the air and cannot be reduced easily. It can be dissolved in acids and absorbs CO_2. It is used in Nerst lambs as filament by alloyed with zirconia and thoria in small quantities. When added to zirconia, it stabilizes the material in cubic structure. Primary yttria minerals are gadolinite, xenotime and fergusonite. Its structure is cubic very refractory.

1.1.3 Magnesia (MgO)

Magnesia is the most abundant one in refractory oxides and its melting point is 2800 ^{0}C. Its thermal expansion rate is very high. It can be reduced easily at high temperatures and evaporate at 2300-2400 ^{0}C. At high temperature levels, magnesia has resistance to mineral acids, acid gases, neutral salts and moisture. When contacted to carbon, it is stable up to 1800 ^{0}C. It rapidly reacts with carbons and carbides over 2000 ^{0}C. The most important minerals of magnesia are magnesite, asbestos, talc, dolomite and spinel.

1.1.4 Alumina (Al_2O_3)

Melting point of alumina is about 2000 ^{0}C. It is the most durable refractory material to mechanical loads and chemical materials at middle temperature levels. Relatively low melting point limits its application. It doesn't dissolve in water and mineral acids and basis if adequately calcined. Raw alumina can be found as corundum with silicates as well as compounds of bauxide, diaspore, cryolit, silimanite, kyanite, nephelite and many other minerals. As its purety rises, it becomes resistant to temperature, wear and electricity.

1.1.5 Beryllia

Beryllia has a high resistance to reduction and thermal stability and its melting point is 2550 ^{0}C. It is the most resistant oxide to reduction with carbon at higher temperatures. Thermal resistance is very high though its electrical conductivity is very low. Mechanical properties of beryllia are steady till 1600 ^{0}C and it is one of the oxides that has high compression strength at this temperature. An important amount of beryllium oxide acquired from beryl. It is a favourable refractory material for molten metals owing to its resistance to chemical materials (Geçkinli, 1992).

2. Ceramic coating applications in internal combustion engines

Ceramic coatings which are applied to reduce heat transfer are divided into two groups. Generally, up to 0,5 mm coatings named as thin coatings and thick coatings are up to 5-6 mm. Thin ceramic coatings are used in gas turbines, piston tops, cylinder heads and valves of otto and diesel engines. At the beginning of ceramic coatings to low heat rejection engines, thick monolithic ceramic coatings were applied to engine parts. Later, it was understood that these coatings are not appropriate for diesel engine operation conditions. Thus, new approaches were started to develop (Yaşar, 1997; Kamo et. al., 1989).

There are a lot of types and system for ceramic and other material coatings. Most important ones are;

- Thermal spray coating: Plasma spray, wire flame spray and powder flame spray, electrical arc spray, detonation gun technique and high speed oxy fuel system
- Chemical ceramic coating: Sole-gel, slurry, chemical vapour sedimentation, physical vapour sedimentation, hard coating
- Laser coating
- Arc spark alloying
- Ion enrichment method (Yaşar, 1997; Kamo et. al., 1989)

Material conglomerations can be avoided by reducing erosion-corrosion, friction-wear, using ceramics as well as improving heat insulation. Non the less, these methods are proper for very thin coatings except thermal spray coatings. Thin layer coatings are successfully used in gas turbine industry, coating turbine and stator blades and combustion rooms. For thick layer coatings like diesel engines, plasma spray and flame spray coatings are generally utilized (Kamo et. al., 1989).

2.1 Flame spray coatings

Oxy-hydrogen and oxy-acetylene systems are preferred in flame spray coatings and usually refractory oxides which have lower melting point than 2760 ^0C are used in coating with these systems. Before ceramic coatings, a binding layer resistant to high temperature like nickel-chromium should be applied to material surface for preventing oxidation as can be seen in Fig. 3. Otherwise, ceramic coating can't adhere to the surface properly. Coating speed in flame spray method is relatively slow and it changes between $4.4x10^{-5}$ and $1.13x10^{-3}$ m/s. There are two flame spray method which are wire flame spray method and powder flame spray method (Geçkinli, 1992).

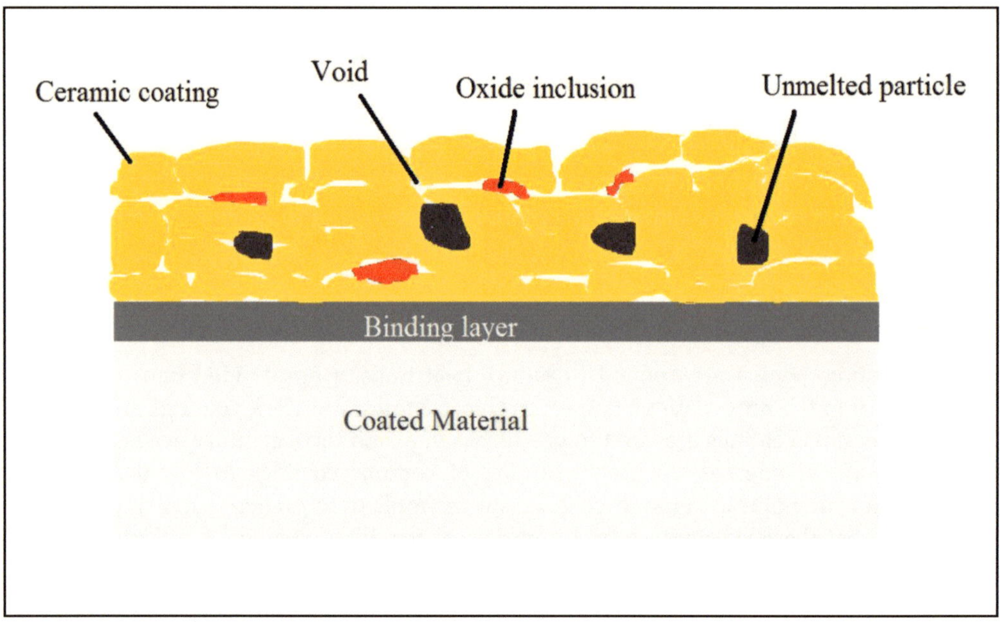

Fig. 3. Ceramic coated material surface, binding layer and coating layer

2.1.1 Powder flame spray coatings

In this method, micro-pulverized powder alloys are sprayed to target surface in oxy-acetylene flame by oxygen vacuum. It is called cold coating because flame temperature is about 3300 ^0C and target surface is about 200 ^0C during coating process. Adherence is mechanical. Coating layer thickness is changed 0,5 to 2,5 mm according to shape of work piece. Using highly alloyed and self lubricant NiCrBSi materials as coating powder and making materials which are not produced in rod or wire shapes possible for coating are the main advantages of this method. Powder flame spray systems are proper for spraying primarily ceramics and metals and cermets (metals and ceramic oxide alloys) as coating materials. Bearing supports, axle and shaft pivots, compressor pistons, cam shafts, bushes, rings and sleeves, hydraulic cylinders and pistons can be coated by this very method (Yaşar, 1997; Anonymous, 2004).

2.1.2 Wire flame spray coatings

Wire flame spray coating method is applied by spraying a wire shaped metal which has a melting point below flame temperature to coated surface. It can be used for metal spray materials and metal surfaces. Coating material wire is molten by oxygen and gas fuel flame after passing from the coating gun nozzle. Acetylene, propane and hydrogen are used for gas fuel. Relatively low equipment costs, high spray speeds and adjustment property according to wire diameters are the advantages of this system. Lower coating intensity and adherence strength comparing with other methods can be told as disadvantages of the method. Bearing supports, hydraulic piston pins, various bearings, shafts, wearing surfaces of axles, piston segments, synchromesh, crank shafts, clutch pressure plates can be coated with wire flame spray coating systems (Yaşar, 1997; Anonymous, 2004).

2.2 Plasma spray coating

Plasma is a dense gas which has equal number of electron and positive ion and generally named as fourth state of the matter. This method has two primary priorities; It can provide very high temperatures that can melt all known materials and provides better heat transfer than other materials. High operating temperature of plasma spray coating, gives opportunity to operate with metals and alloys having high melting points. Also using plasma spray coating in inert surroundings is another positive side of the method. Oxidation problem of the subject material is reduced due to inert gas usage in plasma spray such as argon, hydrogen and nitrogen. All materials that are produced in powder form and having a specific grain size can be used in this method (Yaşar, 1997; Geçkinli, 1992).

The main objective in plasma spraying is to constitute a thin layer that has high protection value over a non expensive surface. The process is applied as spraying coating material in powder form molten in ionized gas rapidly to coated surface. Plasma spray coating system is shown in Fig. 4. The spraying gun is illustrated in Fig. 5. The system primarily consists of power unit, powder supply unit, gas supply unit, cooling system, spraying gun and control unit.

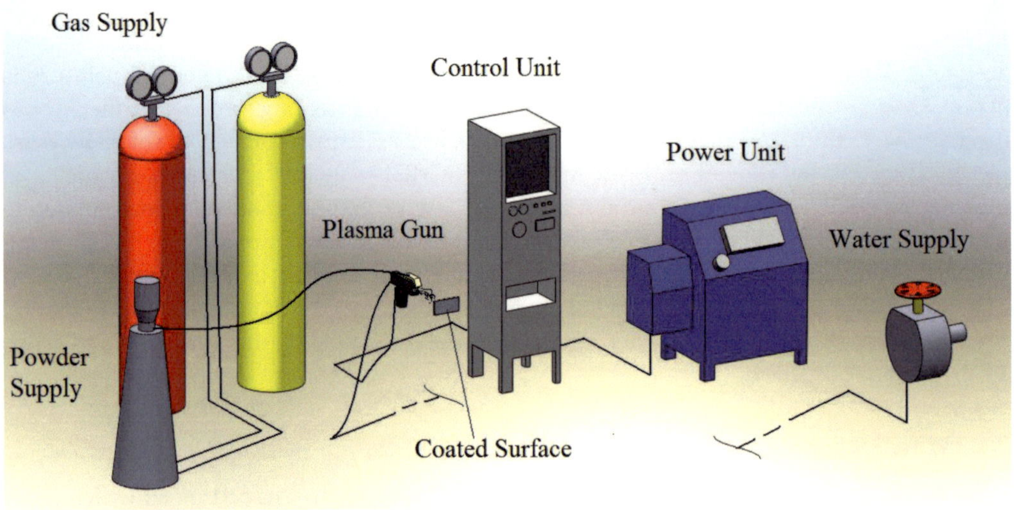

Fig. 4. Plasma spray coating system

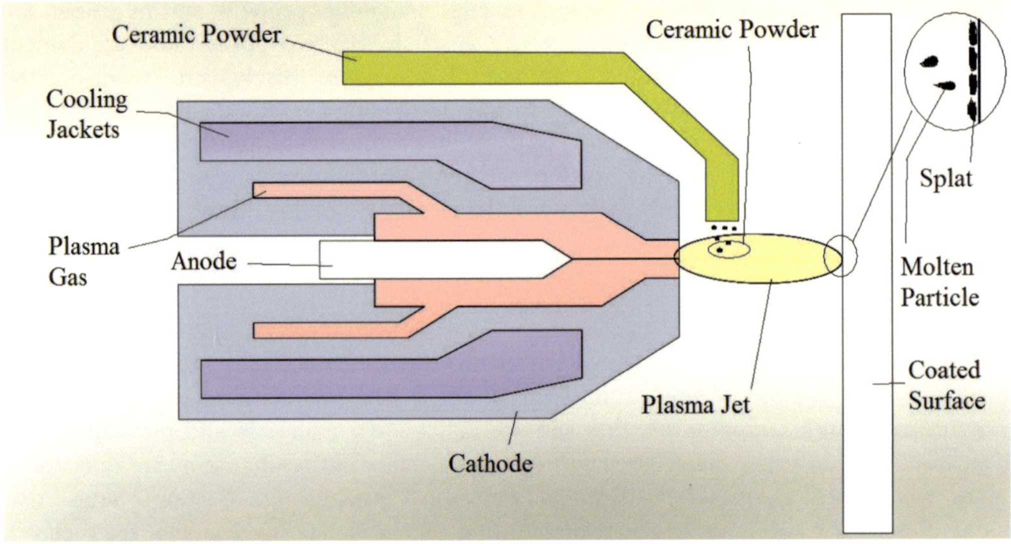

Fig. 5. Plasma spray gun

Direct current electrical arc is formed between electrode and nozzle in plasma spray coating gun. The inert gas (usually argon) and a little amount of hydrogen gas which is used to empower inert gas mixtures are sent to arc area of plasma gun and heated with electrical arc. Gas mixture temperature reaches to 8300 ^{0}C and it becomes ionized. Hence, high temperature plasma beam leaves from gun nozzle. In this system, ceramic grains are supplied to plasma beam as dispersible form. Grains molten by the hot gases are piled up on target surface and hardened. Argon/helium gas mixture increase gas flow and hence ceramic grains speed. Coating layer structure by the plasma spray coating contains equal axial thin solid grains. In some layers, an amorphous structure is attained because of fast solidification (Geçkinli, 1992).

Porosity is a property and a structural indicator of plasma spray coating. By utilizing high viscosity grains and high power plasma units, an intensive coating layer can be attained. Coating layers consisted from brittle and hard ceramic materials have high porosity rates. High porosity negatively affects material hardness which is a mechanical property of the material. While the least porosity layers have about 700 Vickers hardness, porous coating layers have about 300 Vickers hardness. 10 percent of the porosity after plasma spray coating is closed while rest of the porosity is open ones which combined with other defects in the structure because of insufficient fillings of blank areas among settled ceramic grains. Open porosities spoil mechanical properties of substance material by enabling corrosive sediments and gases to diffuse coating layer. On the other hand, spaces parallel to substance surface between layers negatively affect coating adhesion (Yaşar, 1997; Geçkinli, 1992).

Target surface must be rough, cleaned from oxides, oil, dirt and dust for making coating adhere to target surface. Surface roughness usually acquired by spraying an abrasive powder such as dust or alumina to target surface by a pressurized air. By coating base material having its surface prepared with a special binding material, target surface has a proper ground for ceramic coating. In addition to its binding property, binding layers can be used for reducing thermal expansion, protecting base material from effects of corrosion, gases and high temperatures. The most preferred binding material is NiAl. Work pieces which have their surface prepared for coating are placed perpendicular to plasma flame and fixed. Spray powders must hit to target surface perpendicularly to obtain an intensive and good quality ceramic coating (Yaşar, 1997; Geçkinli, 1992).

Another important factor is powder size distribution in the spray. Very small grains in the plasma flame can easily reach plasma flame temperature, big grains however, adhere to target surface without being properly molten and make structure to be porous. Researches show that grain sizes between 60 ±10 μm give good results.

Plasma spray coating can be conducted either in atmospheric conditions or in vacuum conditions. When it is done in vacuum conditions, plasma flame can expand to 20 cm and more intensive coatings can be obtained (Geçkinli, 1992). Fundamental elements and parameters affecting them in plasma spray coating are given in Table 3. One part of the process parameters which are determined for a specific coating application are depended to operator. To eliminate these parameters effecting coating quality, operating plasma gun with a robot arm or making plasma gun to move vertically and horizontally are proposed as solutions and applied.

3. Effects of ceramic coatings to internal combustion engine performance

To reduce damages occurring from high cycle temperatures, high cycle forces, sliding, erosion and corrosion on engine parts, several special techniques have been developed. Water cooling and thick combustion chamber walls had been utilized up to the end of Second World War to transfer excessive heat which material properties of combustion chamber construction materials such as cast iron can't bear. Later on, using low thermal conductivity materials such as glass and its' derivatives were considered. Despite low thermal conductivity, cost and low expansion rate, glass couldn't be used in internal combustion engines due to its' lacking strength. Using glass ceramic materials in engine parts was first seen at 1950s. In those days, ceramics used in spark plugs although low application numbers. Requirements for ceramic coatings for high temperature applications had been started to increase at 1960s. Especially developing gas turbines leaded that requirement because of metals and various alloys that couldn't resist high temperatures. Ceramic coating technology was initially applied to space and aviation areas and then at 1970s it had been started to apply to internal combustion engines, especially diesel engines. Performance increase and specific fuel consumption decrease of aforementioned ceramic coated systems created an interest to the topic.

SPRAYED POWDER PERAMETERS	**COATING PARAMETERS**	**PROCESS PARAMETERS**
COATING MATERIAL	COATED MATERIAL	PROCESS
Chemical composition	Mechanical properties	Atmospheric plasma spray
Phase stability	Thermal expansion rate	Inert gas plasma spray
Thermal expansion	Oxidation resistance	Vacuum plasma spray
Melting characteristics	Work piece dimensions	Under water plasma spray
Grain size distribution	Surface quality	Sprayed powder
Grain morphology	SERVICE CONDITIONS AT OPERATION CONDITIONS	Plasma gases
Specific surface area	Wear	Plasma temperature
Fluidity	Wear-wet corrosion	Speeds of sprayed powders
	Wear-oxidation	Powder supply speed
	Wear- gas corrosion	Pre heating and cooling of work piece
	Wear-erosion	Surface cleaning
	COMPOSITE COATING	Spraying environment
	Chemical components	
	Adhesion strength	
	Metallurgical reaction	
	Mechanical properties	
	Physical properties	
	Coating thickness	
	Porosity	
	Residual stresses	
	Coating properties under load	
	QUALITY CONTROL	
	TEST	
	PRODUCTION	

Table 3. Plasma spray coating technology; Components and parameters

Thermal barrier coatings used for reducing heat loss from cylinders and converting engines to low heat rejection engines also prevent coated materials from decomposing under high temperatures. ZrO_2 is the most preferred material in thermal barrier coated internal combustion engines due to its' low thermal conductivity and high thermal expansion rate. To avoid negative effects of phase changes of ZrO_2 at higher temperatures, it should be partially or fully stabilized with a stabilizer material. By this procedure, whole structure is formed with one phase, generally cubic phase. As stabilizer, usually MgO, CaO, CeO_2 and Y_2O_3 oxides are used.

There are a vast number of studies investigating effects of thermal barrier coatings and especially ceramic coatings to internal engine performance and exhaust emission behaviours. Investigated parameters can be summarized as coating material, coated material, coating thickness, engine types and operational conditions such as engine load and speed. Obtained results can be different in dimensions and magnitudes such as volumetric efficiency, thermal efficiency, engine torque, engine power, specific fuel consumption, heat rejection from cylinders, exhaust temperature, exhaust energy and exhaust emissions. Investigations of thermal barrier coating in internal combustion engines are mostly focused on diesel engines because of detonation and knocking problems of spark ignition engines at higher in cylinder temperatures. For diesel engines, studies can be divided into two main categories; turbocharged engines and non-turbocharged engines. For non-turbocharged engines, thermal barrier coating application and thus ceramic coatings of internal combustion engine cylinders generally results negatively due to decreasing volumetric efficiency. In the other hand, turbocharged diesel engines exhibit better performance and exhaust emissions according to improved volumetric efficiency and in cylinder temperatures. This phenomenon's main reason is the increased exhaust gas energy which is converted to mechanical energy and later on to air mass flow rate increase in turbocharger. For instance, Leising and Prohit (1978) suggested that desired results by heat rejection insulation could only be achieved by the utilization of turbocharger and intercooler. They also reported that a diesel engine performance could be increased up to 20% by the addition of a turbocharger. When studies about thermal barrier coated engines without turbochargers are considered, it was observed that most of the studies were conducted on a single cylinder, four stroke diesel engines. Miyairi et. al. (1989), Dickey (1989) and Alkidas (1989) are some of these researchers. Prasad et. al. (2000), Charlton et. al. (1991), Chang et. al. (1983) can be given as examples for researchers that studied on natural aspirated multi-cylinder diesel engines. In the other hand, multi-cylinder diesel engines types were mostly preferred for turbocharged thermal barrier coated engine researches. For instance Woods et. al. (1992), Kimura et. al. (1992), Woschni and Spindler (1988), Hay et. al. (1986) and Ciniviz (2005) performed parametric studies on thermal barrier coated turbocharged multi-cylinder diesel engines. Parlak (2000) and Kamo et. al. (1997) are two studies among limited turbocharged single cylinder thermal barrier coated engine investigations.

Coating materials and methods can be divided into two categories for this book; ceramics and non-ceramics. Coating thickness is usually changes between 100-500 μm. A typical thickness for coating materials is 0,15 mm binding layer and 0,35 mm coating material. Parlak et. al. (2003) and Taymaz et. al. (2003) are two of these studies which used the typical coating thickness. For the researchers that preferred ceramic materials, zirconia is the most seen material among other ceramics. NiCrAl is frequently used as binding materials for those studies. Uzun et. al. (1999), Beg et. al. (1997), Taymaz et. al. (2003), Marks and Boehman (1997), Schwarz et. al. (1993) and Hejwowski (2002) can be referred for these studies. Alternatively, Sun et. al. (1994), proposed silicon nitride (HPSN) piston materials and thick coating layers of plasma sprayed zirconia between 2-7 mm for cylinders. Matsuoka and Kawamura (1993) used Si_3N_4 instead of zirconia.

Specific literature survey was resulted that specific fuel consumption, heat rejection from cylinders and NO_x emissions are the most reported results of experimental and numerical studies for ceramic coated engines. Depending on rising in cylinder temperatures, almost all studies expressed an increase in NO_x emissions. This event can be named as the main side effect of ceramic coating or thermal barrier coating of internal combustion engines. The increase in NO_x emissions is observed between 10-40% from the literature. Gataowski (1990), Osawa et. al. (1991) and Kamo et. al. (1999) some of the papers in which these aforementioned results can be found. However there are some suggestions for reducing this increase by changing injection timing or decreasing advance angle. Winkler and Parker (1993) reported 26% decrease in NO_x emissions of thermal barrier coated engine by changing injection timing. Similarly Afify and Klett (1996), stated that 30% decrease in NO_x emissions was achieved by advance adjustment. When specific fuel consumption is considered, results are varying both negatively and positively. This is particularly the result of volumetric and combustion efficiency. Specific fuel consumption decrease can be observed from the literature between 1-30%. Ramaswamy et. al. (2000), reported 1-2% specific fuel consumption decrease while Bruns et. al. (1989), stated specific fuel consumption decrease between 16-37% by means of ceramic thermal barrier coating. On the contrary, Sun et. al. (1993) and Beg et. al. (1997) expressed 8% increase in specific fuel consumption by the utilization of ceramic thermal barrier coating. Similarly Kimura et. al. (1992), specified that thermal barrier coating resulted 10% increase in specific fuel consumption. As desired, ceramic thermal barrier coatings were resulted as a decrease between 5-70% in heat rejection from cylinders to engine block and cooling system. Vittal et. al. (1997) reported 12% decrease in transferred heat from cylinders and Rasihhan and Wallace (1991) informed that heat rejection rate was decreased between 49,2-66,5% after ceramic coating.

There are several more indicators that show effectiveness of ceramic thermal barrier coatings. Further search can be conducted for specific parameters.

4. A case study: The effects of Y_2O_3 with coatings of combustion chamber surface on performance and emissions in a turbocharged diesel engine

In this study, the effects of ceramic coating of combustion chamber of a turbocharged diesel engine to engine performance and exhaust emissions were investigated. Increasing mechanical energy by preventing heat losses to coolant and reducing cooling load, improving combustion by increasing wall temperatures and decreasing ignition delay, more power attaining in turbocharged engines by increasing exhaust gas temperatures and decreasing carbon monoxide and soot are aimed. For this aim, cylinder head, inlet and exhaust valves and pistons of the engine were coated with 0.5 mm zirconia by plasma spray coating. Then, the engine was tested for different brake loads and speeds at standard, ceramic coated engine one and ceramic coated engine 2 conditions. The results gained from the experimental setup were analyzed with a computer software and presented with comparatively graphics. Briefly, specific fuel consumption was decreased 5-9 percent, carbon monoxide emission was decreased 5 percent, and soot was decreased 28 percent for a specific power output value. Considering these positive results nitrogen oxide however was increased about 10 percent. By the development of exhaust catalysers, increase in nitrogen oxide becomes no more a problem for present day. When results are generally investigated, it was concluded that engine performance was clearly improved by zirconia ceramic coating.

4.1 Experimental setup

Appropriate measurement equipment, their calibration and operational conditions have an important effect on experimental results. Engine specifications are given in Table 4. Experiments were conducted in internal combustion engines workshop in Gazi University Technical Education Faculty Mechanical Education Department Turkey. Cross section view of the engine is shown in Fig. 6 and solid model view of experimental setup is illustrated in Fig. 7.

ENGINE	SPECIFICATIONS
Brand, type and model	Mercedes-Benz/OM364A/1985
Cylinder number/diameter/stroke	4/97.5 mm/133 mm
Total cylinder volume (Combustion room + cylinder)	3972 cm^3
Compression rate	17.25
Nominal revolution rate	2800 rev/min
Engine power	66 kW (2800 rev/min)
Maximum torque	266 Nm (1400 rev/min)
Operation principle	4 stroke diesel engine
Injection sequence	1-3-4-2 (cylinder numbers)

Table 4. Specifications of engine used in experiments

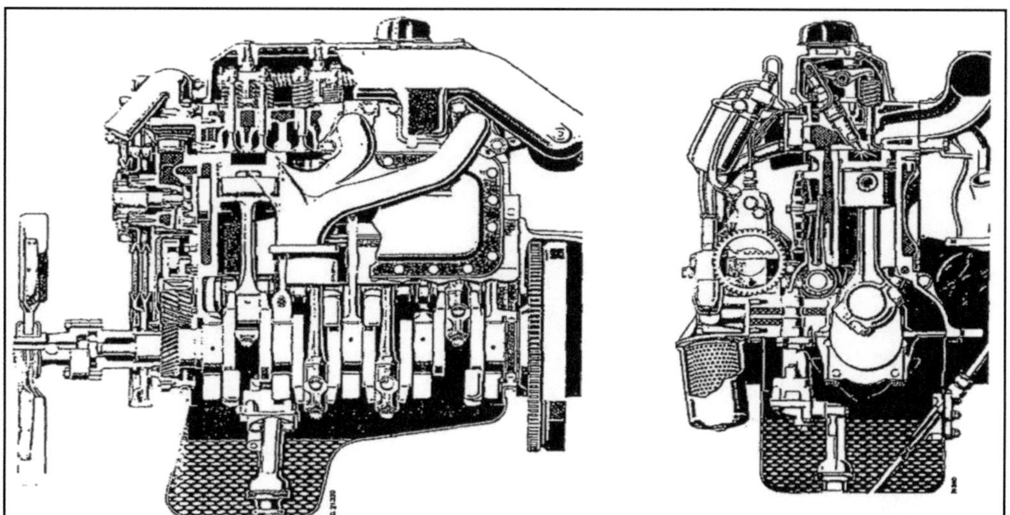

Fig. 6. Cross sectional view of test engine

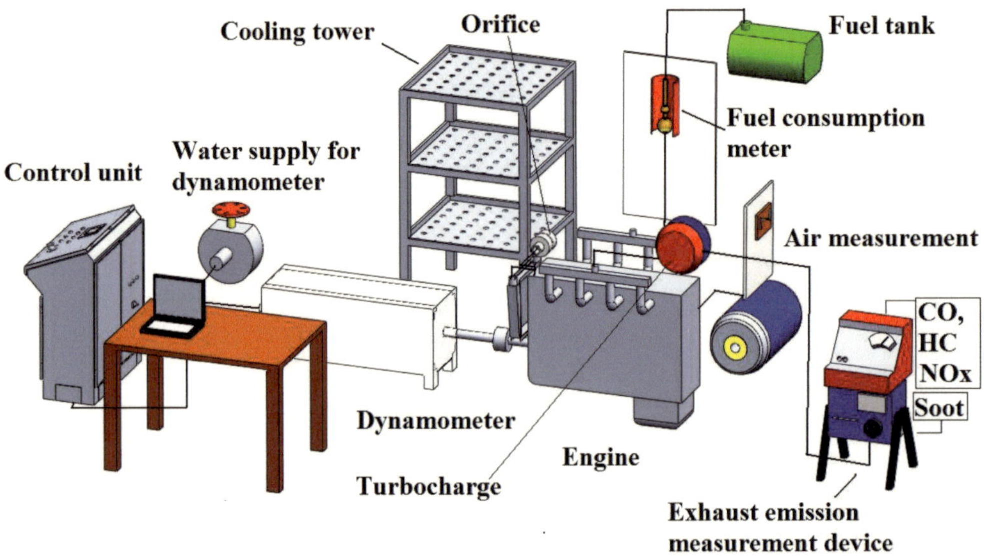

Fig. 7. Solid model view of experimental setup

Measurement devices were used for determining both exhaust emission values and performance characteristics. Photographs of these devices are given in Fig. 8. Experimental setup consist of basic units such as hydraulic brake dynamometer, cooling tower for cooling engine coolant, fuel consumption measurement device, temperature and pressure probes and control panel.

Engine was loaded by hydraulic dynamometer which is connected to engine with a shaft during experiments. Fig. 9 shows test engine in experimental setup. Additionally, flow rate measurement setups were utilized in the system for charge air and coolant.

a) Control panel | b) Charge air flow measurement unit and damper | c) Orifice plate for measuring coolant flow rate

d) Exhaust gas analyser | e) Smoke intensity measurement device | f) Temperature measurement screen

Fig. 8. Measurement devices in experimental setup for determining exhaust emissions and performance characteristics

Air flow measurement device used in the experiments is GO-Power M5000 type. A manometer was placed onto the device and it has a gauge glass of 0-75 mm long. For the conducted experiments, a 2.75 inch nozzle was attached to entrance of damper. Ohaus brand digital mass scale with 0.1 gram sensibility and 8 kg capacity was preferred for determining fuel amount. For exhaust emission, two different exhaust emission measurement devices were used during experiments as it can be seen from Fig. 7 and Fig. 8. For measuring carbon monoxide, carbon dioxide, nitrogen oxides, oxygen and sulphur oxides as ppm (particle per million) and mg/m^3, Gaco-SN branded exhaust gas analyser device was used. It can also calculate combustion efficiency and excess air coefficient. For determining smoke intensity, OVLT-2600 type diesel emission measurement device was used. This device can measure smoke amount as *k* factor and percentage. Measurement range and accuracy of OVLT-2600 are given in Table 5.

Fig. 9. Three different views of the test engine

Measured parameter	Measurement range	Accuracy
k factor	0-10 (m^{-1})	±0.01
Smoke intensity	0-99 (%)	±0.01
Engine revolution	0-9999 rev/min	1 rev/min

Table 5. OVLT-2600 measurement ranges and accuracies

1 ^{0}C accuracy thermometer which have 130 ^{0}C gauge and Precision branded barometer which has measurement range of 710-800 mmHg were used during experiments. A chronometer with 0.01 second resolution was employed while fuel consumption rate was measuring.

4.2. Experimental method

Determining ceramic coating effects on performance and exhaust emissions of turbocharger diesel engine requires standard values for performance indicators. For this purpose, test engine was operated without ceramic coatings according to 1231 numbered Turkish Standards (TS) experimental essentials and results were recorded. Ceramic coatings were applied after those standard tests. Cylinder heads, piston tops and intake exhaust valves were machined at 0.5 mm depth. Machining was done for achieving same compression rate with conventional combustion chamber after ceramic coating. Ceramic coating was applied by plasma spray coating system in Metal & Seramik Kaplama Ltd. Sti. in Turkey.

The most critical coated engine part is pistons due to its thermal expansion rate which is very different from selected ceramic material. In literature, ZrO_2 stabilized with Y_2O_3 and Si_3N_4 ceramic coating materials are told as positive result giving materials. At cylinder heads and intake exhaust valves, ZrO_2 stabilized with MgO can be utilized safely. Another important point in ceramic coatings is the binding layer composition. Coating durability is increased when NiCrAlY is used as binding layer.

Surfaces to be coated were cleaned from lubricants and other unwanted dirt after machining before roughed by sandblasting and prepared for ceramic coating. When surface preparation was done, surface was first coated with binding layer at 0.15mm thickness and then coated with 0.35 mm thick ceramic material layer. Reduction of thermal instability (high heat conduction difference) between coating layer and target surface is aimed by this way. Hence, the failure risk for coating layer is lowered. In Fig. 10, coated piston tops can be seen. Fig. 11 contains two different figures which are illustrating cylinder head and valves before coating and after coating respectively.

Ceramic materials used for coating are;

- Coating sequence for inlet exhaust valves and cylinder head was selected as base material + 0.15 mm thick NiCrAl + 0.35 mm thick Y_2O_3 – ZrO_2.
- Coating sequence for piston heads was selected as base material + 0.15 mm NiCrAlY + 0.35 Y_2O_3 - ZrO_2.

After coating process was done, coated engine parts were mounted to engine. Same circumstances with standard engine test were applied to coated engine tests. Experimental measurements were evaluated via MS Excel and Matlab v6.5 software.

In diesel engines, power output, torque and fuel consumption values according to engine speeds are named as engine characteristics. Differences in these characteristics at different load and engine speeds are illustrated with graphical curves. These curves are called as characteristic curves. Engine characteristic curves provide important information about engine performance at real time operational circumstances. Experimental measurements not always give directly the desired data. These data should be calculated using experimental measurements. Experimental measurements generally consist of torque, engine revolution rate, fuel consumption, charge air flow rate, coolant flow rate, ambient temperature, pressure and humidity, exhaust gases temperatures, coolant entrance and exit temperatures. The most important performance characteristics calculated from these measurements are effective power, torque, mean effective pressure and specific fuel consumption (Ciniviz, 2005).

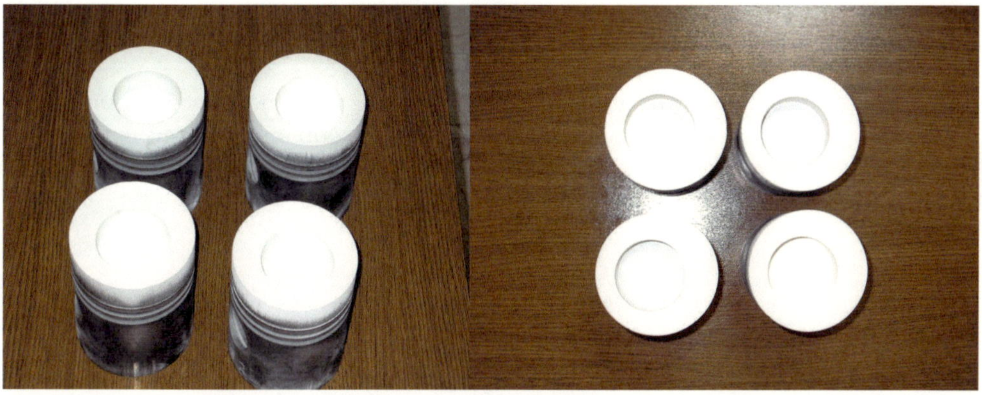

Fig. 10. Ceramic coated piston tops

a) Cylinder heads and valves without coating b) Ceramic coated cylinder heads and valves

Fig. 11. Cylinder head and valves before coating and after coating

During experiments, intake and exhaust valve adjustments were made according to engine catalogue values and injectors were tested at 200 bar injection pressure. Piston rings were renewed. To measure exhaust gas composition, exhaust pipe was drilled after one meter distance from exhaust pipe entrance and measurement probe was fitted to the hole. Experiments were conducted at ten different engine speeds changing between 1100 rev/min and 2800 rev/min and seven different brake loads changing between 40 Nm and full load.

Measurement points are 1100-1200-1400-1600-1800-2000-2200-2400-2600-2800 rev/min and 40-80-120-160-200-240 Nm and full load. Due to vast number of experimental results, only 40, 120, 200 Nm and full load points are presented in this study.

Two different ceramic coated combustion chambers were compared with standard combustion chamber. In the first one, only cylinder heads and intake exhaust valves were coated. This configuration is represented by SKM1 in graphics. In second one, piston tops also coated with selected ceramic material. So, whole combustion chamber was coated in second configuration. Second configuration is represented as SKM2 in graphics. Three dimensional performance curves obtained in experimental study were evaluated and provided in four different regions. These regions are;

1. Low load, low speed
2. High load, low speed
3. Low load, high speed
4. High load, high speed

An example graphic layout was given in Fig. 12 for previously mentioned regions. In two dimensional graphics, results are provided for 40, 120, 200 Nm and full load points. Before experiments, engine was heated by operating low and medium loads thus steady state was acquired.

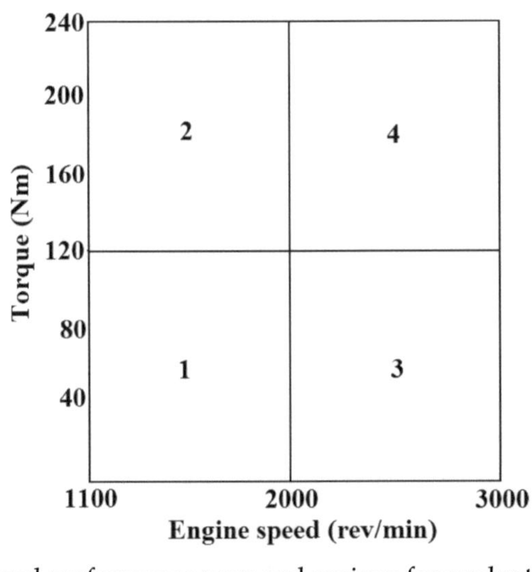

Fig. 12. Three dimensional performance map and regions for evaluation

4.3 Experimental results

In Fig. 13 and Fig. 14, specific fuel consumption comparison of SKM1 and SKM2 with standard engine are provided respectively. Specific fuel consumption changing with engine speed at full load for all engine configurations are given in Fig. 15. For partial load measurement points which are 40, 120 and 200 Nm, similar specific fuel consumption comparison graphics are given for all engine configurations at Fig. 16, 17 and 18 respectively. At first region in three dimensional performance map for specific fuel consumption, SKM1 exhibits 4.5 percent and SKM2 9 percent low specific fuel consumption comparing with standard engine. These figures indicate that there is an important decrease in specific fuel consumption by the utilisation of ceramic thermal barrier coating. This decrease presents continuity at low and medium engine torques. At high torque and high engine speeds, in the other hand, specific fuel consumption decrease continues with a declining trend for ceramic coated engine.

For specific fuel consumption rate, especially second region gives better results. At 1100-1800 rev/min engine speed and 160-200 Nm torque range, standard engine specific fuel consumption is 220 g/kWh while SKM1 has 210 g/kWh and SKM2 has 200 g/kWh.

Fig. 19 and 20 are presented for comparing exhaust gas temperature increase in SKM1 and SKM2 with standard engine respectively. Figures are clearly indicating high exhaust temperatures in ceramic coated engines. In third region, the difference between standard engine exhaust temperatures and ceramic coated engine exhaust temperatures are relatively strong.

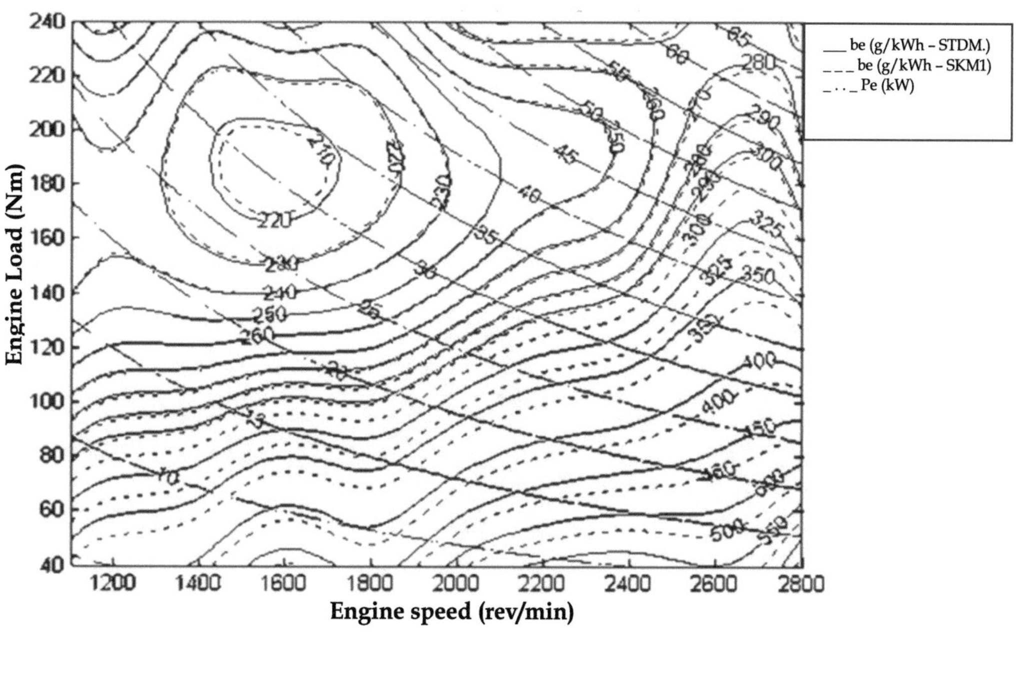

Fig. 13. Three dimensional specific fuel consumption map for SKM1 and standard engine configuration

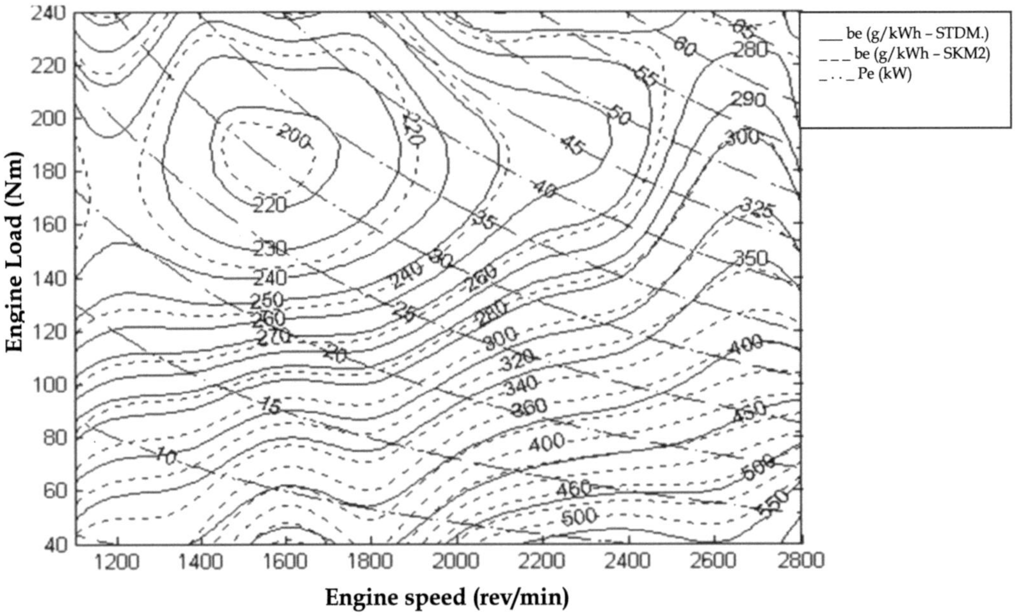

Fig. 14. Three dimensional specific fuel consumption map for SKM2 and standard engine configuration

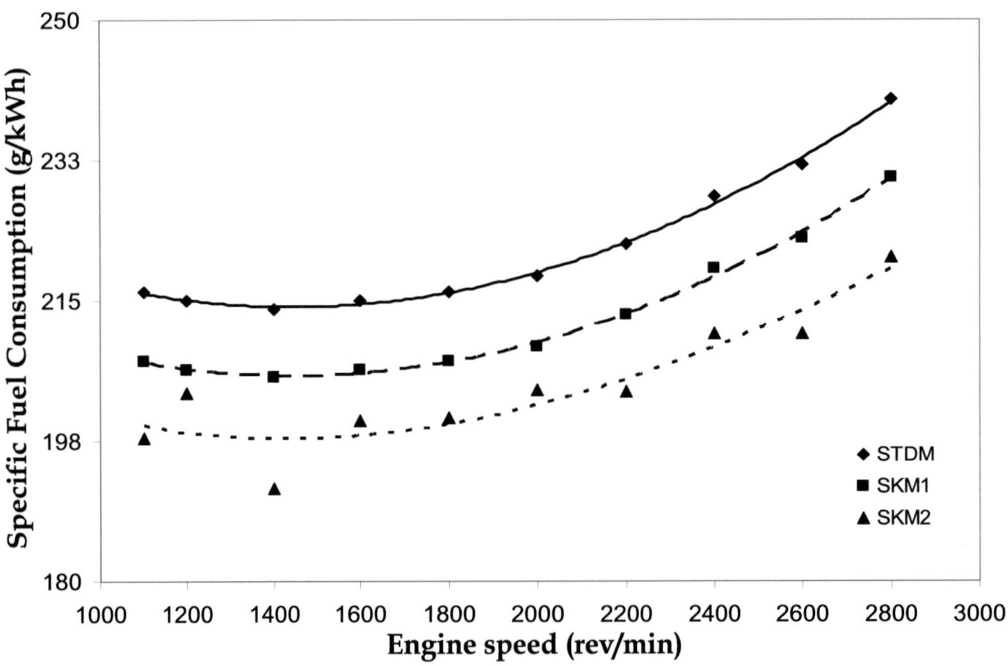

Fig. 15. Specific fuel consumption rate at full load for all engine configurations

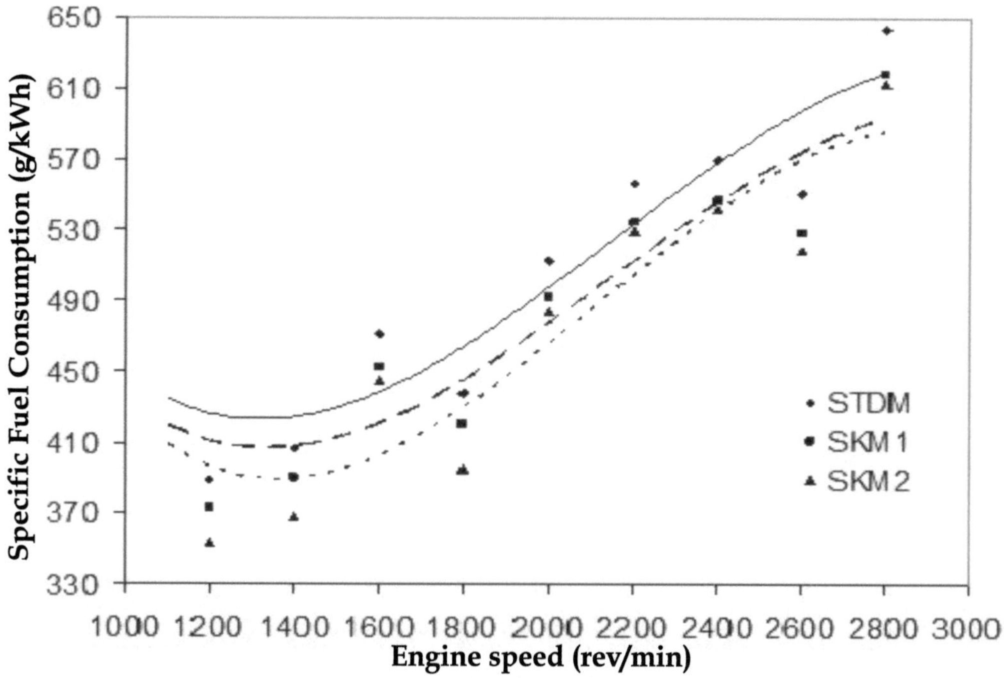

Fig. 16. Specific fuel consumption rate at 40 Nm load for all engine configurations

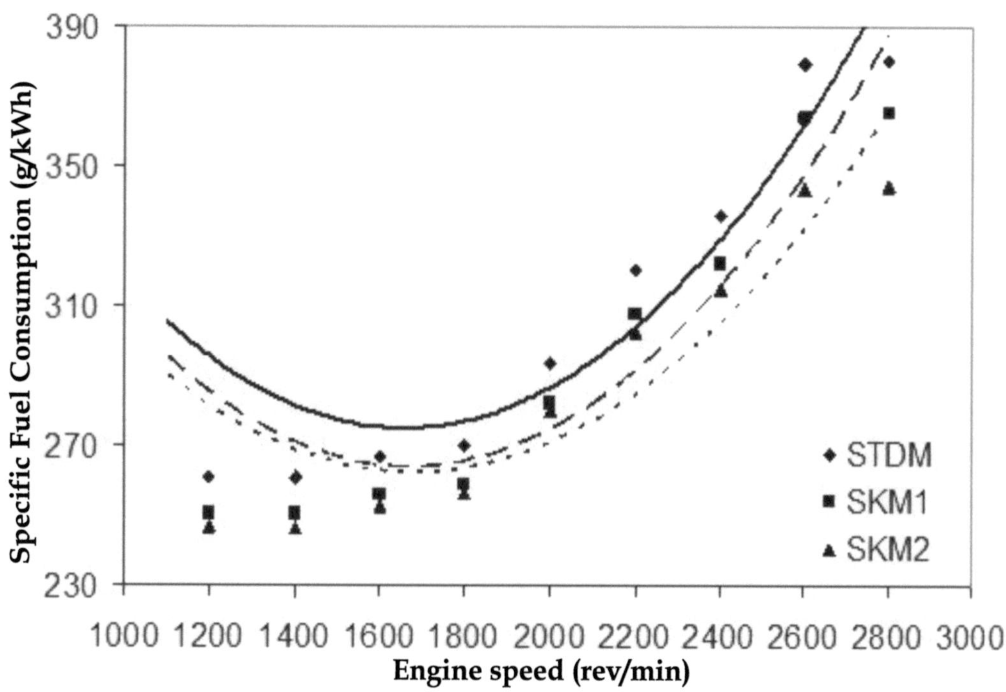

Fig. 17. Specific fuel consumption rate at 120 Nm load for all engine configurations

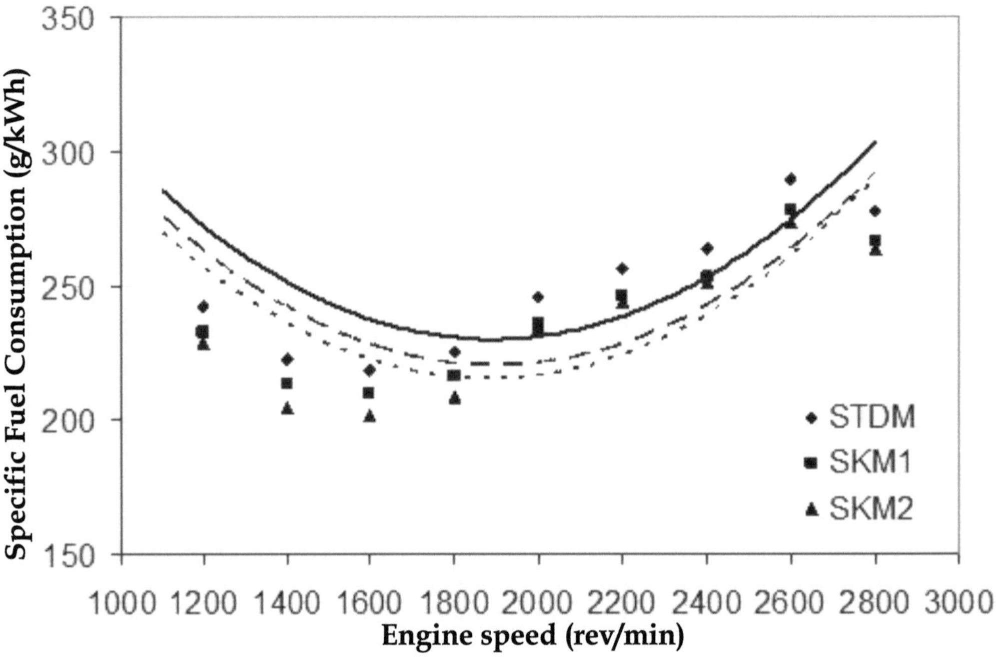

Fig. 18. Specific fuel consumption rate at 200 Nm load for all engine configurations

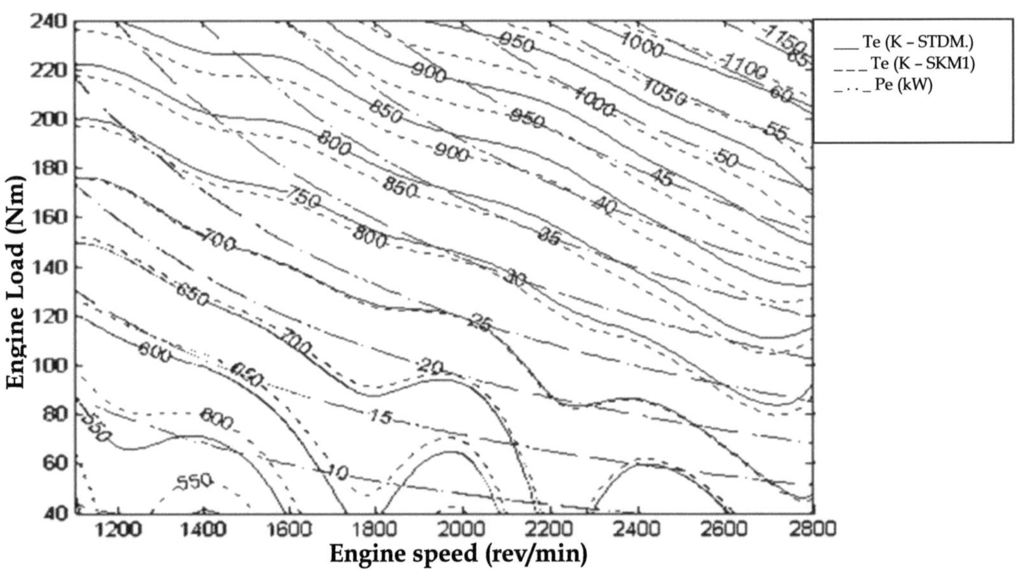

Fig. 19. Three dimensional exhaust temperatures map for SKM1 and standard engine configuration

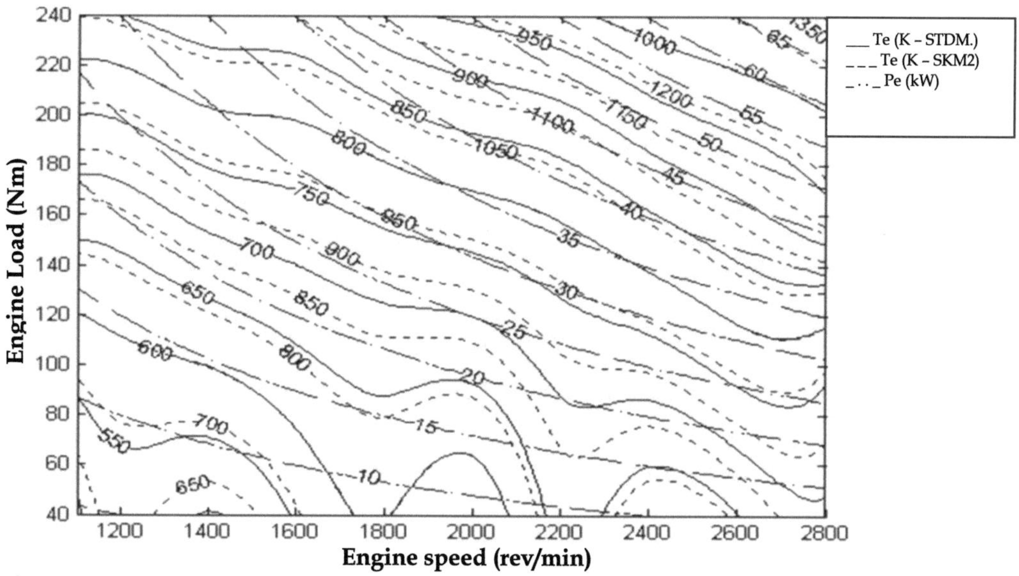

Fig. 20. Three dimensional exhaust temperatures map for SKM2 and standard engine configuration

One can expect that ceramic thermal barrier coating may decrease volumetric efficiency due to increased in-cylinder temperatures. Although exhaust gases and cylinder wall temperatures are high enough to make such effect, turbocharger causes an opposite effect in this study. Fig. 21 illustrates volumetric efficiency change of engine configurations with engine speed at full load. In a same way, Fig. 22, 23 and 24 are presented for 40, 120 and 200 Nm brake loads respectively.

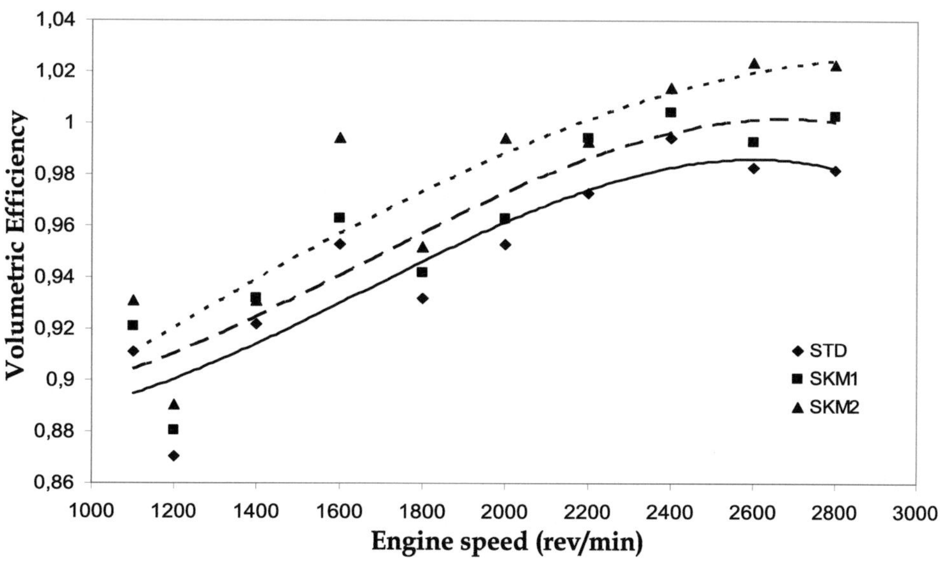

Fig. 21. Volumetric efficiency change at full load for all engine configurations

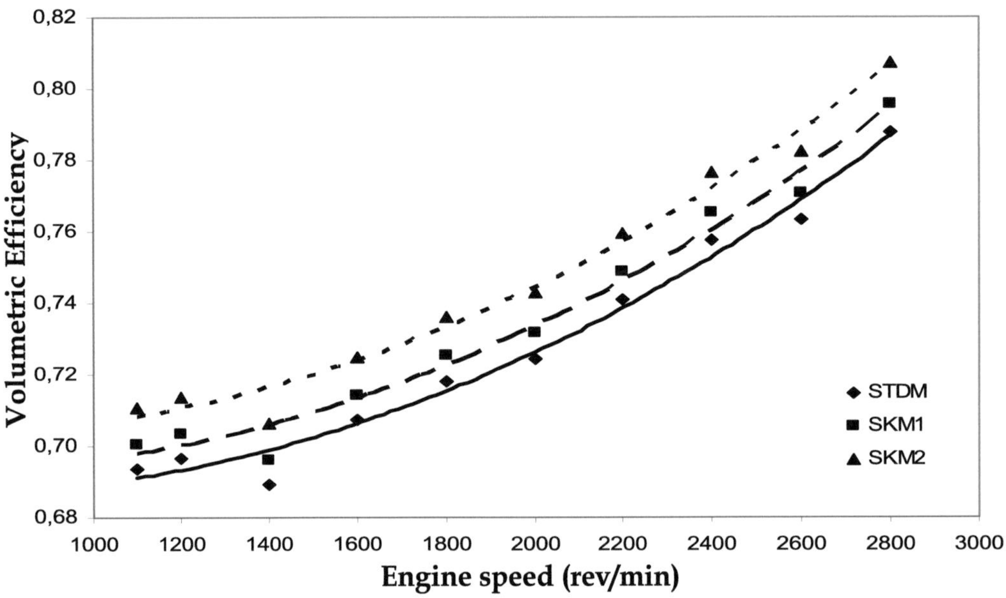

Fig. 22. Volumetric efficiency change at 40 Nm load for all engine configurations

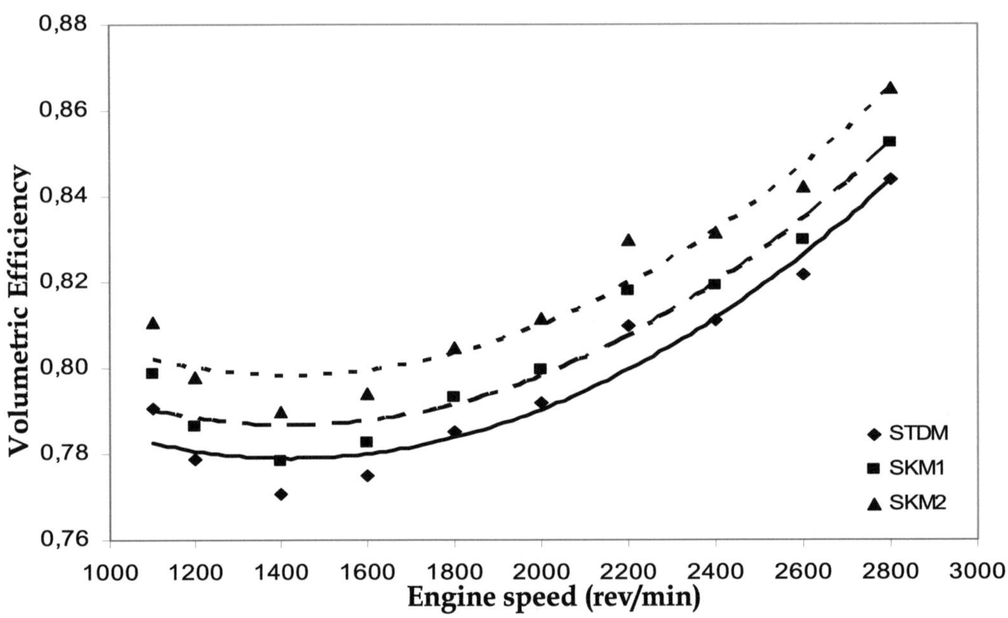

Fig. 23. Volumetric efficiency change at 120 Nm load for all engine configurations

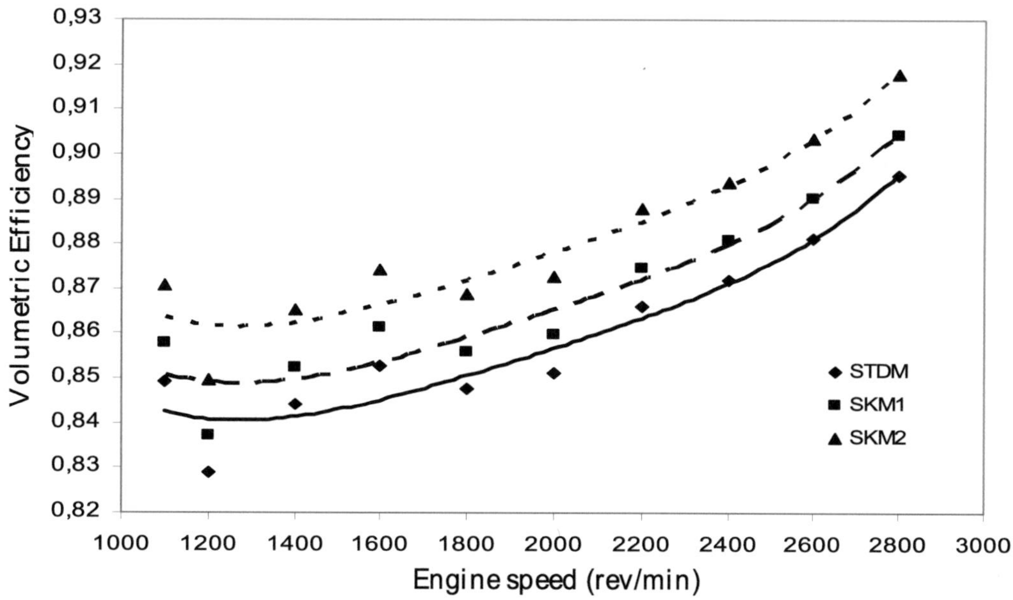

Fig. 24. Volumetric efficiency change at 200 Nm load for all engine configurations

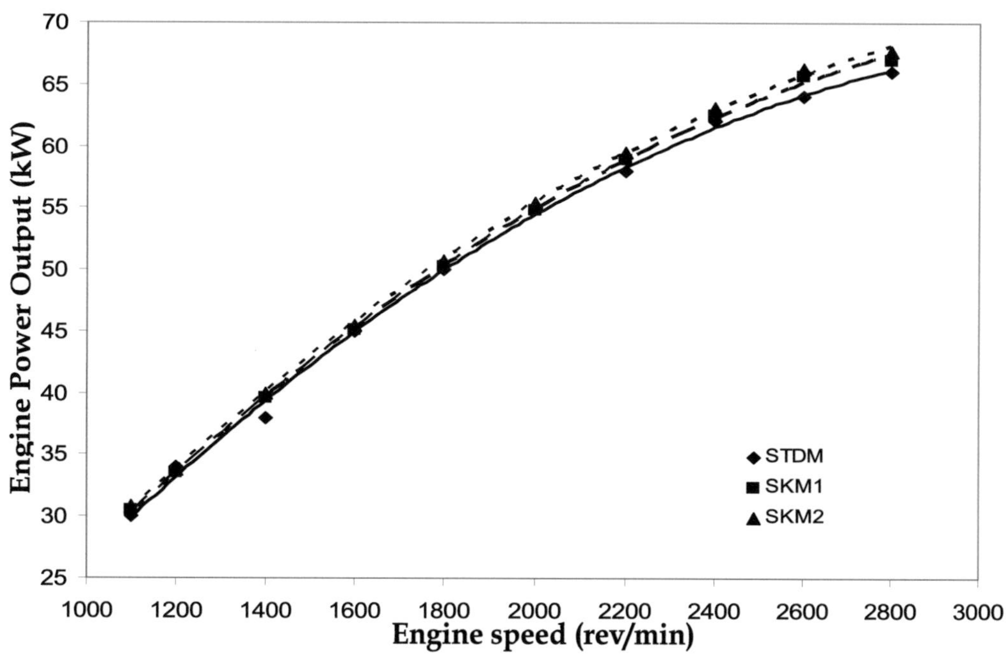

Fig. 25. Engine power output change at full load for all engine configurations

Engine power output is increased between 1-3% and torque increased between 1,5-2,5% by ceramic coating comparing with standard diesel engine. These observations can be chased in Fig. 25 for engine power and in Fig. 26 for torque at full load.

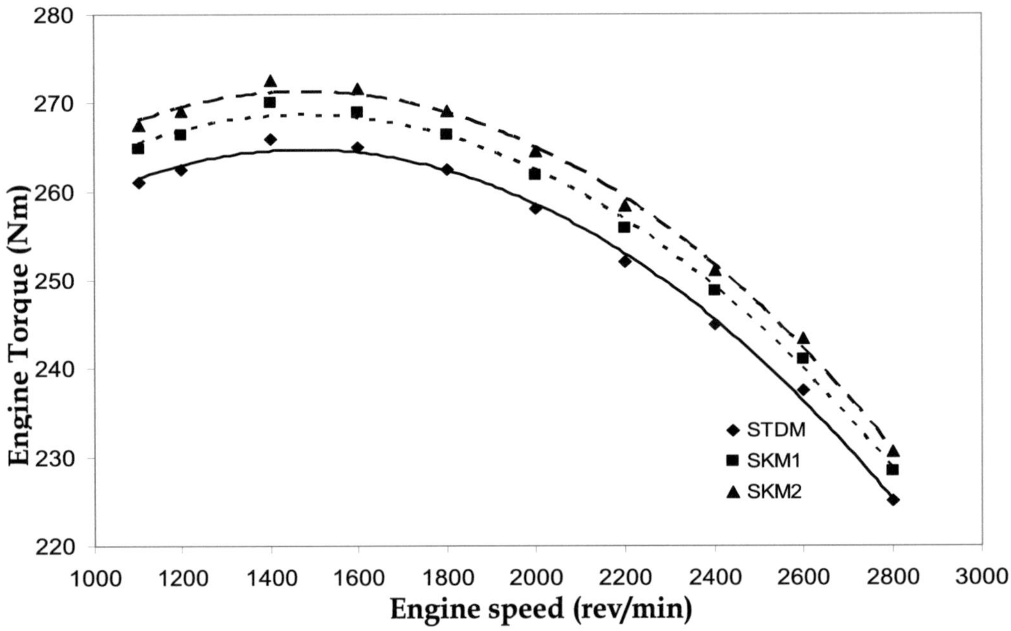

Fig. 26. Engine torque change at full load for all engine configurations

In Fig. 27, heat flux transferred to engine coolant changing with engine speed can be seen at full load for all engine configurations. In Fig. 28, 29 and 30, same graphic was drawn for 40, 120 and 200 Nm loads. In both coated engine configurations and standard engine, heat flux to coolant increase with increasing engine speed however its percentage to total heat is decreasing. These results are compatible with Wallace et. al. (1979; 1984). Experimental results show that heat flux was reduced at a rate of 19 percent by ceramic coating.

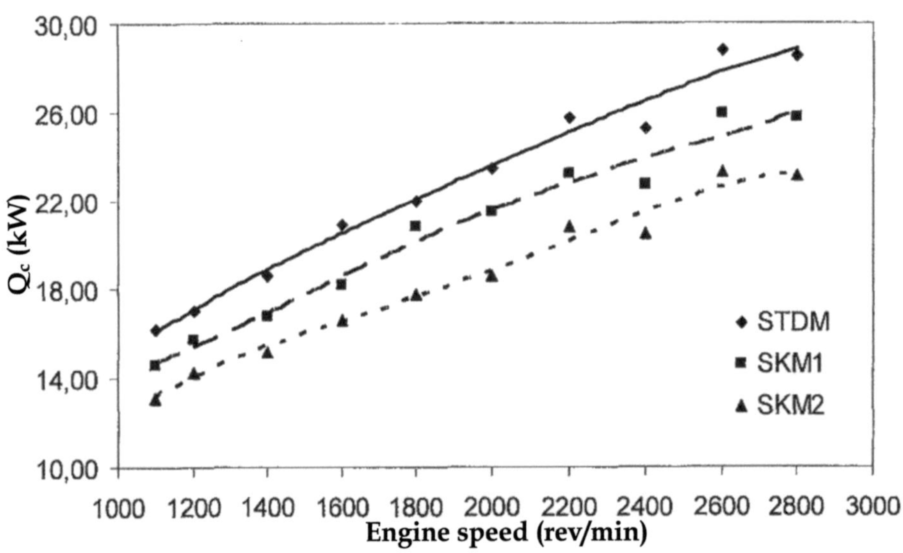

Fig. 27. Heat transfer rate to coolant at full load for all engine configurations

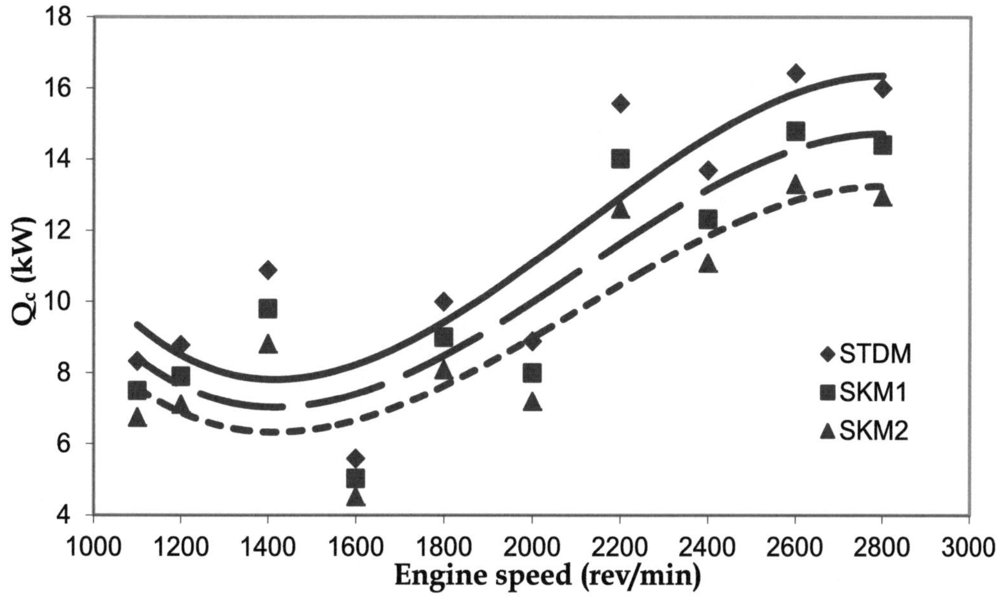

Fig. 28. Heat transfer rate to coolant at 40 Nm load for all engine configurations

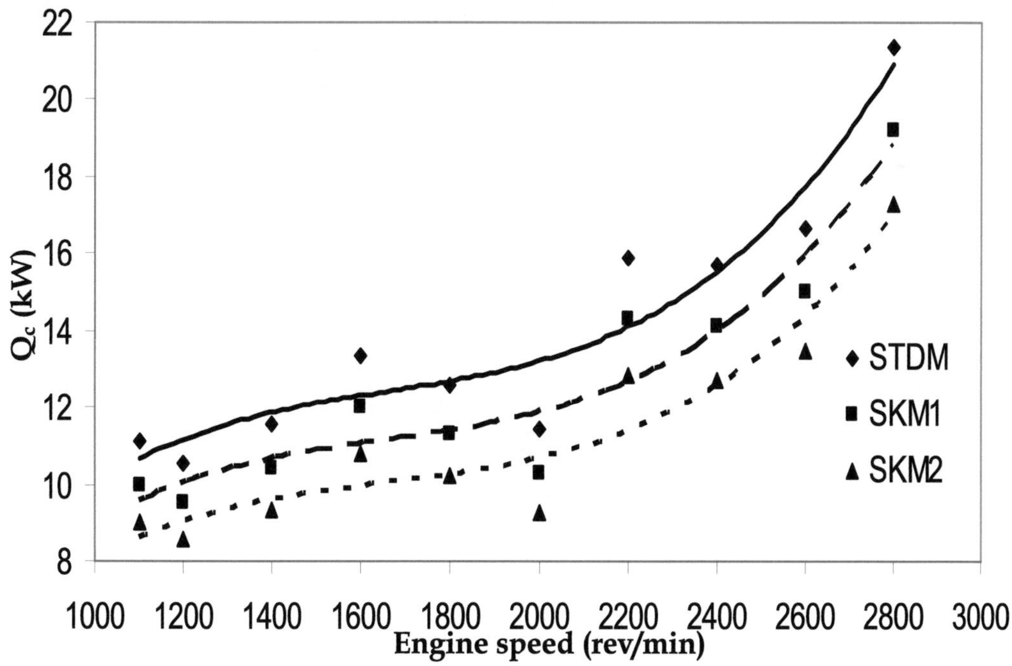

Fig. 29. Heat transfer rate to coolant at 120 Nm load for all engine configurations

Some of the heat after combustion can't be converted into mechanical energy and also it can't be transferred to coolant. This heat portion is carried with exhaust gases. Percentage rate increase of exhaust gas energy is inversely proportional with heat flux to coolant. According to experimental results, about 17.5% increase was observed in the heat energy that passes to exhaust gases. Exhaust heat energy changing with engine speed at full, 40 Nm, 120 Nm and 200 Nm loads for all engine configurations are given in Fig. 31, 32, 33 and 34 respectively.

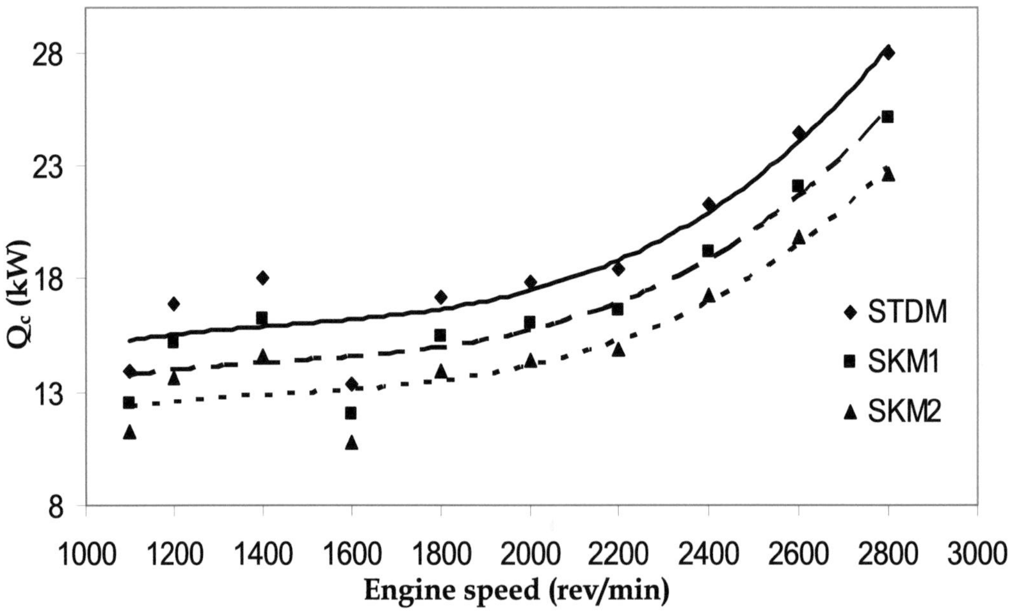

Fig. 30. Heat transfer rate to coolant at 200 Nm load for all engine configurations

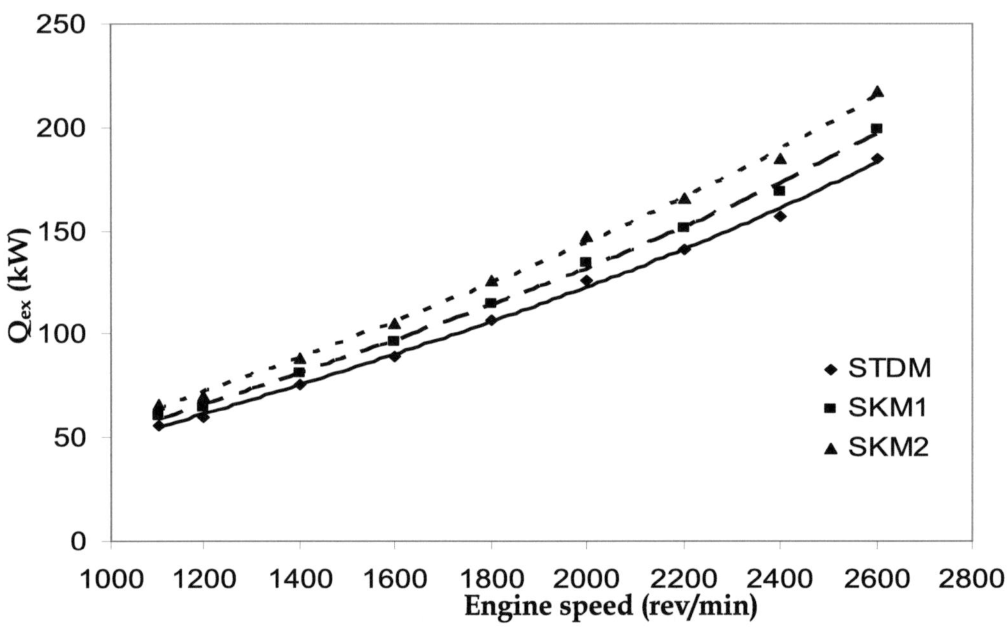

Fig. 31. Heat carried with exhaust gases at full load for all engine configurations

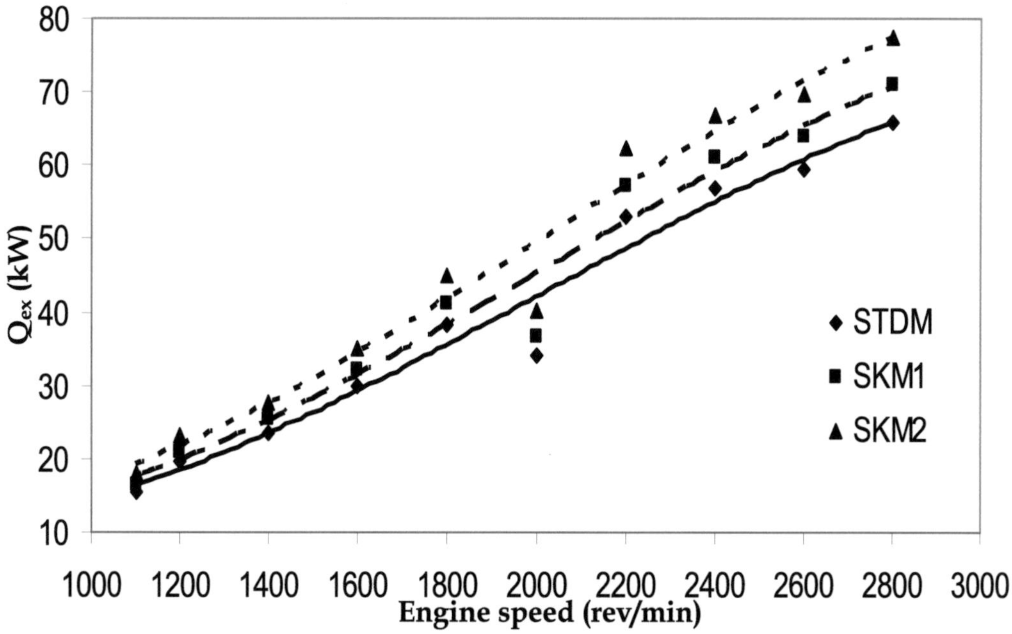

Fig. 32. Heat carried with exhaust gases at 40 Nm load for all engine configurations

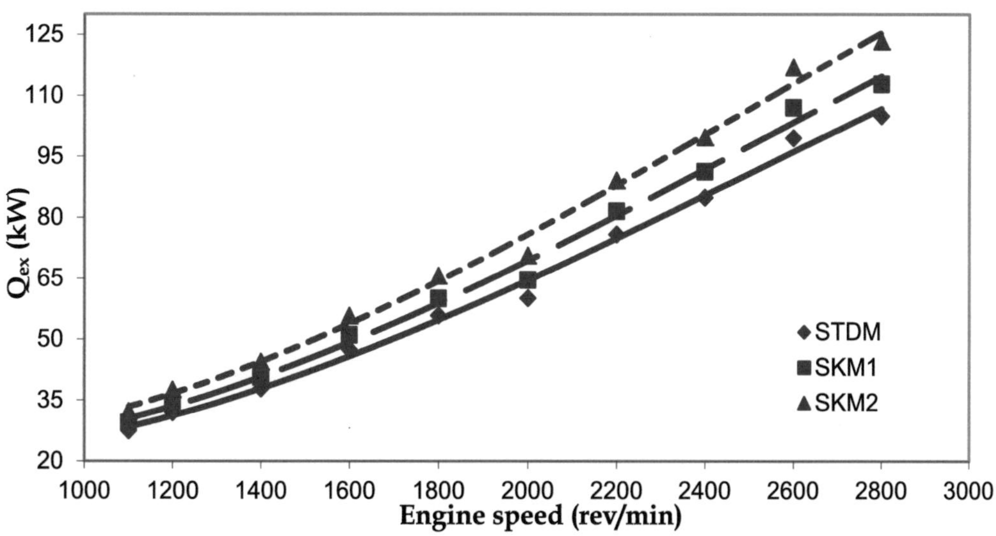

Fig. 33. Heat carried with exhaust gases at 120 Nm load for all engine configurations

One of the most dangerous exhaust emissions is nitrogen oxides in diesel engines. Nitrogen oxide emissions are generally generated over 1800 ^{0}C. Top temperature value during combustion can increase about 150-200 ^{0}C in ceramic thermal barrier coated engines. High in-cylinder temperatures cause an increase in nitrogen oxides emissions about 10% comparing with standard engine operation. Fig. 35 and 36 illustrates nitrogen oxides emissions for SKM1-standard engine and SKM2-standard engine comparisons.

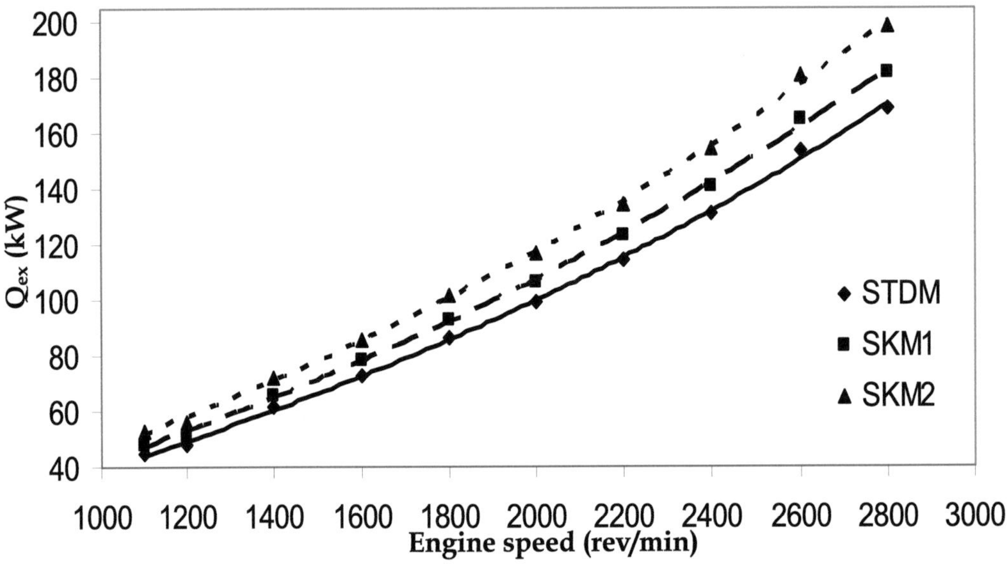

Fig. 34. Heat carried with exhaust gases at 200 Nm load for all engine configurations

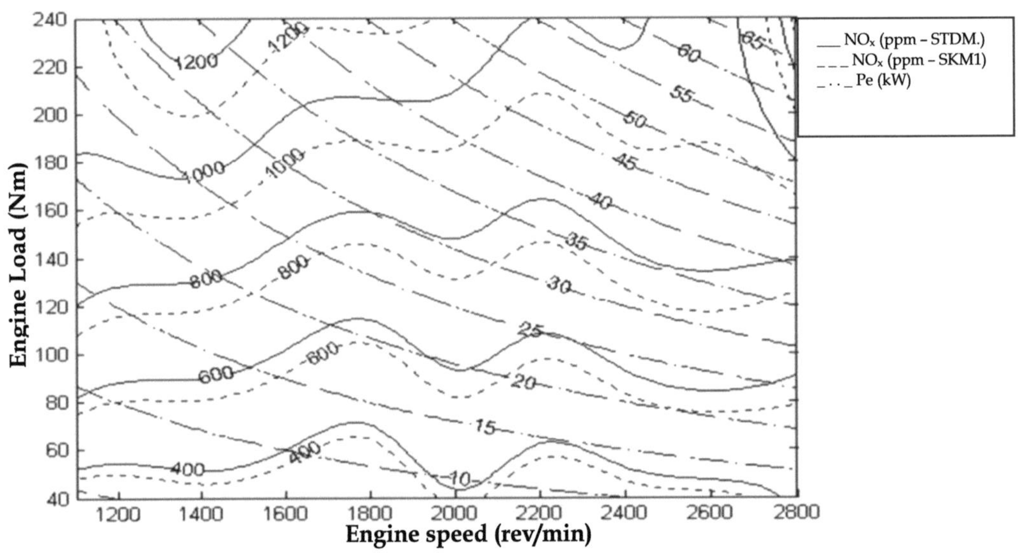

Fig. 35. Three dimensional nitrogen oxides emissions map for SKM1 and standard engine configurations

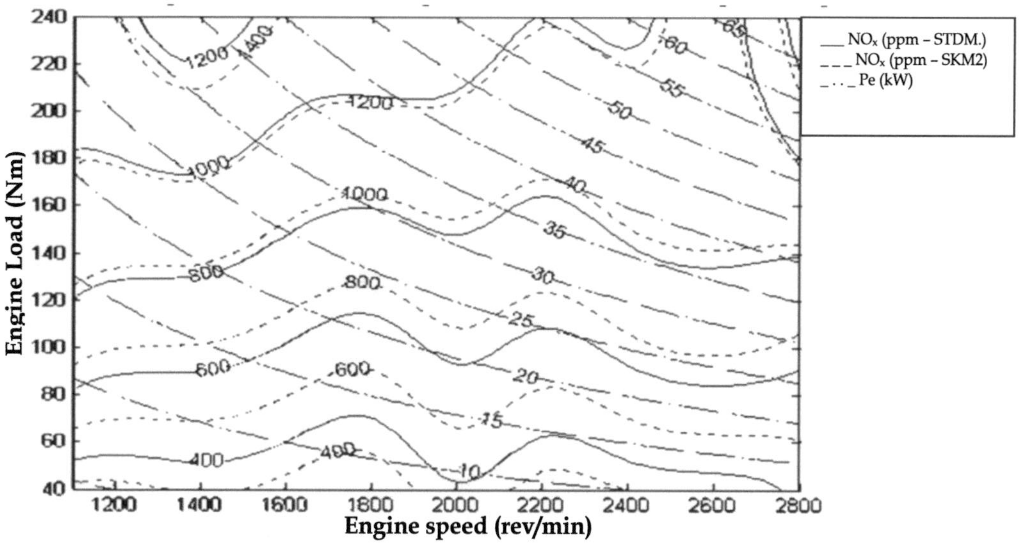

Fig. 36. Three dimensional nitrogen oxides emissions map for SKM2 and standard engine configurations

In standard diesel engines, fuel air mixture ratio is changing with load condition and revolution rate of engine and usually engines are operated at lean fuel air mixture. In this situation, carbon monoxide is converted to carbon dioxide due to sufficient oxygen existence in combustion chamber. However, low combustion temperature, short combustion period and low oxygen content may lead to high carbon monoxide emissions. In ceramic coated engine configurations, carbon monoxide emissions reduced at a rate of 5 to 10% by the increased exhaust temperature. Fig. 37 to 40 show changes in carbon monoxide emissions according to engine speed at full load, 40 Nm load, 120 Nm load and 200 Nm load respectively.

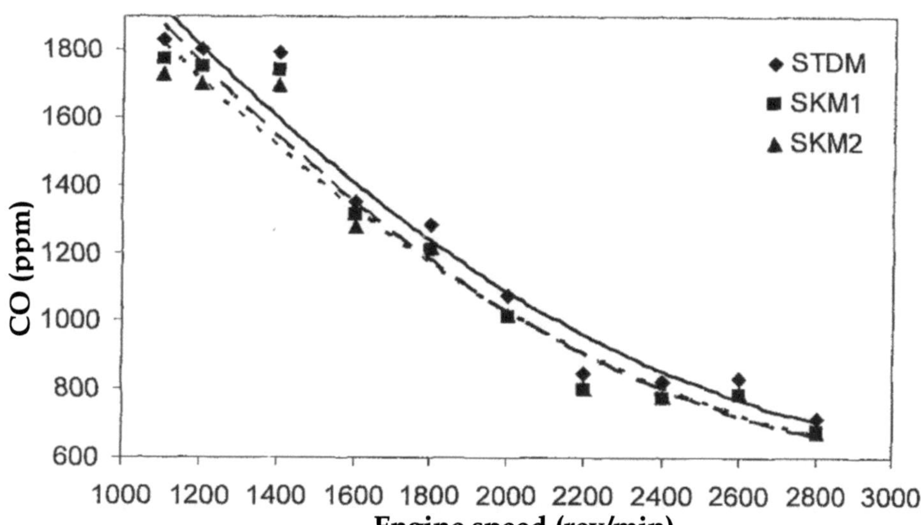

Fig. 37. Carbon monoxide change at full load for all engine configurations

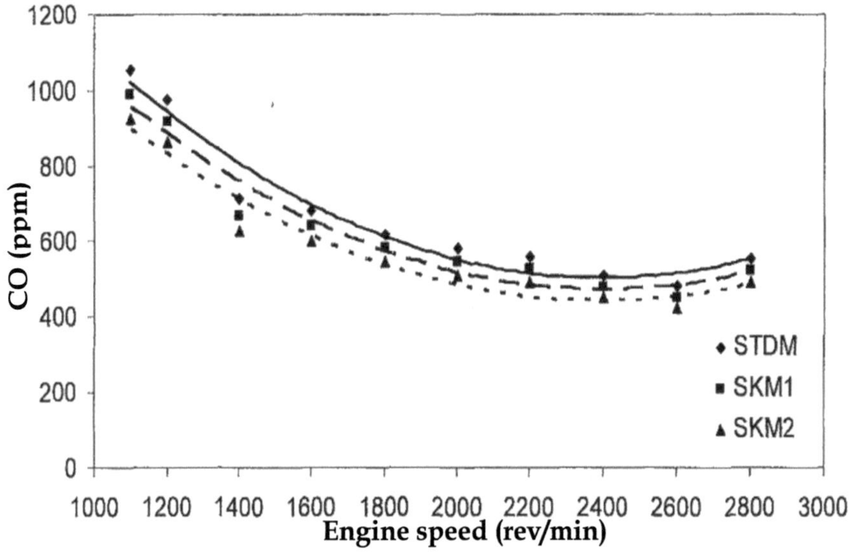

Fig. 38. Carbon monoxide change at 40 Nm load for all engine configurations

Smoke intensity can be evaluated by *k* factor in internal combustion engines. Since diesel engines have a smoke emission problem, the effects of ceramic thermal barrier coating to smoke emissions should be evaluated. Similarly to previous exhaust emission graphics, Fig. 41 to 44 show changes in *k* factor according to engine speed at full load, 40 Nm load, 120 Nm load and 200 Nm load respectively. When figures are investigated, it can be observed that *k* factor decreasing with increasing engine speed. This is due to improved combustion in cylinders owing to increasing temperature. Hence, ceramic coated engine configurations exhibit 18% better smoke emissions.

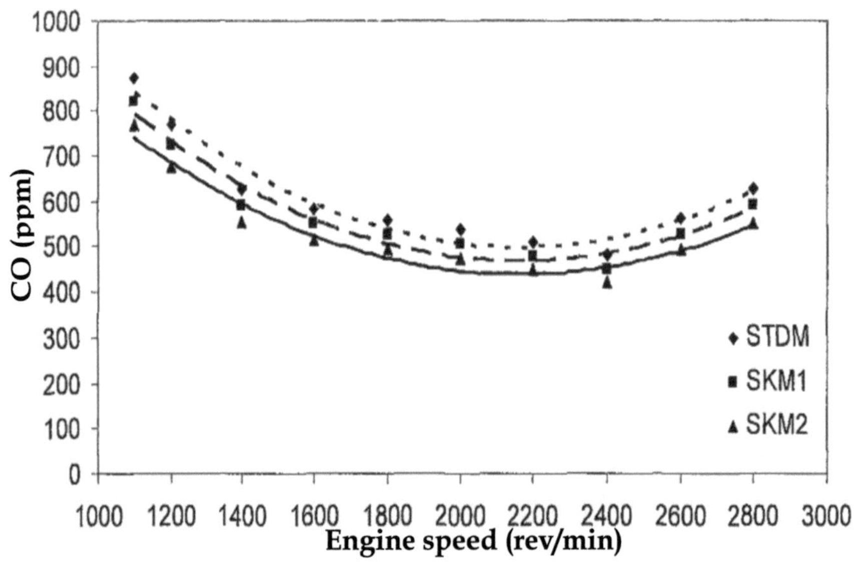

Fig. 39. Carbon monoxide change at 120 Nm load for all engine configurations

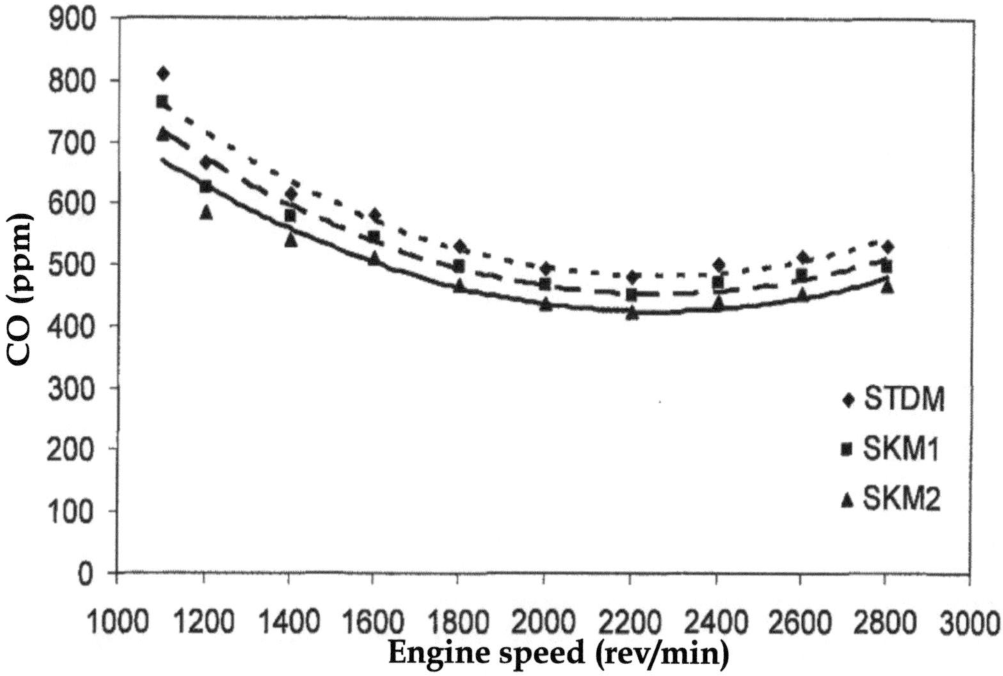

Fig. 40. Carbon monoxide change at 200 Nm load for all engine configurations

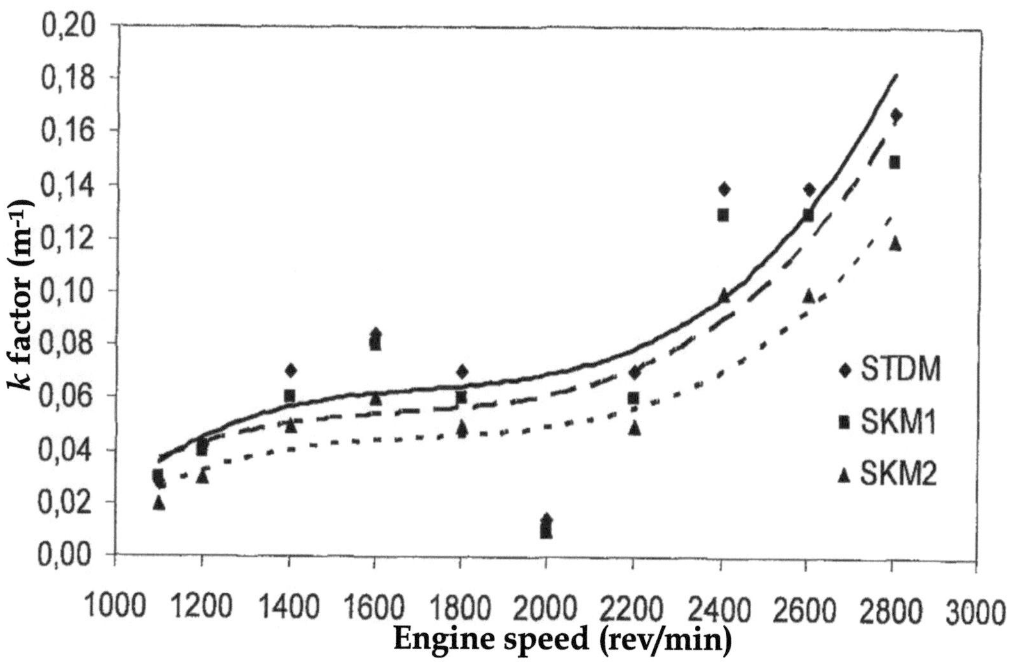

Fig. 41. *k* factor change at full load for all engine configurations

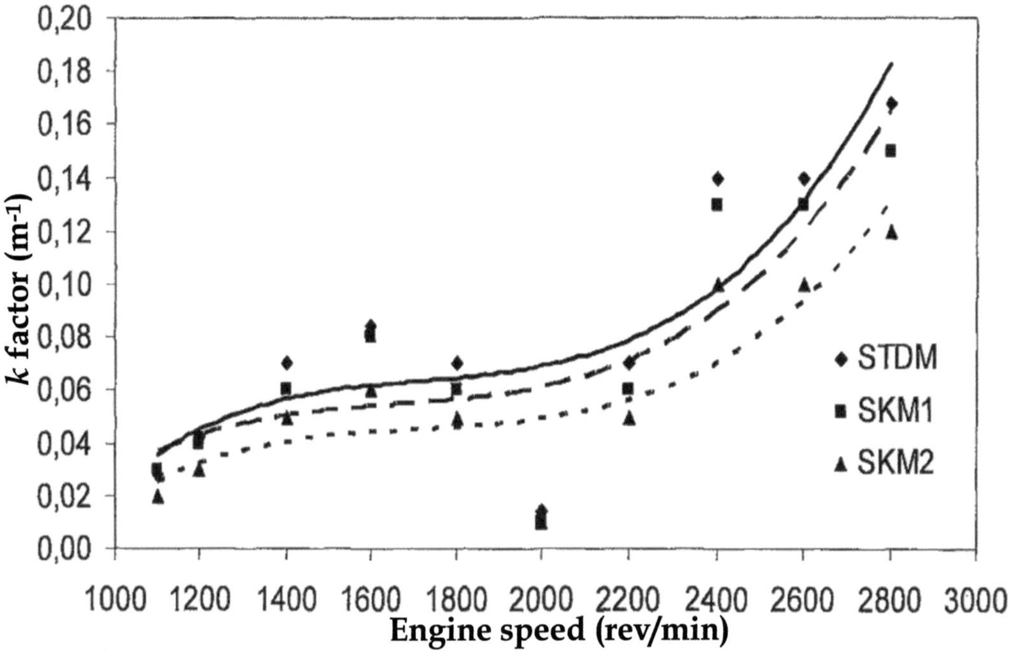

Fig. 42. *k* factor change at 40 Nm load for all engine configurations

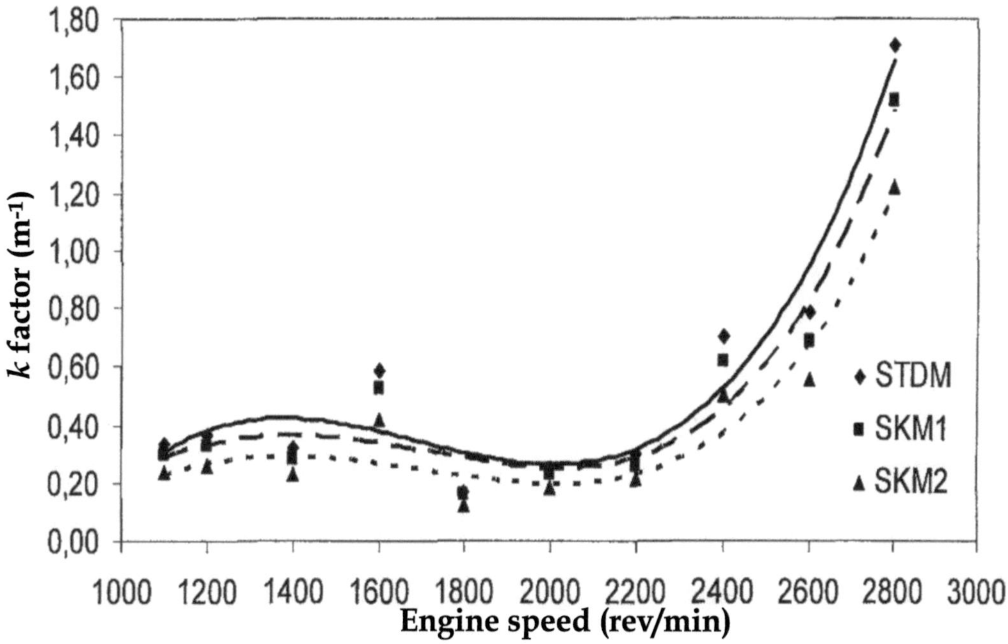

Fig. 43. *k* factor change at 120 Nm load for all engine configurations

4.4 Discussion

In this work, changes in engine performance of a four stroke direct injection four cylinder turbocharged diesel engine were investigated after it was ceramic thermal barrier coated with plasma spray coating method. For study, a specific experimental setup was utilized.

There are some significant problems between coating and coated materials due to thermal expansion ratios in aluminium-silisium alloyed pistons. To avoid these problems, 0.15 mm thick NiCrAlY was coated to base material as binding layer. Zirconia which was stabilized with yttria was used as ceramic coating material for all engine parts.

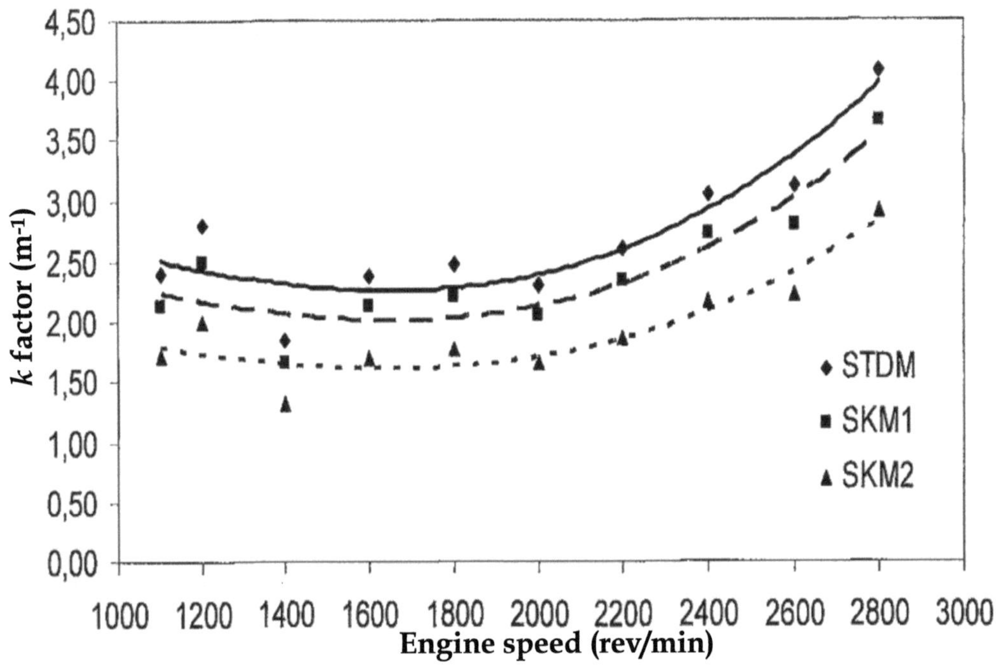

Fig. 44. *k* factor change at 200 Nm load for all engine configurations

A reduction between 4.5 to 9 percent in specific fuel consumption was achieved by ceramic coating in the study. These findings are in accordance with specific literature about ceramic coatings in diesel engines. For instance Coers et. al. (1984) reported 14%, Badgley et. al. (1990) reported 5%, Havstad et. al. (1986) reported 4-9% and Leising et. al. (1978) reported 6% specific fuel consumption reduction in thermal barrier coated engines.

Present experimental study shows that volumetric efficiency was slightly increased at low loads and engine speeds while it was increasing significantly at medium loads and engine speeds. At latter conditions, volumetric efficiency increase reached to 1-2.4%.

Ceramic coating increased exhaust gases temperatures at every operational condition. Exhaust gases temperatures were increased 150 to 200 ^{0}C according to standard engine configuration. This increase corresponds to 7 to 20 percent of standard engine exhaust gases temperatures. When a turbine is combined to the system, aforementioned excess of exhaust energy can be converted to useful mechanical energy.

Heat flux to coolant is also decreased at a rate of 19 percent in present work. This is an important result owing to the possibility of downsizing of cooling system. Reducing sizes of cooling system would be returned as low mechanical energy consume to pumping mechanisms and low weight.

Carbon monoxide emission was decreased 12%, and soot was decreased about 28% in present experimental work. However nitrogen oxides were increased at a rate of 20%. In thermal barrier coating literature for internal combustion engines, reduction of carbon monoxide and soot was emphasized by a lot of researchers. Sudhakar (1984), Toyama et. al. (1989), Assanis et. al. (1991), Amann (1988), Bryzik et. al. (1983) and Matsuoka et. al. (1993) are some of these researchers. Assanis et. al. (1991) reported 30-60% reduction in carbon monoxide emission.

According to present study;

- ZrO_2 stabilized with Y_2O_3 over NiCrAlY binding layer as a coating material gives good results for aluminium alloyed pistons.
- As ceramic coating material, ZrO_2 stabilized with Y_2O_3 is expensive for practical usage. More research should be performed to reduce its cost.
- Cylinder walls also can be coated to reduce heat rejection.
- Injection systems may be tuned for a proper operation in ceramic coated engines. Thus, improvements can be enhanced.
- Alternative fuels can be tested in ceramic coated engines since combustion temperature is increased. Some fuels react positively to this temperature increase as they can be burned more efficiently.

5. References

Afify, E.M. & Klett, D.E. (1996), *The Effect Of Selective Insulation On The Performance, Combustion, And NO Emissions Of A DI Diesel Engine*, International & Congress and Exposition, Detroit, Michigan February 26-29

Alkidas A.C. (1989), *Performance and Emissions Achievements with an Uncooled Heavy Duty, Single Cylinder Diesel Engine*, SAE, Paper 890144, USA

Amann, C.A. (1988), *Promises and Challenges of the Low-Heat Rejection Diesel*, Journal of Engineering for Gas Turbines and Power, Vol. 110

Anonymous, (2004), *Coating Methods*, Senkron Metal A.S., 02.10.2004, Available from www.senkronmetal.com.tr (in Turkish)

Assanis, D., Wiese, K., Schwarz, E. & Bryzik, W. (1991), *The Effects Of Ceramic Coatings On Diesel Engine Performance And Exhaust Emissions*, International & Congress and Exposition, Detroit, Michigan, USA

Badgley, P., Kamo, R., Bryzik, W. & Schwarz, E. (1990), *Nato Durability Test of an Adiabatic Truck Engine*, SAE, Paper 900621, USA

Balcı, M. (1983). *Heat Release Characteristics of a Diesel Type Combustion Chamber*, Msc Thesis, The Univesrsity of Bath, England

Beg, R.A., Bose, P.K., Ghosh, B.B., Banerjee, T.Kr. & Ghosh, A. Kr. (1997), *Experimental Investigation On Some Performance Parameters Of A Diesel Engine Using Ceramic Coating On The Top Of The Piston*, International Congress & Exposition Detroit, Michigan, February 24-27

Büyükkaya, E. (1994). *Ceramic Coating Application and Performance Analysis in a Diesel Engine,* Msc Thesis, Istanbul Technical University, Istanbul, Turkey (in Turkish)

Büyükkaya, E.; Yaşar, H.; Çelik, V. & Ekmekci, M. (1997). *Effects of Thermal Barrier Coationg to Exhaust Emissions of a Turbocharged Diesel Engine,* pp. 21-23, 5. International Combustion Symposium, Bursa, Turkey, 21-23 July 1997

Bruns, L., Bryzik, W. & Kamo, R. (1989), *Performance Assessment of U.S. Army Truck with Adiabatic Diesel Engine,* SAE, Paper 890142

Bryzik, K. & Kamo, R. (1983). *TACOM/Cummins Adiabatic Engine Program,* SAE, Paper 830314, USA

Ciniviz, M. (2005), *The effects of Y2O3 with coatings of combustion chamber surface on performance and emissions in a turbocharged diesel engine,* PhD Thesis, Selçuk University, Turkey (in Turkish)

Chang, S.I. & Rhee, K.I. (1983), *Computation of Radiation Heat Transfer in Diesel Combustion,* SAE International of Highway Meating, Wisconsin; 327-341, USA

Charlton, S.J., Campbell, N.A., Shephard, W.J., Cook, G. & Watt, M. (1991), *An Investigation of Thermal Insulation of IDI diesel Engine Swirl Chamber,* Proc. Instn. Mech. Engrs., Vol.205

Coers, R.B., Fox, L.D. & Jones, D.J. (1984), *Cummins Uncooled 250 Engine,* SAE International Congress & Exposition, Michigan, USA

Çevik, İ. (1992), *Changing Physical and Chemical Properties of Zirconia Based Ceramic Coatings,* PhD Thesis, Istanbul Technical University, Turkey (in Turkish)

Dickey, D.W. (1989), *The Effect of Insulated Combustion Chamber Surfaces on Direct-Injected Diesel Engine Performance, Emissions and Combustion,* SAE, Paper 890292, USA

Gataowski, J.A. (1990), *Evaluation of a Selectively Cooled Single Cylinder 0.5 L Diesel Engine,* SAE, Paper 900693,USA

Geçkinli, A. (1992), *Advanced Technology Materials,* Teknik Üniversite Matbaası Publishing House, Istanbul, Turkey (in Turkish)

Hay, N., Watt, P.M., Ormerod, M.J., Burnett, G.P., Beesley, P.W. & French, B.A. (1986), *Desing Study for a Low Heat Loss Version of the Dover Engine,* Proc. Instn. Mech. Engrs. Vol.200, No.DI

Hejwowski, T. & Weronsk., A. (2002), *The Effect Of Thermal Barrier Coatings On Diesel Engine Performance,* Vacuum Surface Engineering, Vacuum 65, 427-432

Hocking, M.G., Vasatasree, V. & Sidky, P.S. (1989). *Metallic and Ceramic Coatings,* High Temperature and Applications, London, UK

Schwarz, E.; Reid, M.; Bryzik W.; & Danielson E. (1993). *Combustion and Performance Characteristics of a Low Heat Rejection Engine,* SAE, Paper 930988, USA

Kamo, R.; Assanis, D. & Bryzik, W. (1989), *Thin Thermal Barrier Coatings for Engines,* SAE, Paper 890143, USA

Kamo, R., Bryzik, W., Reid, M., & Woods, M. (1997), *Coatings For Improving Engine Performance,* International & Congress and Exposition, Detroit, Michigan February 24-27, USA

Kimura, S., Matsui, Y. & Ltoh, T. (1992), *Effects of combustion chamber insulation on the heat rejection and thermal efficiency of diesel engines,* International & Congress and Exposition, Detroit, Michigan February pp, 24-28, USA

Leising, C.J. & Purohit, G.P. (1978), *Waste Heat Recovery in Truck Engines,* SAE National West Coast Meeting, California, USA

Marks, D.A. & Boehman, A.L. (1997), *The Influence Of Thermal Barrier Coatings On Morphology And Composition Of Diesel Particulates*, International Congress & Exposition Detroit, Michigan, February 24-27, p51-60, USA

Matsuoka, H. & Kawamura, H. (1993), *Structure of Heat-Insulation Ceramic engine and Heat Insulating Performance*, JSAE, review vol. 14, No.2

Miyairi, Y., Matsumsa, T., Ozawa, T, Odcawa, H. & Nakashima, N. (1989), *Selective Heat Insulation of Combustion Chamber Wall, for a DI Diesel Engine with Monolithic Ceramics*, SAE, Paper 890141, USA

Osawa, K., Kamo, R. & Valdmanis, E. (1991), *Performance Of Thin Thermal Barrier Coating On Small Aluminum Block Diesel Engine*, International & Congress and Exposition, Detroit, Michigan February 25- March 1

Parlak, A. (2000), *Experimental Investigation of Injection Advance and Compression Ratio of a Supercharged Ceramic Coated Diesel Engine*, PhD Thesis, Sakarya University, Turkey (in Turkish)

Parlak, A., Yaşar, H. & Şahin, B. (2003), *Performance And Exhaust Emission Characteristics Of A Lower Compression Ratio LHR Diesel Engine*, Energy Conversion and Management, 44, 163-175

Prasad, C.M.V., Krishna, M.V., Reddy, C.P. & Mohan, K. R. (2000), *Performance Evaluation Of Non-Edible Vegetable Oils As Substitute Fuels In Low Heat Rejection Diesel Engines*, Proceedings of the institution of mechanical Engineers, 214,2; ProQuest Science Journals.

Ramaswamy, P., Seetharamu, S., Varma, K.B. & Rao, K.J. (2000), *Thermo Mechanical Fatigue Characterization Of Zirconium (8% Y_2O_3 - ZrO_2) And Mullite Thermal Barrier Coatings On Diesel Engine Components: Effect Of Coatings On Diesel Engine Performance*, Proceedings of the institution of mechanical Engineers, 214,5; ProQuest Science Journals, pg. 729

Rasihhan, Y. & Wallce F.J. (1991), *Piston-Liner Thermal Resistance Model for Diesel Engine Simultion*, Proc. Inst. Mech. Engrs. Vol.205

Schwarz, E., Reid, M., Bryzik, W. & Danieison, E. (1993), *Combustion and performance characteristics of a low heat rejection engine*, International & Congress and Exposition , Detroit, Michigan March 1-5, USA

Sudhakar, V. (1984), *Performance Analysis of Adiabatic Engine*, SAE International Congress & Exposition, Detroit, Michigan, USA

Sun, X., Wang, W.G., Lyons D.W. & Gao, X. (1993), *Experimental Analysis And Performance Improvement Of A Single Cylinder Direct Injection Turbocharged Low Heat Rejection Engine*, International & Congress and Exposition, Detroit, Michigan March 1-5

Sun, X., Wang, W.G., Bata, R.M. & Gao, X. (1994), *Performance Evaluation of Low Heat Rejections Engines*, Transactions of the ASME, Vol. 116

Taymaz, I., Çakır, K. & Mimaroğlu, A. (2003), *Experimental Investigation Of Heat Losses In A Ceramic Coated Diesel Engine*, Surface and coatings technology, 169-170

Toyama, K., Cheng, W.K., Wong, V.W. & Cao, F. (1989), *Heat Transfer Measurement Comparations in Insulated and Non-Insulated Diesel Engines*, SAE, Paper 890570, USA

Uzun, A., Çevik, İ. & Akçil, M. (1999), *Effects Of Thermal Barrier Coating On A Turbocharged Diesel Engine Performance*, Surface and coatings technology, 116-119, 505-507

Vittal, M., Borek, J.A., Boehman, A.L. & Okrent, D. A. (1997), *The Influence Of Thermal Barrier Coatings On The Composition Of Diesel Particulate Emissions,* International Fall Fuels & Lubricants Meeting & Exposition Tulsa, Oklahama, October 13-16

Wallace, F.J., Way, Rj.B. & Vollmert, H. (1979), *Effect of Partial Suppression of Heat Loss to the Coolant on the High Output Diesel Engine Cycle*, SAE, Paper 790823, USA

Wallace, F.J., Kao, T.K., Tarabad, M., Alexander, W.D. & Cole, A. (1984), *Thermally Insulated Diesel Engines,* Proc. Instn. Mech. Engrs., Vol. 198A, N5, USA

Winkler, M.F. & Parker D.W. (1993), *The Role Of Diesel Ceramic Coatings In Reducing Automotive Emissions And Improving Combustion Efficiency,* SAE, Paper 930158, International & Congress and Exposition , Detroit, Michigan March 1-5

Woods, M., Bryzık, W. & Schwarz, E. (1992), *Heat Rejection from High Output Adiabatic Diesel Engine,* SAE, Paper 920541, USA

Woschni, G. & Spindler, W. (1988), *Heat Transfer with Insulated Combustion Chamber Walls and its Influence on the Performance of Diesel Engines,* Transaction of the ASME, Vol. 110, July

Yaşar, H. (1997), *Effects of Thermal Barrier Coating to Performance of a Turbocharged Diesel Engine,* PhD Thesis, Istanbul Technical University, Turkey (in Turkish)

Part 4

Pigment

8

Ceramic Coatings for Pigments

A.R. Mirhabibi
[1]Institute for Materials Research (IMR), Leeds University,
[2]Center of Excellence for Advanced Materials and Processes (IUST)
[3]Iran University of Science and Technology (IUST)
[1]UK
[2,3]Iran

1. Introduction

Special effect pigments, which can be natural or synthetic, show outstanding qualities of luster, brilliance and iridescent colour effects based upon optically thin layers [1.1–1.4]. This visual impression develops by reflection and scattering of light on thin multiple layers. In nature this is not limited to pearls and mussel shells alone; there are a multitude of birds, fish, precious stones and minerals, even insects, that demonstrate a luster effect. Experiments to understand the optical principles of natural pearl luster demonstrate that the brilliant colours are based upon structured biopolymers and upon layered structures, which are developed by biomineralization. Figure 1.1 illustrates the various optical principles of conventional pigments (A) (absorption pigments), metal effect pigments (B), and pearls (C) and pearl luster pigments (D), the most important group of special effect pigments. In the case of absorption pigments, the interaction with light is based upon absorption and/or diffuse scattering. A completely different optical behavior can be observed with the group of effect pigments including pearl luster and metal effect pigments. Metal effect pigments consist of small metal platelets (for example aluminum, titanium, copper), which operate like little mirrors and almost completely reflect the incident light.

Pearl luster pigments simulate the luster of natural pearls. They consist of alternating transparent layers with differing refractive indices. The layers consist of $CaCO_3$ (high refractive index) and proteins (low refractive index). This difference in refractive indices, arising equally on the interface between an air/oil film or oil film/water, is a prerequisite for the well-known iridescent colour images in these media. Small highly refractive platelets of pearl luster pigments align themselves parallel in optically thin systems such as paints, printing inks, or plastics. Interference effects develop when the distances of the various layers or the thicknesses of the platelets have the right values. Synthetic pearl luster pigments are either transparent or light-absorbing platelet shaped crystals. They can be monocrystalline, as in $Pb(OH)_2$. $2PbCO_3$ and BiOCl, or possess a multi-layered structure in which the layers have differing refractive indices and light absorption properties. The use of pearls and nacreous shells for decorative purposes goes back to ancient times (e.g., in Chinese wood intarsia). The history of pearl pigments dates back to 1656, when French rosary maker Jaquin isolated a silky lustrous suspension from fish scales (pearl essence) and applied this to small beads to create artificial pearls.

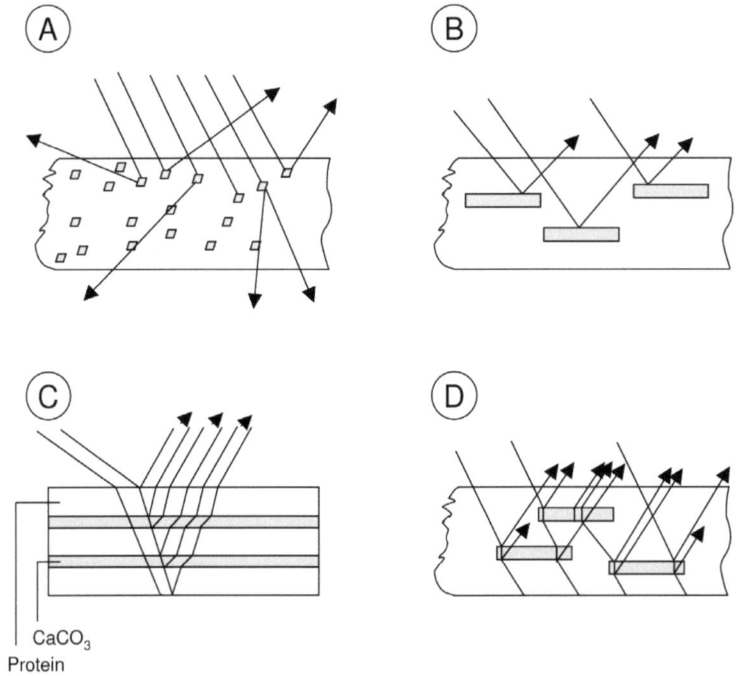

A) conventional pigment; B) metal effect pigment; C) natural pearl; D) pearl luster pigment (1.12).

Fig. 1.1. Optical properties of absorption pigments, effect pigments and natural pearls.

It took more than 250 years to isolate the pearl essence material (guanine platelets) and understand the pearl effect. Attempts were made to create synthetic pearl colours as organic or inorganic, transparent, highly refractive coatings and pearl pigments as crystalline platelets. From 1920 onwards, hydroxides, halides, phosphates, carbonates, and arsenates of zinc, calcium, barium, mercury, bismuth, lead, and other cations were produced for this purpose. Only the traditional natural pearl essence, basic lead carbonate and bismuth oxychloride, is still of importance. The strong demand for pearl effects came from the growing coatings and plastics industries, which wanted to improve the acceptance and popularity of their products. Furthermore, pearl luster pigments also allowed artists and designers to create new visual effects similar to those found in nature. The breakthrough for pearl luster pigments came with the invention of mica coated with metal oxides. Mica-based pearl luster pigments now account for >90% of the world market. Important manufacturers of pearl luster pigments are Merck KGaA, Germany (with overseas subsidiaries EMD Chemicals Inc., USA and Merck Ltd., Japan) and Engelhard Corp., USA.

Table 1.1 shows an overview of inorganic pigments with luster effects. Effect pigments can be classified with regard to their composition as metal platelets, oxide-coated metal platelets, oxide-coated mica platelets, oxide-coated silica, alumina and borosilicate flakes, platelet-like monocrystals, comminuted PVD films (PVD = physical vapor deposition), and liquid crystal polymer platelets (LCP-pigments, the only industrially relevant organic effect pigment type) [1.3–1.5]. The aims of new developments are new effects and colours, improvement of hiding power, more intense interference colours, increased light and weather stability, and improved dispersibility.

Of special interest are pigments which are toxicologically safe and which can be produced by ecologically acceptable processes. The total market for effect pigments can be estimated to be about 50,000 tons per year. Half of this amount can be calculated to be special effect pigments, the other half to be metal effect pigments.

1.1 Optical principles of pearl luster and interference pigments

The physical background of optical interference effects has been the subject of many publications [1.1-1.4, 1.6–1.9]. The optical principles of pearl luster (interference) pigments are shown in Figure 1.2 for a simplified case of nearly normal incidence without multiple reflection and absorption. At the interface P_1 between two materials with refractive indices n_1 and n_2, part of the beam light L_1 is reflected (L_1) and part is transmitted (i.e., refracted) (L_2). The intensity ratios depend on n_1 and n_2. In a multilayer arrangement, as found in pearl or pearl luster and iridescent materials (Figure 1.1D), each interference produces partial reflection. After penetration through several layers, depending on the size of and difference between n_1 and n_2, virtually complete reflection is obtained, provided that the materials are sufficiently transparent.

Pigment type	Examples
Metallic platelets	Al, Zn/Cu, Cu, Ni, Au, Ag, Fe (steel), C (graphite)
Oxide-coated metallic platelets	Surface oxidized Cu-, Zn/Cu-platelets, Fe_2O_3 coated Al- platelets
Coated mica platelets*	non-absorbing coating: TiO_2 (rutile), TiO_2 (anatase), ZrO_2, SnO_2, SiO_2 selectively absorbing coating: FeOOH, Fe_2O_3, Cr_2O_3, TiO_{2-x}, TiO_xN_y, $CrPO_4$, $KFe[Fe(CN)_6]$, colorants totally absorbing coating: Fe_3O_4, TiO, TiN, $FeTiO_3$, C, Ag, Au, Fe, Mo, Cr, W
Platelet-like monocrystals	BiOCl, $Pb(OH)_2 \cdot 2\,PbCO_3$, α-Fe_2O_3, α-$Fe_2O_3 \times n\,SiO_2$, $Al_xFe_{2-x}O_3$, $Mn_yFe_{2-y}O_3$, $Al_xMn_yFe_{2-x-y}O_3$, Fe_3O_4, reduced mixed phases, Cu-phthalocyanine
Comminuted thin PVD-films	Al, Cr (semitransp.)/SiO_2/Al/SiO_2/Cr (semitransp.)

*Instead of mica other platelets such as silica, alumina, or borosilicate can be used.

Table 1.1. Overview of inorganic effect pigments (1.12)

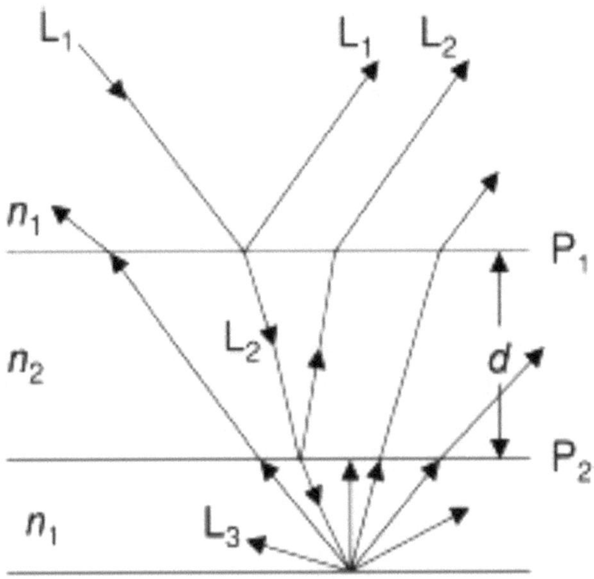

Fig. 1.2. Simplified diagram showing nearly normal incidence of a beam of light (L_1) from an optical medium with refractive index n_1 through a thin solid film of thickness d with refractive index n_2. L_1 and L_2 are regular reflections from phase boundaries P1 and P_2. L_3 represents diffuse scattered reflections from the transmitted light.

In pigments that simulate natural pearl effects, the simplest case is a platelet shaped particle with two phase boundaries P_1 and P_2 at the upper and lower surfaces of the particles, i.e., a single, thin, transparent layer of a material with a higher refractive index than its surroundings. For small flakes with a thickness of ca. 100 nm, the physical laws of thin, solid, optical films apply.

Multiple reflection of light on a thin solid film with a high refractive index causes interference effects in the reflected light and in the complementary transmitted light. For the simple case of nearly perpendicular incidence, the intensity of the reflectance depends on the refractive indices (n_1, n_2), the layer thickness (d), and the wavelength (λ):

$$I = \frac{A^2 + B^2 + 2AB\cos\Theta}{1 + A^2B^2 + 2AB\cos\Theta}$$

$$\text{Where } A = \frac{n_1 - n_2}{n_2 + n_1},\quad B = \frac{n_2 - n_1}{n_2 + n_1},\quad \Theta = 4\pi\frac{n_2 d}{\lambda}$$

With given n_1 and n_2 the maximum and minimum intensities of the reflected light, seen as interference colours, can be calculated and agree well with experimental results. Values for the refractive indices of the most important materials for pearl luster pigments are shown in Table 1.2.

Material	Refractive index
Vacuum/Air	1.0
Water	1.33
Proteins	1.4
Organic polymers (plastics, Lacquers, etc.)	1.4-1.7
Mica	1.5
$CaCO_3$ (aragonite)	1.68
Natural pearl (guanine, hypoxanthine)	1.85
$Pb(OH)_2 . 2PbCO_3$	2.0
BiOCl	2.15
TiO_2 (anatase)	2.5
TiO_2 (rutile)	2.7
Fe_2O_3 (hematite)	2.9

Table 1.2. Refractive indices of materials.

In practice, platelet crystals are synthesized with a layer thickness *d* calculated to produce the desired interference colours (iridescence). Most pearl luster pigments now consist of at least three layers of two materials with different refractive indices.

Thin flakes (thickness ca. 500 nm) of a material with a low refractive index (mica, silica, alumina, glass) are coated with a highly refractive metal oxide (TiO_2, Fe_2O_3, layer thickness ca. 50–150 nm). This results in particles with four interfaces that constitute a more complicated but still predictable thin film system. The behavior of more complex multilayer pigments containing additional, thin, light-absorbing films can also be calculated if appropriate optical parameters are known.

Colour effects depend on the viewing angle. Pearl luster pigment platelets split white light into two complementary colours that depend on the platelet thickness. The reflected (interference) colour dominates under regular (maximum) reflection, i.e., when the object is observed at the angle of regular reflection. The transmitted part dominates at other viewing angles under diffuse viewing conditions, provided that there is a non-absorbing (white) or reflecting background.

Variation of the viewing angle therefore produces a sharp gloss (reflectance) peak, and the colour changes between two extreme complementary colours. The resulting complex interplay of luster and colour is measured goniophotometrically in reflection and at different angles. A pearl luster pigment is characterized by a minimum of three $L^*a^*b^*$ data sets (CIE $L^*a^*b^*$-system) measured under different conditions (e.g., 0°/45° black background, 22.5°/22.5° black background, 0°/45° white background). An analysis of these data specifies a pigment on the basis of its hiding power, luster, and hue [1.1, 1.10, 1.11, 1.12].

2. Application of the Taguchi method to develop a robust design for the synthesis of mica-SnO_2 gold pearlescent pigment

2.1 Introduction

For a long time beautiful and deep pearlescent pigments have attracted human attention and have been used in many cases [2.1]. These pigments consist of thin transparent small flat surfaces with high reflective index. They reflect most of the radiant light and transmit a bit. Simultaneous reflection of light from small parallel surface layers of pigments causes the effect of deepness and brightness, such as exhibited by a pearl [2.2]. Each layer regularly reflects a part of the light and transmits the rest. The transmissible light is again reflected by other layers. Therefore, the manifestation of lustrous and interfering colours are revealed due to reflection from the interface between layers. This phenomenon is observed in natural pearl, fish scale, pearl body, birds feather, the butterfly wing, etc .

Pearlescent pigments are synthesized via two main methods. In the first, single crystals such as BiOCl or polycrystals like TiO_2 are considered as bright and light materials due to their special structure. In the second, the pearlescent state is formed through the coating of materials with a high refractive index, (mainly metal oxide) on a transparent substrate like mica. This group is more important because of higher mechanical stability and brightness [2.3].

One of the most important kinds of these pigments is TiO_2 coated mica which due to its high refractive index of TiO_2, has high light resistance, low cost, good chemical and heat resistance and nontoxicity has many applications in different industries.

However, a new kind of these pigments is gold pearlescent pigment of mica-tin dioxide which is obtained through coating mica platelets by tin dioxide. This chapter section examines mica-tin dioxide and uses muscovite mica as the substrate and is prepared by laying a particle layer of tin dioxide on its surface. Due to the absorption and reflection of light in this layer, the pigment is termed a colouring pearlescent pigment. This pigment is widely used in many fields such as glass, glaze, automotive, plastics, cosmetics, etc.

The main purpose of this study was to prepare pearlescent pigment and to find the optimum values of process parameters which affect its properties using the Taguchi statistical method.

2.2 Taguchi techniques

The Taguchi technique is a powerful tool for the design of high quality systems developed by Taguchi between 1950 to 1960 [2.4-2.6]. It provides a simple, efficient and systematic approach to optimize designs for performance, quality and cost. The methodology is valuable when design parameters are qualitative and discrete. Taguchi parameter design can optimize the performance characteristics through the setting of design parameters and reduce the sensitivity of the system performance to source of variation [2.6-2.7]. This technique is multi-step process, which follow a certain sequence for the experiments to yield an improved understanding of product or process performance. This design of experiments process is made up of three main phases: the planning phase, the conducting phase and analysis interpretation phase. The planning phase is the most important phase and one must give a maximum importance to this phase. The data collected from all the experiments in the set are analysed to determine the effect of various design parameters. This approach is to use a fractional factorial approach and this may be accomplished with the aid of orthogonal arrays. Analysis of variance is a mathematical technique, which is based on a least square approach. The treatment of the experimental results is based on the analysis of average and analysis of variance [2.8-2.9].

2.3 Experimental

2.3.1 Raw materials and reagents

Muscovite mica in bulk state was milled for 1 hour and then sieved and was used as the base. $SnCl_2$ (analytical reagent) was used to produce the metal oxide hydrate on mica flakes and the precipitation was performed by the presence of $KClO_3$ (analytical reagent) as the oxidizing agent. Sorbitan mono-oleate (analytical reagent) was used as the non-ionic surfactant. Also, NaOH (chemical reagent) and HCl (chemical reagent) were used for adjustment of pH.

2.3.2 Design of experiments

The experiments were conducted by standard orthogonal array. The selection of the orthogonal array is based on the condition that the degrees of freedom for the orthogonal array should be greater than or equal to sum of those lightness and reflective percent parameters [2.6-2.9]. In the present investigation, an L16 orthogonal array was chosen, which has 5 rows and 16 columns as shown in Table 2.1. Table 2.2 indicates the factors and their level. The experiment consists of 16 tests (each row in the L16 orthogonal array) and the columns were assigned with parameters. The first row was pH, second row was reaction temperature (T), third row was concentration of $SnCl_2$ solution (C), fourth row was reaction time (t) and fifth row was stirring rate (R). The response to be studied was the lightness and reflective percent with the objective of the bigger the better. The experiments were conducted by orthogonal array with level of parameters given in each array row. The l test results were subject to the analysis of variance.

Experiment number	1	2	3	4	5	6	7	8	9	10	11	12	13	14	15	16
pH	1	1	1	1	2	2	2	2	3	3	3	3	4	4	4	4
T	1	2	3	4	1	2	3	4	1	2	3	4	1	2	3	4
C	1	2	3	4	2	1	4	3	3	4	1	2	4	3	2	1
t	1	2	3	4	3	4	1	2	4	3	2	1	2	1	4	3
R	1	2	3	4	4	3	2	1	2	1	4	3	3	4	1	2

Table 2.1. Orthogonal array $L_{16}(4^5)$ of Taguchi [2.4]

Parameters	Levels			
	1	2	3	4
pH	1	2	2.5	3
T (°C)	60	65	70	75
C (g/lit)	10	20	40	60
t (h)	5	7	8.5	10
R (r.p.m)	100	200	300	400

Table 2.2. Process parameters with their values at four levels

2.3.3 Analysis

The spectral reflectance of the pigment was measured using a Colour Quest Goniospectrophotometer (CE-741-GL-Gretag Macbet Co). The SEM photomicrographs were obtained using an electron microscope (LEO-1455VP) and SEM studies were performed using this microscope equipped with EDX (to determine chemical composition of the observed objects). Appearance and colour of the covered mica flakes were observed by reflectance light microscope (Laborlux 11 pol, Leitz matallux 3). Finally, the phase analysis of the resultant pigment powders was determined by X-ray diffractometer (XRD-Philips-Xpert). Density and particle size of resultant pigment were measured by helium pycnometer (Accupyc 1330) and Master sizer 2000 (Mal 100229) respectively.

2.3.4 Preparation method

Firstly, a suitable suspension of 2 g milled and sieved mica (20-60 μm) in 40 ml de-ionized water was prepared. The suspension pH was adjusted with dilute hydrochloric acid (5%) to pH which is shown in table 2.1 and then heated to desired temperature according to table 2.1 experiment plane. The mixture was heated up to desired temperature, then 0.0001 % (based on mica wt%) of nonionic surfactant (sorbitan mono-oleate) was added to this suspension. Then, potassium chlorate aqueous solution ($KClO_3$) was added to this suspension as an oxidizing agent.

According to the conditions shown in table 2.1, stirring rate, suspension temperature and pH were controlled. In order to coat the mica flake, $SnCl_2$ solution and aluminium chloride ($AlCl_3$) with a constant proportion $SnCl_2/AlCl_3 = 9/1$, were added to the mica suspension. The pH of the suspension was kept at desired level by addition of 3% aqueous sodium hydroxide solution. The coating process was stopped after the desired time. Then, the suspension was filtered and the prepared pigments were washed with distilled water until neutral pH was reached and dried at 100ºC. Finally the powders were calcined at 950ºC for 1h to get the pigment powders with pearlescent luster.

2.4 Result and discussion

2.4.1 The colour characteristics of pigments

Lightness (l*) and reflective percent (R) of pigments were measured using goniospectrophotometer colour analysis apparatus and D65 illuminant in different view angles (20º, 45º, 75º, 110º). The results are shown in table 2.3.

Experiment number	1	2	3	4	5	6	7	8
reflection(R)%	45	75	82	58	46	57	74	79
lightness(l*)%	79.1	88.01	93.84	84.76	78.05	84.59	90.02	94.64
Experiment number	9	10	11	12	13	14	15	16
reflection(R)%	82	85	74	54	62	79	59	61
lightness(l*)%	93.20	95.48	88.94	83.25	84.64	90.12	82.30	83.04

Table 2.3. Results of lightness and reflective percent of pigments

Doing different analyses such as analysis of variance by using Winrobust software, important parameters were identified. This showed that some of these parameters have less effect in synthesis of the pigments than others. The amount of the effects is shown in figure 2.1.

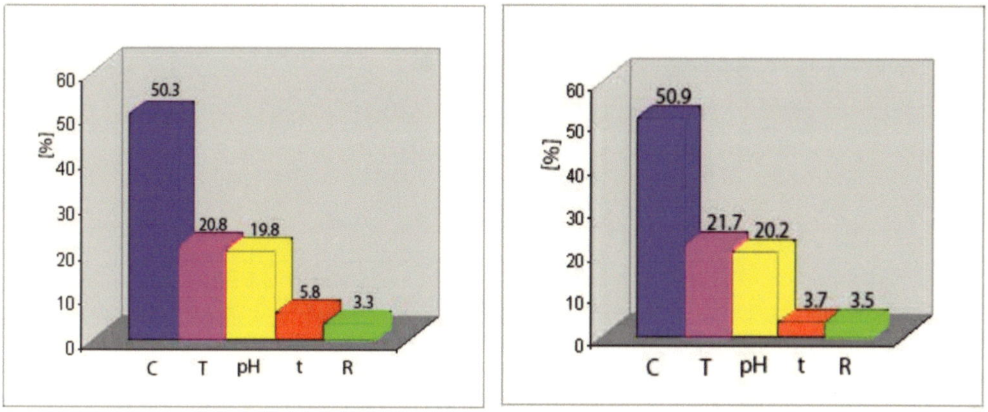

a b

Fig. 2.1. Effect of synthesis parameters on the a) lightness and b) reflective percent of pigments

Since in the Taguchi method it is possible that an optimum does not exist in the performed experiments in the designed array [2.10-2.11], some operations were then done by Winrobust software to determine the optimized point. Optimized levels of identified parameters for lightness and reflective percent of pigments is shown in table 2.4.

Parameters	Optimize level of lightness	Optimize level of reflection
pH	3	3
T	2	2
C	3	3
t	2	2
R	2	2

Table 2.4. Optimized levels of parameters determined with Winrobust software

To confirm this statistical method, synthesis is performed at the optimized conditions. If the experimental results equal the results with Winrobust software, it shows that the method is correct. In figure 2.2, normal probability distribution by Winrobust software is shown for the residual amount in which the continuity of the points in this figure is another confirmation for this method [2.10].

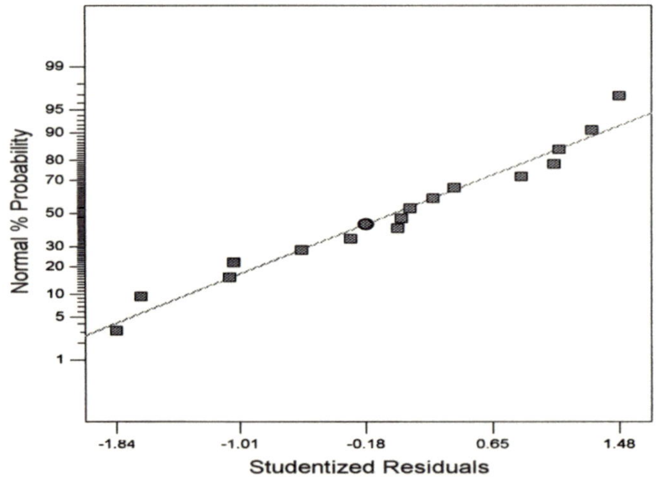

Fig. 2.2. Normal probability distribution of residual amounts

In the following section, according to the presented method and results of experiments, the effect of the parameters has been studied.

2.4.2 The factors that affect the lightness and reflection of pigment

2.4.2.1 pH

By plotting brightness and reflectance versus pH based on Taguchi method and using of Winrobust software it can be observed that maximum brightness and reflectance is related to pH of about level 3 (figure 2.3).

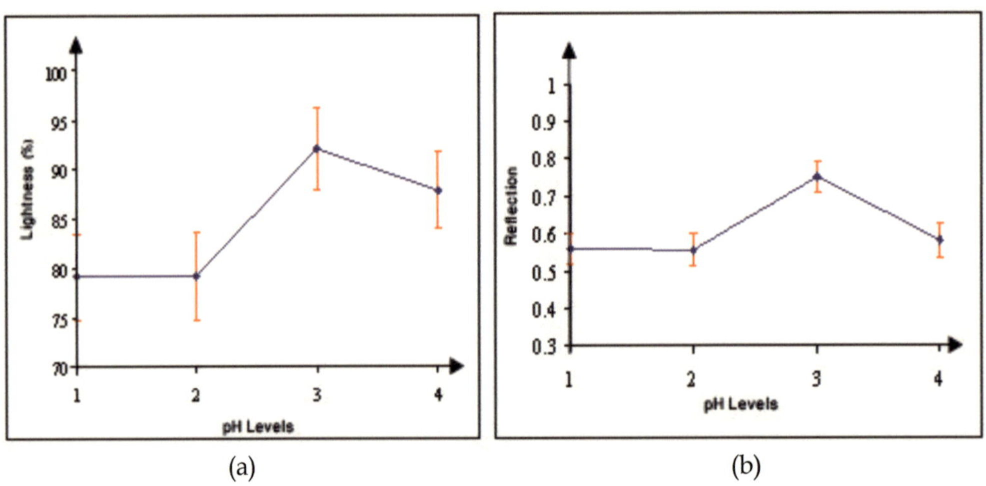

Fig. 2.3. Effect of parameter pH on the a) lightness and b) reflection , in different levels

To follow and conform the effect of these changes, SEM and EDS analysis were performed on pigments synthesized at different levels of pH. The results are shown in figures 2.4, 2.5 and 2.6. According to figure 2.4-a, SnO_2 particles have precipitated on mica flakes uniformly in sample 9 which its pH is optimum (pH = level of 3). Figure 4-b (EDS analysis of sample 3) shows a great amount of tin on the mica flakes. In sample 6, only a small quantity of SnO_2 particles have been precipitated on mica flakes that its pH is less than optimum pH because hydrolysis process has not been performed completely . The results are shown in figures of 2.5-a and 2.5-b but in sample 15 a great amount of SnO_2 particles are not agglomerated on the mica flakes uniformly. The results are illustrated in figures 2.6-a and 2.6-b. It seems that due to the progress of hydrolysis the homogeneity of the SnO_2 film can vary and the hydrolysis rate is very fast.

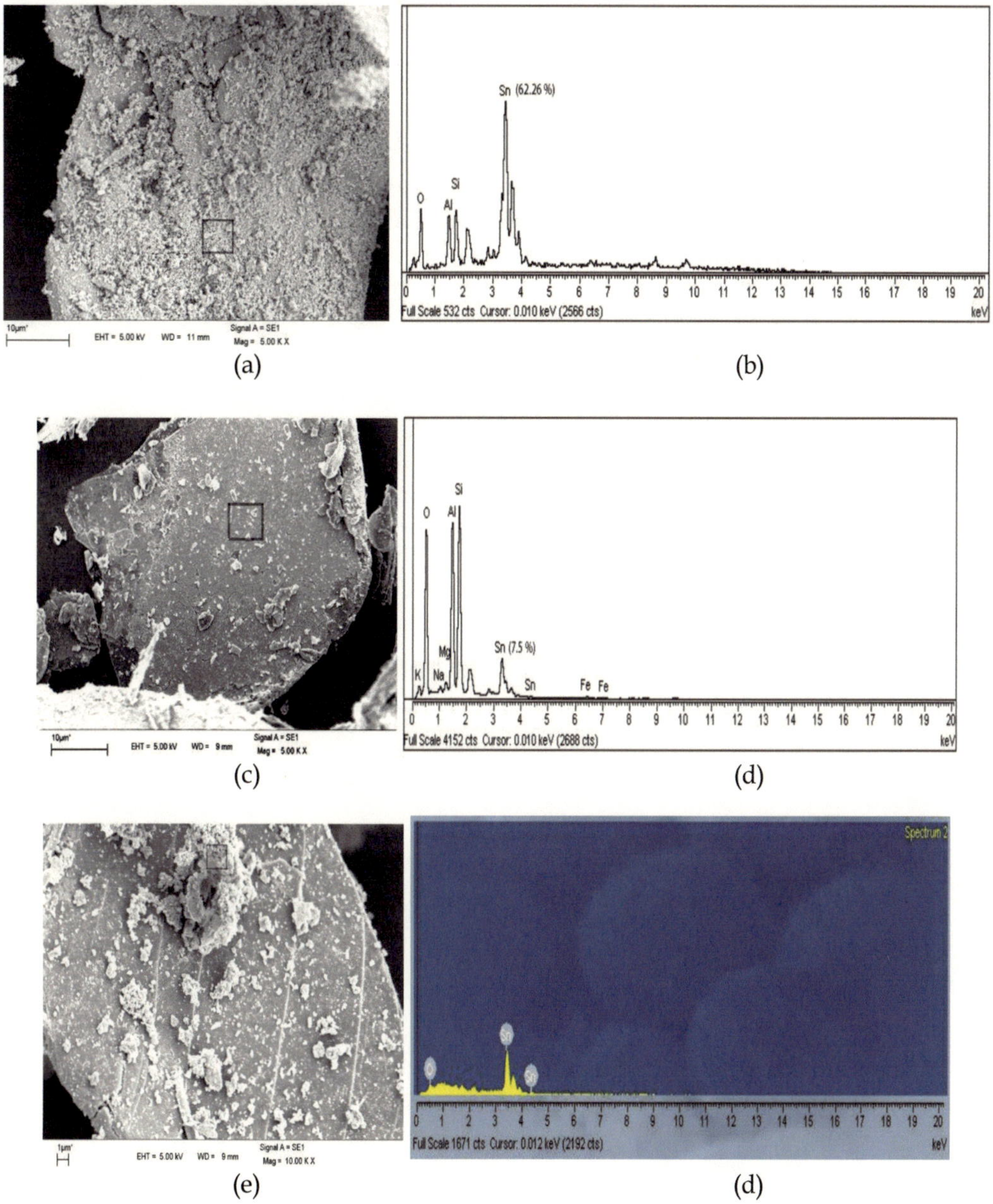

Fig. 2.4. SEM and EDS Images of pigments synthesized at different levels of pH

Therefore, the uniformity of SnO_2 coated on mica flakes depends on the hydrolysis rate of $SnCl_2$ solution. The equation of hydrolysis (Equation 1) and rate of hydrolysis (Equation 2) can be considered as follows:

$$\left[M\left(OH_2\right)_N\right]^{Z+} + hH_2O \rightarrow \left[M\left(OH\right)_h\left(OH_2\right)_{N-h}\right]^{(Z-h)^+} + hH^+{}_{Solvated} \quad (1)$$

$$h = \left[\frac{1}{1+0.41pH}\right]\left[\left(1.36Z-N\right)\left(0.236-0.08pH\right) - \frac{2.621-0.02pH-X_m^*}{\sqrt{X_m^*}}\right] \quad (2)$$

Where "h" is the hydrolysis rate, "Z" is the charge of M cation, "N" is the coordination number of M, "X^*_m" is electronegativity of M. According to the equation (2), if the pH level is less than the optimum level, the hydrolysis process is not performed and SnO_2 particles do not precipitated, because the hydrolysis rate is negative. Therefore, brightness and reflectance will decrease. On the other hand, if the pH level is more than the optimum level, the hydrolysis rate increases with pH increasing. Therefore, agglomerated particles are initially formed in suspension and then on the mica flakes. These agglomerated particles cause irregular scattering of light which in turn decrease the brightness and reflectance [2.12, 2.13, 2.14].

Furthermore, deposition of hydrolysed particles on the mica flakes depends on the electrical charges of hydrolysed particles and mica flakes. On the other hand, the electrical charge of particles in the suspension depends on pH of the media. Therefore SnO_2 particles can precipitate on the mica flakes in a special pH range. As the electrical charge of mica flakes is negative in the suspension, most of the deposition is performed at a pH less than 3.5. This is because the charge of SnO_2 particles are positive at this pH. The variation of electrical charges can be seen in figure 2.5.

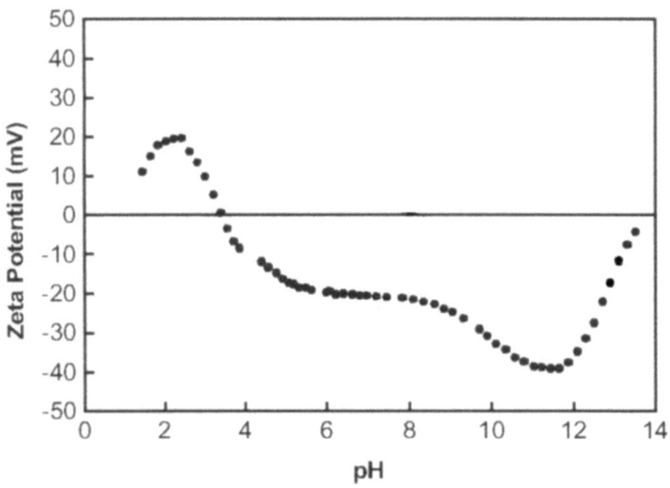

Fig. 2.5. Variation of zeta potential with pH for SnO2. Isoelectric point of SnO_2 is 3.5 and optimum amount of positive charge of SnO_2 particles is pH 2.5 [2.15]

2.4.2.2 Reaction temperature

By observing brightness and reflectance versus reaction temperature based on Taguchi method and the use of Winrobust software it can be observed that the maximum of brightness and reflectance is related to reaction temperature of about level 2 (figure 2.6).

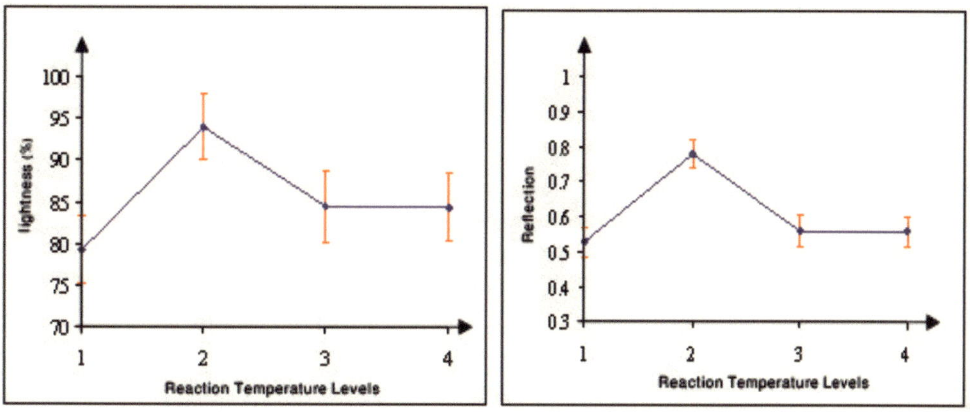

(a) (b)

Fig. 2.6. Effect of parameter temperature on the a) lightness and b) reflection , at different levels

According to SEM analysis observed in figures of 7-a, 7-b and 7-c, when the reaction temperature is low (T = level 1), approximately no SnO2 coating is formed on the mica flakes, thus lightness and reflectance are small (figure 2.7-a). Lightness and reflectance increase with temperature increasing until level of 2. SnO_2 particles coated on the mica flakes are uniform at temperature of level 2 (figure 2.7-b). Increasing temperature causes the flocculation of the colloid particles, the particles become larger, and the membrane on substrate surface becomes loosen and SnO_2 particles irregularly deposit on the mica flakes (figure 2.7-c).

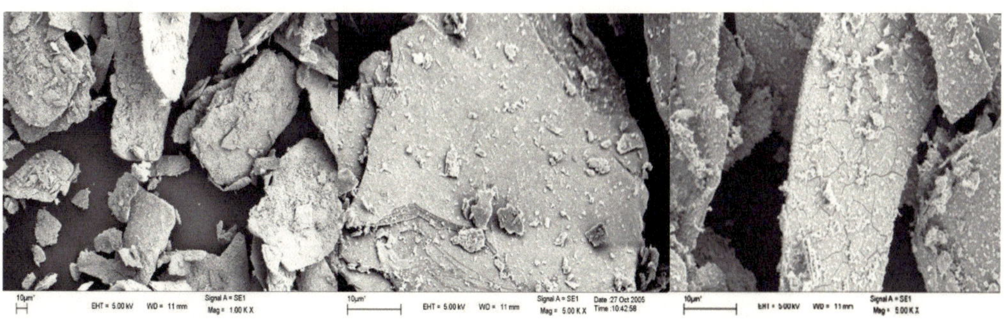

(a) (b) (c)

Fig. 2.7. SEM images of pigments synthesized at different levels of reaction temperature

As the hydrolysis process is a precipitation reaction, precipitation decreases with temperature decreasing. This seems to be due to the solubility of tin hydroxide increasing with temperature increasing. So, the precipitate reduces and the coating becomes uneven.

Also increasing temperature causes the weakening of hydrogen bonding between SnO_2 particles and mica flakes [2.13].

2.4.2.3 Concentration of $SnCl_2$ solution

By observing brightness and reflectance versus concentration of $SnCl_2$ solution, based on Taguchi method and the use of Winrobust software it can be observed that the maximum of brightness and reflectance is related to concentration of about level 3 (figure 2.8). It can be seen that with the increase in the concentration of $SnCl_2$ solution, to a level higher than 3 causes the lightness and reflectance of the pigment to decrease.

The relationship between the intensities of reflected light, transmitted light, scattered light, absorbed light and the intensity of incident light is I = S+T+D+A [2.16], where S, T, D and A represent the strength of reflecting light, transmitting light, scattering light and absorbing light respectively. These are function of tin chloride concentration. When I is fixed, the thicker the membrane layer, the larger the value of A, while the smaller the value of S and T. As T decreasing, there is less incident light traveling through the coating layer and arriving at the surface of the mica substrate. Thus, the intensity of reflecting light generated at the surface decreases, the reflection and interference action becomes weaker, and the lightness decreases.

Figure 2.9 shows SEM images of a great thickness of SnO_2 layer at a great concentration of $SnCl_2$ solution. This figure is related to the SEM photo of sample of 2.7 in that the lightness and reflectance are small, because the thickness of SnO_2 layer is so high.

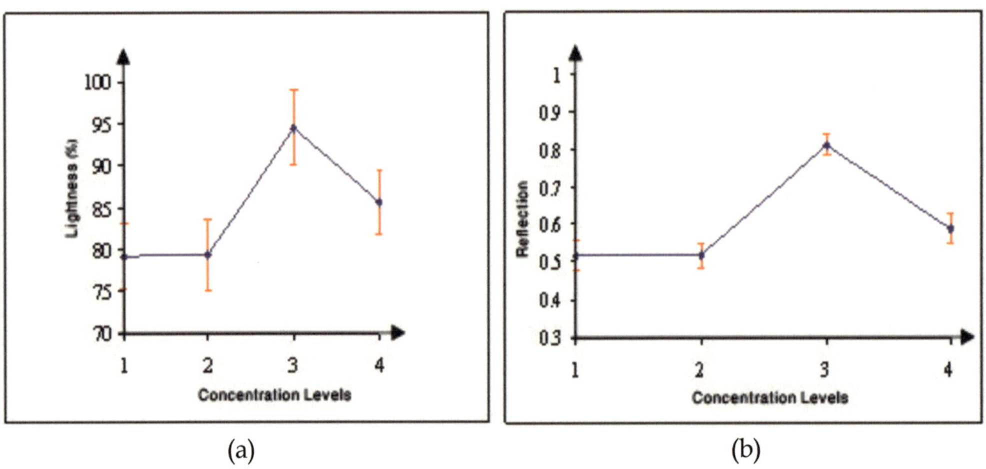

Fig. 2.8. Effect of concentration parameter on the a) lightness and b) reflection , at different levels

Fig. 2.9. SEM image of pigment synthesized at great concentration

2.4.2.4 Reaction time

By observing brightness and reflectance versus reaction time based on the Taguchi method, it can be observed that the maximum of lightness and reflectance are related to reaction time of about level 2 (Fig. 2.10).

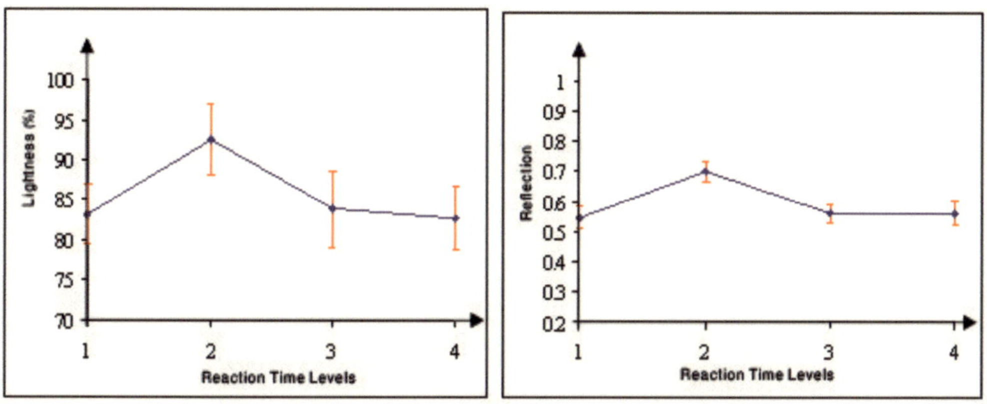

(a) (b)

Fig. 2.10. Effect of time parameter on the a) lightness and b) reflection, at different levels

As SEM analysis show in figures 2.11-a and 2.11-b, when reaction time is low a few coating of SnO_2 are formed on the mica flakes, thus lightness and reflectance are decreased (figure 2.11-a). While increase of reaction time, until level of 2, a uniform SnO_2 coating is formed on the mica flakes and thus, lightness and reflectance increase as shown in figure 2.13-b. However, increasing time more than level 2 causes decreasing of lightness and reflectance, because the SnO_2 coatings formed on the mica flakes are separated gradually.

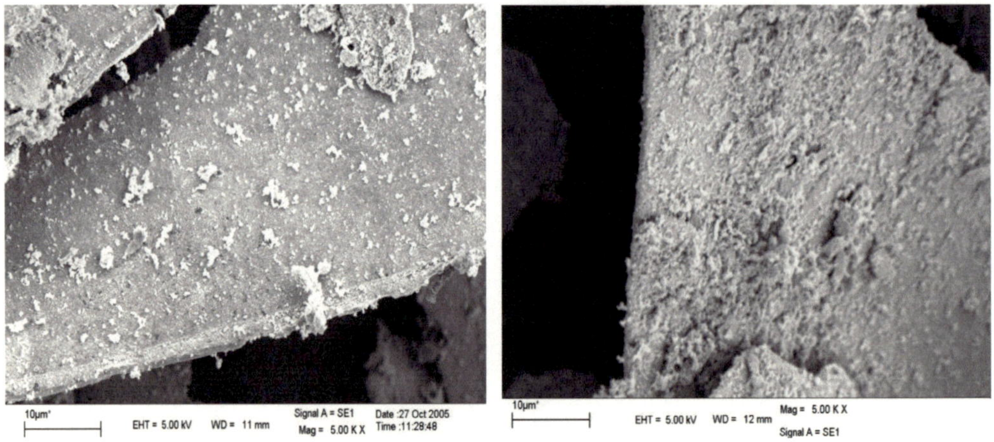

(a) (b)

Fig. 2.11. SEM images of pigments synthesized at different levels of reaction time

2.4.2.5 Stirring rate

By observing brightness and reflectance versus stirring time based on (Figures 2.12-a and 2.12-b) Taguchi method, it can be seen that the maximum of lightness and reflectance are related to a stirring rate of about level 2.

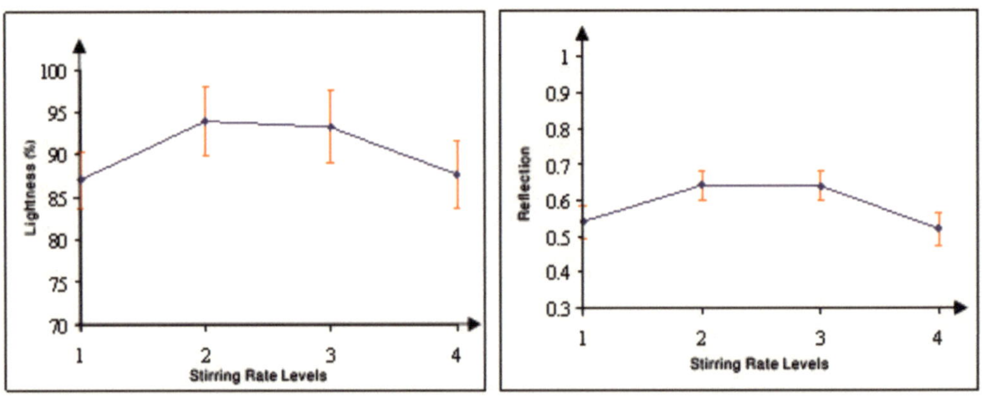

(a) (b)

Fig. 2.12. Effect of stirring rate parameter R on the a) lightness and b) reflection at different levels

The stirring rate affects the densification of the membrane layer directly. If the stirring rate is too low, the reaction solution can not form sufficient turbulence [2.17], then the microscopic mixture is uneven, and the generated crystal particles of SnO_2 are of various sizes. On the other hand, if the stirring rate is too high, it will affect the rate of growing of the crystal nucleus, with the result that some tiny colloidal micro-particles enter the solution through the filter paper rather than depositing on the surface of mica flakes, which will cause the light to scatter. Therefore, the stirring rate in this reaction should be of level 2.

2.4.2.6 Surfactant effect

Figure 2.13 shows the SEM images of optimized sample with surfactant and without it. As figure 2.13-a shows in the sample containing surfactant, SnO_2 particles coat mica flakes uniformly without any agglomeration. In addition, these particles agglomerate on mica flakes only and are not formed freely among mica flakes, but in the sample without surfactant, SnO_2 particles formed irregularly and are agglomerated on mica flakes and also freely among them (Figure 2.13-b).

(a) (b)

Fig. 2.13. SEM images of coated mica flakes in optimized sample a) with surfactant in magnification of 1 KX and b) without surfactant in magnification of 500 X

Surfactant molecules are able to surround a small volume of suspension including seed and due to repulsive forces between electrical charged boundaries of mica and SnOH particles, prevent the over sticking of particles and growth of agglomerates [2.13, 2.18, 2.19]. In addition the mica boundary and hydrated metal oxide formed by hydrolysis is activated with increasing amount of surfactant making for a smooth surface and stable adsorption [2.20].

2.4.3 Optimized sample

After performing experiments designed by Taguchi method and analysing results with Winrobust software the optimized sample was provided using the following condition: pH = 2.5, T = 65°C, C = 40 g/lit, t = 7 h, R = 200 r.p.m .

The XRD pattern of this sample shows the hydroxide phase before calcination and Casiterite SnO_2 phase after that.

According to SEM results, SnO_2 particles are coated on mica flakes almost regularly and uniformly (figure 2.13-a). According to the laser beam diffraction technique the mean particle size of pigments is about 65 micron. The density of this pigment is 3.3 g/cm^3 and the lightness and reflective percent are 97% and 89% respectively in which case these amounts of l^* and R are equal to the Winborust software results. Consequently choosing to use the Taguchi method and L_{16} algorithm is correct for this work.

Table 2.5 shows the results of Goniospectrophotometery of the optimized sample. Figure 2.14 shows the amount of optimized sample reflection from different angle views.

Colorimetery parameters	Angle view			
	20°	45°	75°	110°
a^*	0.81	0.74	0.73	0.67
b^*	13.04	14.20	14.34	13.78
l^*	96.76	87.57	87.03	86.99
$R_{570\ nm}$	0.89	0.64	0.56	0.55

Table 2.5. R results of goniospectrophotometery of optimized sample

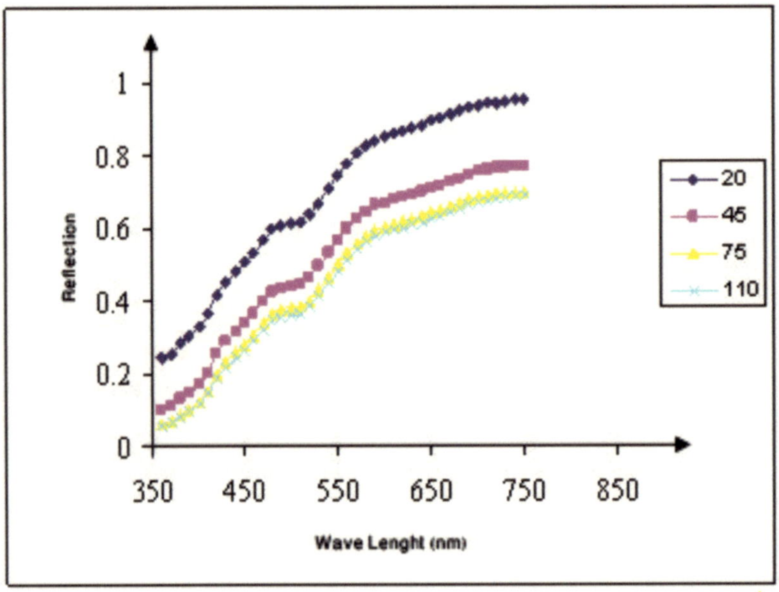

Fig. 2.14. Amount of optimized sample reflection from different angle views

According to table 2.5 and figure 2.14 parameters l^* and R in the different angle views have different values, and also these values are higher in smaller angle views. Therefore the optimized sample has lightness and reflection [2.21, 2.22]. In addition since the maximum value of reflection is in the range of yellow light wavelength (570-600 nm), the pigment is yellow. So it is possible to say that the optimized pigment is gold pearlescent pigment.

Figure 2.15 shows the microscopic image of the optimized sample. As can be seen from Fig. 2.14 and Fig. 2.15, most of mica flakes coated with SnO_2 have a golden colour. However, in this sample, flakes with other colors can be seen as well, which might be due to different thickness of coating on the mica flakes.

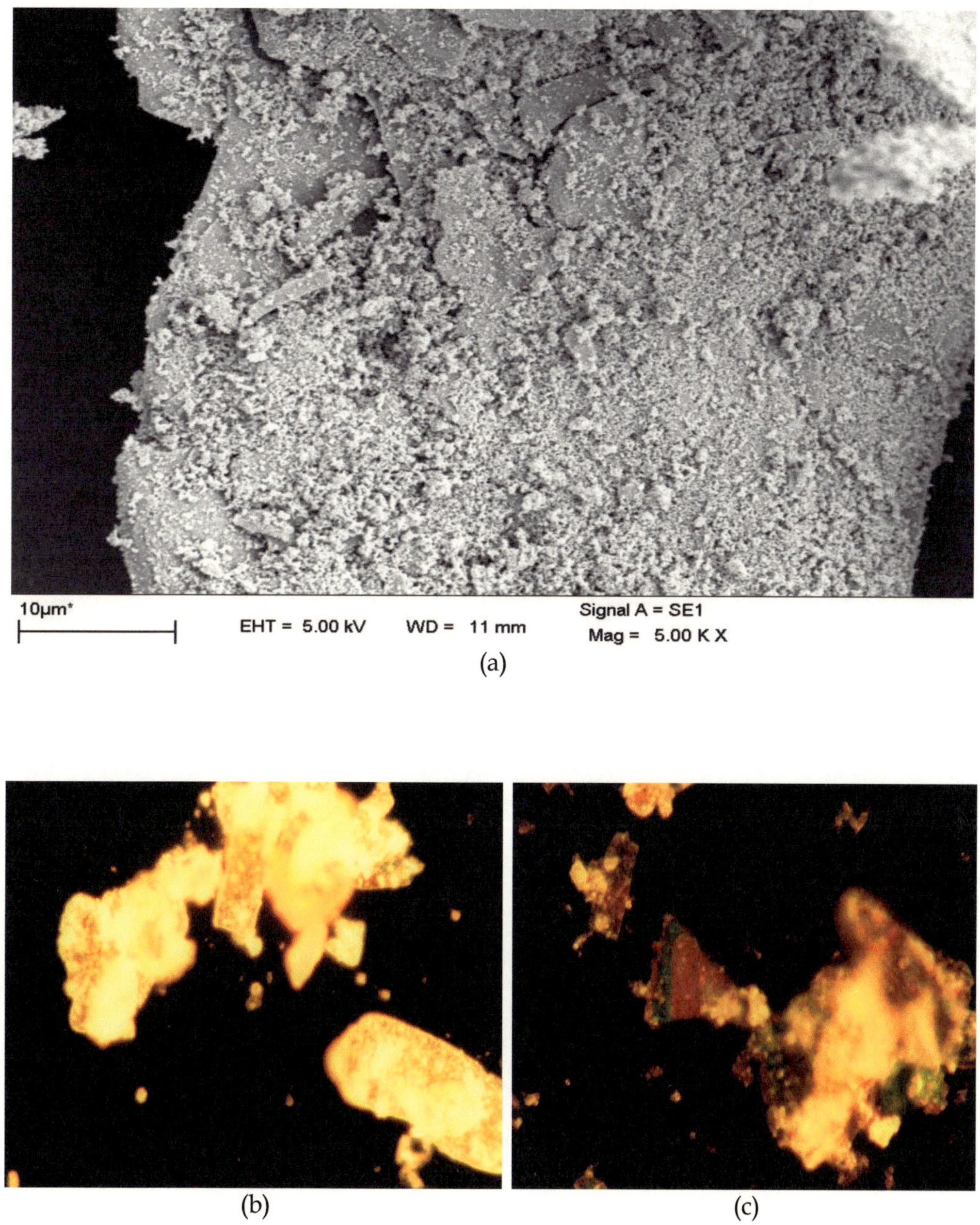

Fig. 2.15. Microscopic image of optimized sample, (a) SEM, (b) and (c) light microscopy

2.5 Chapter conclusions

1. The results obtained in this research according to the Taguchi method, provide optimum values for pH, temperature, concentration of $SnCl_2$ solution, time and stirring velocity of 2.5, 65 °C, 40 g/lit, 7h and 200 rpm respectively.
2. According to SEM investigation, a relatively good and homogenously flat coating is formed on the mica platelet at the optimum conditions.
3. The effects of using surfactant are: formation of a homogenous thickness of coating, specifically on the flakes surfaces and absence of intra particles agglomerations of SnO_2. This is not the case without using the surfactant.
4. XRD analysis of the optimum sample before and after calcination show the amorphous and crystalline phase of SnO_2 and Casiterite respectively.
5. Colourimetry measurements show a bright golden appearance of the optimum pigment.
6. Light microscopic observation of the coated flakes using a polarized filter shows that the majority of the flakes have the golden colour, implying the final interference pigment colour of the optimum synthesized pigment is a golden one [2.23].

3. Acknowledgement

Thanks are due to Mrs M. Esfidari who helped for preparation of second part of this chapter.

4. References

4.1 Part one references

[1.1] Glausch, R., Kieser, M., Maisch, R., Pfaff, G., Weitzel, J., Special Effect Pigments, Vincentz Verlag, Hannover, 1998.
[1.2] Pfaff, G., Franz, K.-D., Emmert, R., Nitta, K., Ullmann's Encyclopedia of Industrial Chemistry: Pigments, Inorganic, see Section 4.3, 6th edn. (electronic release), VCH Verlagsgesellschaft, Weinheim, 1998.
[1.3] Pfaff, G., Chem. Unserer Zeit 31, (1997), p. 6.
[1.4] Pfaff, G., High Performance Pigments (Ed.: Smith, H. M., Wiley-VCH, Weinheim, 2002, p. 77.
[1.5] Ostertag, W., Nachr. Chem. Tech. Lab. 42 (1994), p. 849.
[1.6] Bäumer, W., Farbe + Lack 79 (1973), p. 747.
[1.7] Greenstein, L. M., Pigment Handbook, Part 1, Wiley, New York, 1988, p. 829.
[1.8] Schmidt, C., Friz, M., Kontakte (Darmstadt) 2 (1992), p. 15.
[1.9] Pfaff, G., Reynders, P., Chem. Rev. 99 (1999), p. 1963.
[1.10] Hofmeister, F., Eur. Coat. J. 3 (1990), p. 80.
[1.11] Hofmeister, F., Pieper, H., Farbe + Lack, 95 (1989), p. 557.
[1.12] Buxbaum, G. and Pfaff, G., Industrial Inorganic Pigments, Third, Completely Revised, _c WILEY-VCH Verlag GmbH & Co KGaA, Weinheim, 2005.

4.2 Part two references

[2.1] Buxbaum, G. and Pfaff, G., Industrial Inorganic Pigments, Third, Completely Revised, _c WILEY-VCH Verlag GmbH & Co KGaA, Weinheim, 2005.

[2.2] S. Hanchisu, "Nacreou Pigments", Prog. Org. Coat. 3:191-220 (1975).
[2.3] K.Othmer, "Encyclopedia of Chemical Technology", 3rd ed., 17, J.Wiley & sons, NewYork, 1978-1984, PP. 78-838.
[2.4] G. Taguchi and S. Konishi, Taguchi methods, orthogonal arrays and linear graphs, tools for quality engineering, American Supplier Institute, Dearborn, MI (1987) p. 35-38.
[2.5] G. Taguchi, Taguchi on robust technology development methods, ASME Press, New York, NY (1993) p. 1-40.
[2.6] Phillip J. Ross, Taguchi Technique for quality engineering, McGraw-Hill, New York (1988).
[2.7] K. Roy Ranjit, A Primer on Taguchi method, Van Nostrad Rainhold, New York (1990).
[2.8] J. Paulo Davim, An experimental study of tribological behaviour of the brass/steel pair, J Mater Process Technol 100 (2000), p273-279.
[2.9] J. Paulo Davim, Design optimization of cutting parameters for turning metal matrix composites based on the orthogonal arrays, J Mater Process Technol 132 (2003), p. 340-344.
[2.10] Genechi Taguchi, "System of Experimental Design" Vol. 1, 1987, KRAUS International Publication.
[2.11] Yeow Nam Ng, Don Black, Khanh Luu, "Taguchi Methods", 1995, Curtin University Handout Notes for Computer Aided Engineering.
[2.12] T. Junru, Sh. Lazhen, F. Xian song, H. Wenxiang, "Preparation of Nanometer - Sized (1-x) SnO_2. xsb_2O_3 Conductive Pigment Powders and the Hydrolysis Behavior of Urea", Dyes and pigments 61 (2004) 31-38.
[2.13] J. P. Jolivet, "Metal Oxide Chemistry and Synthesis", John Wiley & Sons, (2002).
[2.14] M. Yamamoto, A. Ando, "Pearlescent Pigment, and Paint Composition, Cosmetic Material, Ink and Plastics Blended with the New Pearlescent Pigment", U.S. Patent 5741 355 (1998).
[2.15] H.R.Castro, B.S.Murad, D.Gouvea, "Influence of the Acid-Basic Character of Oxide Surfaces in Dispersants Effectiveness" Ceramics International 30 (2004) 2215-2221.
[2.16] Miller HA. Optical property of pearl pigment GFR. Farbe lack 1987; 12:93.
[2.17] Xiong Y, Zhou X, Hu L. Study on the process of the ultrafine α-FeOOH synthesis by dripping method. Journal of East china university of science and technology 1996; 22(5): 541-7.
[2.18] K. Matsui, M. Ohgai, "Formation Mechanism of Hydrous Zirconia Particle Produced by Hydrolysis of $ZrCL_2$ Solution, Kinetics Study for Nucleation and Crystal Growth Processes of Primary Particle", J. Am. Ceram. Soc. 84: 2203-2313 (2001).
[2.19] M.R. Porten, "Hand book of Surfactan", Chapter 4, Chapter 8, Hall (1994).
[2.20] K.Chang, etal, "A Method for Preparing a Pearlescency Pigment", U.S. Patent. APPL. 20040096579 (2004).
[2.21] H.M.Smith, "High Performance Pigment", Chapter 7, Wiley- VCH, Weinheim, Germany, (2002).
[2.22] L.M. Greenteain, "Pigment Handbook", Vol. 1, Chapter III-D-d-2, John Wiley & Sons NewYork (1998).

[2.23] M. Esfidari, "Preparation of Pearlescent Pigments for Low Temperature Glazes", MSc Thesis, Ceramic Department, Iran University Of Science and Technology, Tehran, Iran 2005.

Part 5

Application in Foundry

9

Ceramic Coating for Cast House Application

Zagorka Aćimović-Pavlović[1], Aurel Prstić[1],
Ljubiša Andrić[2], Vladan Milošević[2] and Sonja Milićević[2]
[1]*University of Belgrade, Faculty for Technology and Metallurgy,*
[2]*Institute for Technology of Nuclear and Other Mineral Raw Materials*
Republic of Serbia

1. Introduction

The coatings for moulds and cores represent an integral part of castings production. Basic role of ceramic coatings is to form an efficient refractory barrier between the sand substrate and liquid metal flow during the phase of casting, solidification and forming of the castings. This provides a smooth and clean surface of castings, with no adhered sand or defects due to metal penetration into the mould (lumps, dents, rough surface and alike). Dimension accuracy and surface appearance of castings depend both on metal and mould. In casting practice, sand casting technology is widely applied for castings production. Quartz sand mostly used in the composition of mould and core blends has a number of faults - low refractoriness, high heat expand coefficient, causing casting surface faults, especially when metals and alloys with high melting temperatures are concerned. Higher quality mould blends based on zircon, olivine, chromite, sinter magnesite containing much better thermal-physical properties than quartz sand are relatively less applied for their high pricing. Various additions to mould blend are used more frequently, as well application of ceramic coatings on moulds and cores. Application of higher quality ceramic coatings significantly influences either reduction or elimination of expensive cast house cleaning and machining operations for the castings, thus directly reducing production costs of a foundry.

A practice to coat sand moulds and cores in order to improve the surface quality is dated from 19th century, when so called „black coating" was applied, based on graphite, silica-dioxide and chamotte dispersed in water, with molasses as a binding agent. This kind of coating is very simple, but its application was efficient at the time for improvement of the casting surface quality (Svarika 1977, Tomović 1990).

Depending on use, contemporary coatings represent mixtures of ceramic materials in a solvent with suspension agent and binding agent. Coating composition analyses show that these consist of a number of components. Ceramic coatings influence both an improvement of the existing casting methods and a development of the new ones, primarily expandable & meltable pattern casting (EPC process and precision investment casting). A completely new concept of coating development has arisen over the past decades with some producers in a form of electrostatic coating with powder, coatings based on aluminium molecular oxides in a form of fishbone and alike. Coating development should be carried out through systematic researches in order to make an optimum choice of the coating for concrete casting methods, types of castings and types of alloys. At the same time, all relevant quality indicators and economy indicators for the casting production should be monitored. Coating properties are strictly defined by standards; therefore, it is very important to make the right choice of coating, as well as its preparation and application in concrete foundry working conditions (Cho 1989, Cibrik 1977, Brome 1988, Davies 1996, Josipović et al. 1994).

2. Characteristics and division

Ceramic coating includes materials with a composition set in advance, which refractoriness and other properties prevent reactions on the mould-metal contact surface from happening. There are three basic classes of coatings for moulds and cores from the aspect of the basic ingredient as follows:

- The class based on graphite or some other sort of carbon, (graphite coatings),
- The class based on highly refractory materials, (ceramic coatings), and
- The class based on a mixture of carbon and highly refractory ingredients, (blended coatings).

Graphite coatings are agents containing graphite or similar materials as a refractory filler. Ceramic coatings are agents containing fine milled ceramic materials such as zircon, quartz, chamotte, magnesite, olivine, chromite, talc, mica, cordierite and alike as a refractory filler. The coatings not changing casting surface properties are classified into the group of passive coatings. In terms of physical properties, there are the following coating types:

- Liquid coatings based on organic solvents, such as isopropanol, ethanol and alike;
- Liquid coatings with water;
- Semi-liquid coatings (pastes) based on organic solvents;
- Semi-liquid coatings (pastes) with water;
- Dry (powder) coatings.

Liquid and semi-liquid coatings are used for either dry or semi-dry moulds, while powders are used for wet moulds. Growth tendency regarding quality and complexity of casting production, constant requirements for as high quality casting surface as possible and for production cost reduction impose the need to improve the production of foundry coatings and thereby to widen their function range. Nowadays, so called active coatings are increasingly used; their role, different to the one of protective (passive) coatings, is reflected in the change of properties in the surface layer of castings. Active ingredients of this type of coating penetrate into inner metal layers either by merging or diffusion. It has a different effect on casting quality, such as hard structure formation (spots on a casting which are in contact with the coatings containing tellurium, bismuth), cementation (presence of carbon in coatings) or prevention of appearance of scratches on castings (iron (III) oxide, (Fe_2O_3) (Gorni & Marcinkowski 1977, Monroe 1994, Tsai & Chem 1988).

There are different divisions of coatings, often according to the type of casted metal (for non-ferrous metals, steel, cast iron), and according to the chemical character (acid, base, neutral). The choices of coatings are made based on the type of mould and core either. The method of mould production out of different mould blends with different systems to bind the sand grains (cold mould casting, hot mould casting, CO_2 procedure, expendable pattern casting, shell process) influence the choice of the type of coating. For moulds and cores made out of the quartz sand mixed with binding agent, one kind of ceramic coatings is used, while the other kind of ceramic coatings is used for metal moulds and metal cores. It includes the choice of a refractory filler, binding agent, additives and solvents in accordance with notions and processes carried out on the mould-coating-liquid metal contact surface during the phase of inflow, cool down and solidification of castings (Tomović 1990).

3. Physical-chemical reactions on the liquid metal-ceramic coating sand mould boundary

In order to properly choose the ceramic coating required, it is necessary to be familiar with the phenomena related to physical-chemical and thermo-dynamic changes carried out on the liquid metal-mould boundary during the phase of inflow, cool down and solidification of castings. Basic physical and physical-chemical processes carried out on the metal-mould contact surface are the following:

- Physical-chemical processes of interaction between liquid metal and mould determining castings surface properties,
- Metal transfer from liquid state into solid state as a result of structure solidification and formation, and
- The phenomenon of metal shrinkage and related notions.

During sand casting, oxidation atmosphere prevails within the mould cavity, influencing alloy components to oxidize first; afterwards, reactions among the metal oxides and mould blend are carried out. In order to obtain a quality casting surface, it is necessary to examine the interaction between metal oxides and mould blend, as well as pore formation and mould blend maceration with metal. When liquid metal penetrates the mould pores formed under the influence of capillary forces, complex reactions of chemical nature between the mould material and liquid metal are carried out. Depending on the mould blend composition, oxidation atmosphere has different oxidation ability which depends on CO, CO_2, H_2 and O_2 contents. Character of the oxides formed on the metal surface is determined by the ratio of individual components from the gas stage composition, as well as by specific properties of the alloy being casted. It should be stressed that chemical action of metal oxide and mould blend is also determined by oxidation of some elements out of the blend composition. Pore formation is prevented by an appropriate choice of a mould blend either not reacting with metal oxides or forming solid compounds on the liquid metal-mould contact surface. Therefore, for example, manganese steel with 13% Mn gives unsuitable casting surface during sand casting, due to active activity of MnO and SiO_2. Being alloyed with aluminium, the same kind of steel does not have the faults mentioned above. It is explained by the fact that Al_2O_3 oxide skim with melting temperature of 2050°C together with an easily meltable eutecticum $Al_2O_3 \cdot SiO_2$ with melting temperature of 1545 °C forms a layer protecting the metal from further oxidation. When carbon steels are casted in the moulds made out of mould blend, iron oxide (FeO) prevails in the metal oxide composition. At the steel casting temperature, these oxides are highly overheated (1370°C) causing their hastened activity and fluidity. Figure 1 shows a schematic of interaction between the metal oxides and mould material. Metal oxides being formed on the melt surface penetrate the pores and, reacting with quartz sand grains, form silicates of volatile compositions of $(FeO)_m(SiO_2)_n$ type, influencing casting surface quality, Figure 1.a. Silicates formed will penetrate the mould wall if they have a low viscosity, while metal might penetrate the surface pores formed. It has been shown that depending on the metal temperature (1100°, 1300°, 1500°C) and mould material quality, mould is macerated with oxides (in the case of sand-clay, iron chromite, chrome-magnesite), and silicates are intensively formed (with sand-clay blend). Examinations concerning liquid metal casting into the sand mould showed that iron oxide (FeO) reacted with the SiO_2 sand grains and formed an easily meltable silicate. Thus created silicate fills the space between the sand grains and gets suppressed by liquid metal into the mould interior, Figure 1 b. Further, by interaction between iron oxide and silica dioxide SiO_2 , a thin membrane may be formed, disabling metal penetration into the mould pores, (Figure 1.c). Pore formation in the mould blend is influenced by the character of the oxides formed, as well as their actions with the mould surface (Cibrik 1977, Svarika 1977, Tomović 1990).

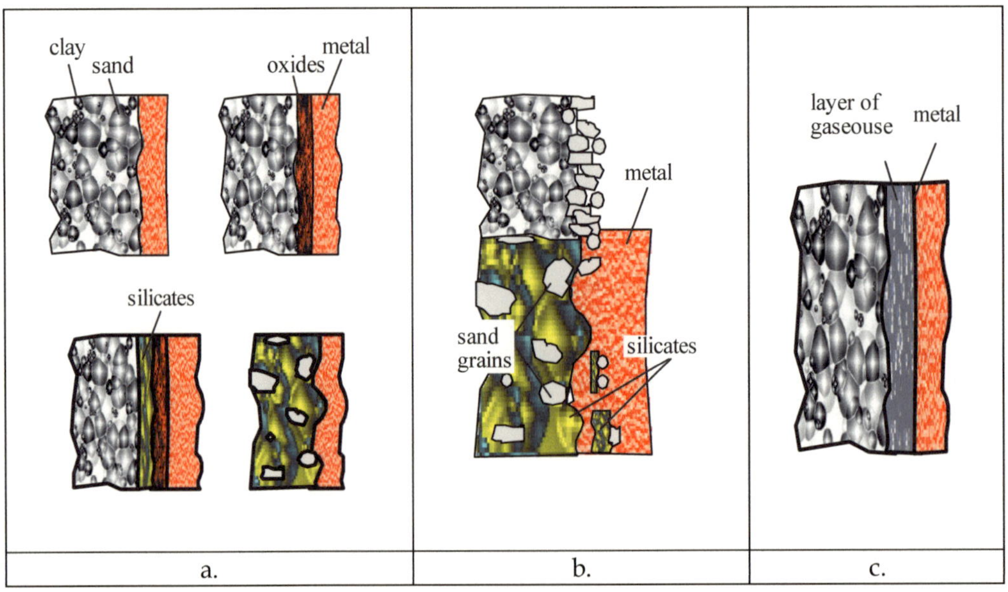

Fig. 1. Schematic view of non-metallic layer formation on the metal-mould boundary (Svarika 1977)

Iron silicate formation on the metal-mould contact boundary is carried out in three stages, Figure2. At the first stage, metal oxidation is carried out through the mould pores (Figure 2.a), second stage is featured by absorption of iron oxide dissolved in metal on the sand grain surface and by its reaction with SiO_2 (Figure 2.b), while silicate with volatile composition $(FeO)_m(SiO_2)_n$, is formed at the third stage (Figure 2.c). As this process is carried out in an excessive SiO_2 , there is no possibility that easily meltable iron silicates will be formed. Absence of these mobile silicates (with a low viscosity), as well as duration of a forming process, make metal penetration either difficult or disabled. In case that the silicates are hard to melt, i.e. if they have a high viscosity, they may represent a barrier for metal penetration into the mould pores, like with sintered material(Cibrik 1977, Svarika 1977, Tomović 1990).

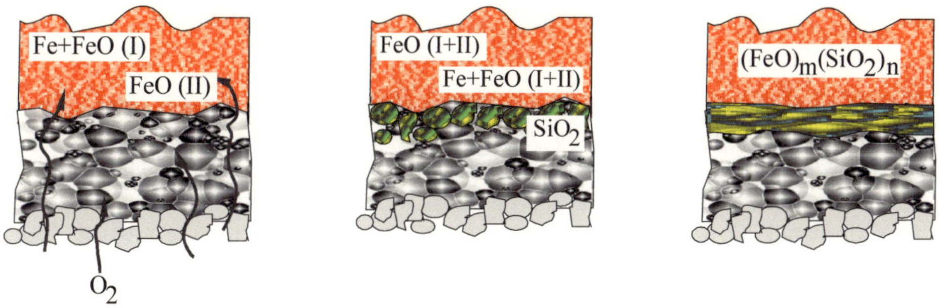

a. metal surface oxidation through pores

b. interaction between metal oxides and mould sand

c. silicate layer with volatile composition

Fig. 2. Schematic view of iron silicate formation on the metal - mould contact boundary (Svarika 1977)

Mould blend maceration with metal has a decisive influence on the casting surface quality. It is known that metal and alloy melt, if not „contaminated" with oxides, do not macerate the mould walls. If a melt is oxidized, i.e. if there is a thin oxide skim on the melt surface, mould wall maceration will depend on the oxide properties. For instance, if a steel cast contains certain amount of iron oxides (if a melt disoxidation process has not been fully completed or if the melt oxidizes due to a casting process) maceration will happen, i.e. send will be burnt on the casting surface, Figure 3.a. Steel cast is most frequently disoxidized by aluminium; then, an aluminium oxide skim is present on the cast surface, preventing the melt from further oxidation, thereby preventing a sand burn fault on the casting surface. If these oxides are not easily dissolved in the basic metal, the metal will then penetrate the mould wall pores only if the pores are sufficiently sized, Figure 3.b (Ballman 1988, Svarika 1977, Tomović 1990).

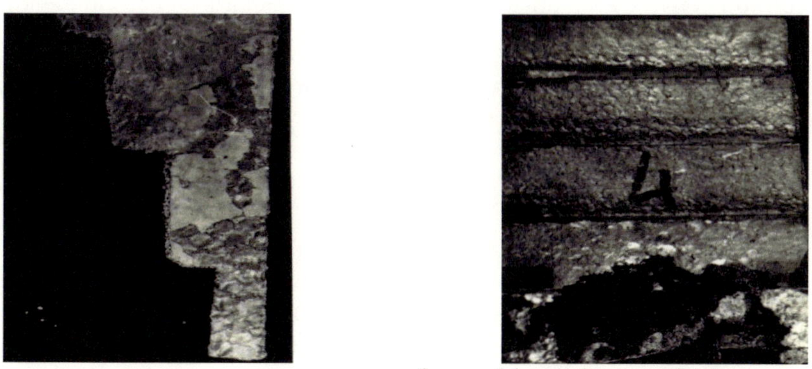

a. sintered sand on the casting surface

b. mould punching on thicker cross sections of castings

Fig. 3. Faults on casting obtaining in quartz sand - based moulds

From the aspect of liquid metal action to the mould or core made of typical mould and core blends (quartz sand or some other sand and a binding agent, various additions or impurities), sand blend features, such as refractoriness and permeability, have a crucial influence at the casting stage.

Examinations of surface faults on the iron alloy castings obtaining into mould and core blends with quartz sand point to the fact that a higher melting temperature, as well as oxidation atmosphere, i.e. an immediate contact between SiO_2 and FeO, cause appearance of a sintered layer on the casting surface. Defects of mould and core blends with quartz sand may be eliminated by replacing this kind of sand with highly refractory sands based on zircon, olivine, sinter magnesite, chromite, corundum and other, or by application of protective ceramic coatings for moulds and cores (Aćimović et al. 1994, Burdit 1988 Clegg 1978 , Shivukumar et al. 1987).

Melted metal often contains large amounts of dissolved gases; they are likely to disappear from the mould cavity mostly through the mould blend. When choosing a blend permeability, one should bear in mind that blends with different sand grain sizes may cause surface defects on castings. In case of high blend permeability with large sand grains, flowing metal, due to its viscosity, can easily get into intermediate areas among the sand grains, forming a very rough surface of castings after solidification. Large sand grains and spacious intermediate areas make melted metal flow more difficult due to high friction, thereby abrupting individual grains, causing a change of mould cavity shape and contaminating liquid metal itself. Surface defects on castings being present as impurities of mould or core blend are mostly originated from a mould not being firm enough but also from the inflow system being improperly sized or due to carelessness during mould and core manufacture. Apart from impurities, certain swollen parts are frequently noticed on castings on the spots where mould or core material is taken off by the melt flow. It does not happen when a very tiny sand grain is concerned, but the metal flowing in would become ruffled due to low friction. Besides, permeability of such sand blend would have an extremely low value (Cibrik 1977, Svarika 1977)

While interpreting the role of ceramic coatings, their effect on sand moulds or cores and the processes being carried out on the sand mould-ceramic coating-liquid metal contact surface, it can be concluded that the coatings increase relative refractoriness of sand blend, since:

- They prevent chemical reactions from happening on the metal-mould contact surface, i.e. they disable formation of iron oxides (FeO) due to presence of reduction atmosphere created by burndown of the coating layer applied,
- There is no direct contact between sand mould and liquid metal due to reduction atmosphere formation, thus avoiding a heat shock in the moment of contact with liquid metal flowing into the mould cavity,
- Temperature gradient is created on the metal-mould contact surface due to presence of the gas film mentioned above, thus avoiding a sintering process, as a consequence of lighter allotropic modifications of quartz (SiO_2) in case of sand mould being immediately heated up to the liquid metal temperature,
- The coatings have a more suitable grain size and a lower quantity of adverse matters than refractory blend.

Ceramic coatings influence a better metal fluidity, because its firm particles get into intermediate areas filling the empty spots among the sand grains on the moulding surface, reducing friction and preventing liquid metal penetration. They thus provide for smooth surfaces of castings. Utilization of ceramic coatings minimizes the risk of mould cavity erosion and liquid metal contamination.

4. Physical-chemical properties on the liquid metal-ceramic coating-expandable pattern-sand mould boundary

Different to sand mould or metal mould casting, where liquid metal flows into the mould cavity, with expendable polymer pattern casting (EPC casting process), patterns and inflow systems made of polymers are retained in the mould until liquid metal flows in ("full mould" casting). In contact with liquid metal, polymer pattern degradation and expansion process is carried out violently, during a relatively short time, and is followed by moulding crystallization. In order to attain a high quality and cost effective production of castings by EPC casting process, it is necessary to achieve the balance of the expandable polymer pattern-liquid metal-ceramic coating-sand mould system at the stage of metal inflow, polymer pattern degradation and expansion and casting formation and solidification, Figure 4.a. It requires a systematic research of both complex notions and processes carried out in the pattern and the notions and processes carried out in the metal-pattern contact zone, as well as in the metal-ceramic coating-sand contact zone (Aćimović-Pavlović et al. 2007,2010,2011, Brome 1988).

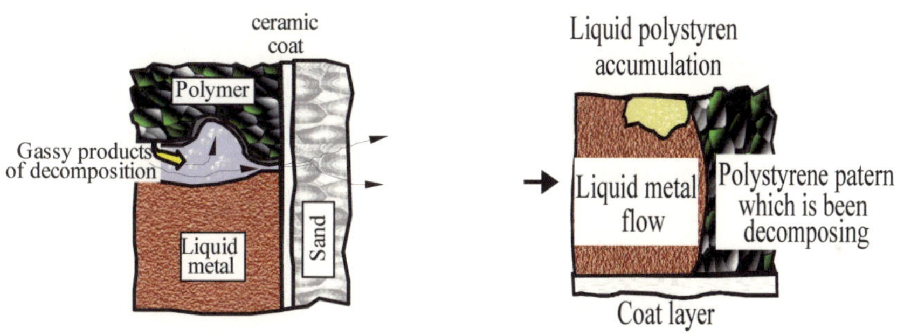

a) System balance: liquid metal-ceramic coat-pattern-sand

b) liquid products accumulation on casting's surface

c) minor permeability: higher coat layers' thickness and smaller sand grains

d) major permeability: less coat layers' thickness and bigger sand grains

Fig. 4. EPC process: The role of ceramic coating

In order to properly understand the role of ceramic coatings for polymer patterns in this process, it is necessary to point out that the polymer pattern degradation is an endothermal process commencing during liquid metal inflow. Kinetics of the pattern degradation is a function of the temperature of liquid metal brought in contact with the pattern. At the inflow stage, while metal is passing through the polymer pattern, 70-90% of pattern degradation products are liquid. Liquid degradation products are pushed toward the upper surface of mould cavity in front of liquid metal front during the process. In case ceramic coatings and mould sand are less permeable, these liquid pattern degradation products are retained in upper parts of castings causing surface, subsurface or volume defects (Figure 4.b). Further degradation of the liquid stage is made by evaporation (formation of the boiling stage); the rest of polymer chain solidifies forming a monomer, as well as benzene and other products of polymer degradation (Aćimović-Pavlović et al. 2011).

Influential factors for the process of pattern degradation and evaporation, apart from the pattern temperature and density, are the type and thickness of the ceramic coating layer lining the expandable pattern, type and size of mould sand grains, i.e. mould sand permeability, casting constructions and inflow systems. Pattern density and permeability of ceramic coating and sand mould determine polymer evaporation speed, Figure 4.c.,d (Aćimović-Pavlović et al. 2003, 2007).

To obtain the castings with the requested quality, critical parameters of EPC process should be determined for both each individual polymer pattern and the alloy type for casting. It requires a long-term research aimed at optimization of this type of casting process to obtain the mouldings with the properties set in advance.

5. Quality requirements and properties

The coatings are featured by general, operative and technological properties, Table 1. (Svarika 1977, Tomović 1990).

General properties	Operative properties	Technological properties
- density - viscosity - sediment suspension stability - granulometric composition - hardness - chemical composition - pH indicator - flammability - toxicity	- burn extent - thermal stability - heat radiation absorption - heat conductivity - maceration ability - erosion resistance - gas permeability - gas forming ability - alloying properties - reduction properties	- ability to be applied on the surface of mould, core, pattern - thixotropy - drying time - endurance - adhesion to mould, - easy adherence to mould surface - hygroscopy

Table 1. Ceramic coating properties

Examinations concerning different physical-chemical properties of foundry coatings showed that there are general conditions which must be met by the coatings, regardless of their type:

- They must have suitable refractoriness,
- Refractory filler must have a low heat expansion coefficient,
- They must not contain materials subject to become softer or melted in contact with liquid metal,
- They must be resistible to metal penetration into the mould or core wall,
- To make a permeable layer for gases,
- With EPC process, it is particularly important to have a suitable coating permeability for the products of degradation and evaporation of polymer pattern formed during contact with liquid metal while flowing into the „full mould“,
- Not to create gases when in contact with liquid metal,
- They must not form the compounds with low melting temperature with metal, its impurities or oxides,
- To be uniformly distributed across the core or mould blend surface, pattern surface, mould surface, to adhere and to be firmly bound to the surface;
- To get dried quickly, not to crack, not to peel off the mould surface during drying or casting time and to be resistible to sudden temperature changes,
- That there is a possibility to control and set the coating layer gauge,
- After drying, they must form a thin visible layer on the mould, core or pattern surface, firmly bound to this surface
- They must not stratification during utilization,
- Intermediate layer (made by metal oxides and coating ingredients) should have a space parameter grid closer to the coating material grid rather than metal oxides ; otherwise, this transitive layer tends to bind the metal oxides to form membranes on the alloy surface leading to blend immersion into castings;
- The coatings should form gases to repulse the metal oxides from the mould (core) walls, especially at sand mould casting with high vertical walls or upright core positions
- The coatings increase strength and abrasion resistance by polymer patterns at EPC process preventing their distortion and break during filling and compression of unbound sand at the mould manufacture stage,
- The coatings should make the separation of mould or core blend from castings easier, reducing the casting cleaning time (Aćimović et al. 2003, Svarika 1977, Tomović 1990, Trumbulović et al. 2004).

The quality of the coating applied depends on its uniformity; it is better if it has a lower precipitation speed. Otherwise, casting surface presents burns from mould or core blend or from the coating itself due to low refractoriness of the filler. During „full mould casting“, expendable pattern degradation products created in contact with liquid metal disappear through the refractory coating layer into unbound sand which the mould is made of, if its permeability is satisfactory. It is primarily attained by choosing the suitable coating type, coating preparation procedure, coating suspension density and the coating dry film thickness on the pattern, Figure 5 (Brome 1988).

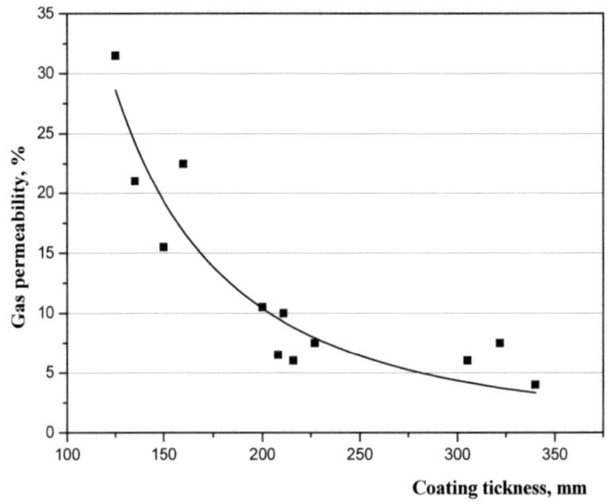

Fig. 5. Gas permeability dependence on the coating dry film thickness

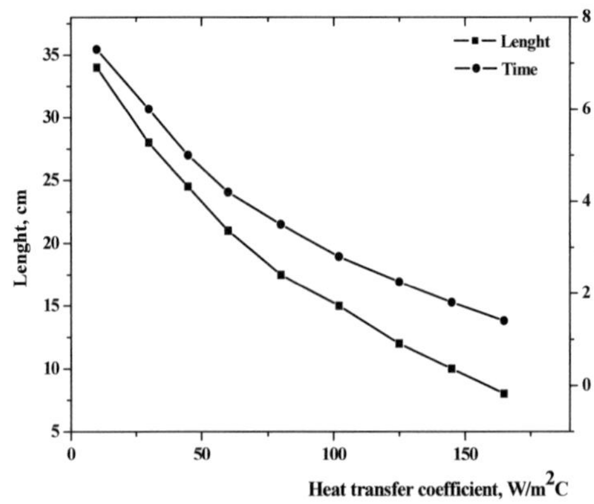

Fig. 6. Coating insulation effect on the metal flow length and mould filling time

Ceramic coatings present an insulation effect influencing liquid metal temperature drop reduction when flowing into a mould; they also influence liquid metal fluidity and the way a mould is filled, certainly affecting the casting quality. Metal fluidity reduction has been noticed while using some types of coatings (based on silica, zircon, graphite); it most frequently resulted in an increment of the coating layer thickness on either a mould or a polymer pattern (Monroe 1994).

With EPC process, when a mould is filled and when polymer pattern degradation and evaporation is carried out, insulation effect of the coating influences liquid metal temperature drop reduction. When the mould is filled up with liquid metal, i.e. when polymer pattern is expended, the coating, due to insulation effect, influences a casting cooling and solidification speed reduction. At the same time, subcooling, created in liquid metal as a consequence of endothermic degradation of polymer pattern in contact with liquid metal, has a significant influence on the casting solidification. If subcooling is huge, fine and tiny-grained casting structure is preferably formed, because such structure provides for the better casting properties. All of that points to the complexity of casting solidification conditions as far as „full mould" is concerned, and to a significant role of ceramic coating in the casting process to form the casting structure and properties. Figure 6 shows the coating insulation effect on the metal flow length and mould filling time with polymer pattern casting. Taking into account the casting conditions and the character of the EPC method („full mould casting"), basic requirement in terms of ceramic coating permeability is more expressed than the one of the coatings for sand moulds and cores (Ballman 1988, Brome 1994).

6. Composition

Research showed that, depending on use, most of contemporary foundry coatings represent very complex blends consisting of over 15 components. However, four basic components are the following:

- Refractory filler,
- Binding agent,
- Suspension maintaining agent,
- Liquid carrier or solvent.

6.1 Refractory filler

Refractory filler is the most important component of ceramic coatings. It determines metal penetration resistance by reducing the permeability of the surface to which it is applied, prevents sand blend erosion and reactions on the metal-mould contact surface. As mentioned above, the choice of the refractory filler depends primarily on the casting alloy type, casting wall thickness and weight, preferable inflow system, i.e. metalostatic pressure in the mould. Main physical-chemical and thermo-physical characteristics of refractory filler are the following:

- High melting temperature providing for the clean casting surface with no sintered sand appearance,
- Low heat expand coefficient and its uniform growth reducing risk of coated layer crack; at the same time, higher dimensional casting stability is ensured,
- Lower hardness of ceramic material enabling eased attainment of suitable filler granularity providing for an eased coating application to the mould or core surface,
- Non-maceration with liquid metal and low reactivity with metal oxides, i.e. resistance to their activity on elevated temperatures, preventing reactions on the metal-mould contact surface, providing for the clean and smooth casting surface,
- Low density value, since high density value requires more intensive rheology of foundry coatings due to an increased precipitation speed,
- Low heat conductivity coefficient value, since elevated values of this parameter cause insufficient thermal stability, i.e. they decrease heat shock resistance, causing a rapid growth of temperature gradient on the metal-mould contact surface,
- Refractory filler must not develop gases when metal flows in, because it causes gas porosity of the moulding (Aćimović-Pavlović et al. 2007, Brome 1994, Svarika 1977).

The choice of the refractory filler highly depends on the metal casting temperature. In the foundry technology, fine grinded mineral raw materials are used for coatings; these mineral raw materials are based on olivine, chromite, zircon, siner magnesite, mica, corundum, cordierite and other refractory materials, Table 2.

In order to use ceramic powder as a filler, it should have the grain size below 40 $\cdot 10^{-6}$ m to attain an appropriate coating rheology and adhesive properties. It has been shown in practice that rounded filler particles with different size (25-40 $\cdot 10^{-6}$ m) are suitable. These contribute to form a uniform, continuous film, i.e. a thinner coating layer, especially when applied for the EPC casting process, due to a better mutual stacking-packing of filler particles. Researches showed that refractory filler particles are more tiny after being fine-milled and mechanically activated (individual particle size is below 15-20 $\cdot 10^{-6}$ m, even below 5 $\cdot 10^{-6}$ m); they are homogenously distributed in the solvent.

Having compared the coatings made either with or without mechanical activation of the filler, it has been noted that the mechanical activation process contributes to coating properties improvement; thus prepared coatings are easily applied and are more efficiently adhered to the sand mould and core surfaces, as well as to the polymer pattern surface. When being applied, they make a fine continuous film of coating on the subject surfaces, which is easily dried, without being either wiped off or broken after drying (Clegg 1978).

Refractory material	Specific weight, kg/m^3	Mos hardness	Melting point, °C	Specific heat, kJ/kgK	Heat cond. W/mK on 827°C	Recommendation
Aluminium oxide	3.30	9.0	2016	1.13	2.59	steel, Mg alloys
Periclase	3.40	5.5-6.0	2800	1.08	3.32	Mn- steel
Mullite	3.20	7.5	1830	1.00	2.02	cast iron, Al & Cu alloys
Quartz	2.65	7.0	1716	1.05	1.73	cast iron, nodular cast iron
Zircon	4.50	7.5	2500	0.54	2.31	cast iron, nodular cast iron, steel, Al & Cu alloys
Zirconium-oxide	5.5-6.1		2700	0.71	0.72	cast iron, nodular cast iron, steel, Al & Cu alloys
Sinter magnesite	2.85-3.5	5.5-6.0	2400	0.95	4.50	steel, Mg alloys
Chromite	4.5-4.8	5.5	2180	0.8-1.1	1.69	cast iron, nodular cast iron, steel, Al & Cu alloys
Talc	2.6-2.8	1-1.5	1547	0.19	2.40	cast iron, nodular cast iron, Al & Cu alloys
Graphite	2.23	1-1.5	3000	0.75-1.4	1.50	cast iron, nodular cast iron, Al & Cu alloys

Table 2. Type and properties of refractory fillers most frequently used

6.2 Binding properties

Coating ability to retain its required properties and to be firmly adhered to the subject mould or core surfaces after a liquid component has been dried or expanded is ensured by addition of a binding agent. Binding agent is chosen in accordance with the refractory filler applied, as well as with ability to be dissolved in the liquid carrier of cast house coatings. The quantity of binding agent depends on the particle size of the refractory filler used. It must be carefully chosen, because gas development from the film of coating depends on it, as it is proportional to the excess of binding agent. Binding agents may be organic and non-organic. Depending on the solidification temperature, binding agents are divided into the binding agents being solidified at the room temperature and binding agents being solidified through drying or curing. There are three mechanisms to form the coating hardness: binding agent drying, solidification after binding agent melting and solidification due to chemical processes. With the aqueous coatings containing bentonite, silica esters or water glass, coating solidification is carried out through the loss of a liquid component during heat treatment. With the coatings containing calophonium, bitumen, phurane, formaldehyde, phenol-formaldehyde binding agents and with the sugar-based or glucose-based binding agents, solidification of applied layer is carried out as a result of binding agent solidification after being heated up to its melting temperature. In case that a binding agent polymerisation is carried out or if various components chemically reacts when the applied layer is dried, the molecules of the binding agents are mutually connected into long chains or nets. That's how a binding agent solidifies. The coatings with a binding agent which is polymerised at the room temperature during a liquid component evapouration are very suitable because they do not require a thermal treatment. The coatings containing this type of binding agent (for example, polyvinilebutirol) easily solidifies once they have been applied on the mould or core surfaces, without heating. Basic requirements for the binding agent quality are the following: thermo-stability, i.e. maximum hardness maintenance at elevated temperatures, during either drying or inflow of liquid metal, in order to prevent coating stratification and crack, minimum gas separation, they must not soak humidity in and alike (Svarika 1977, Tomović 1990)].

6.3 Suspension maintaining agent

Suspension maintaining agent keeps the refractory powder particles in a dispersed state. They are divided into two groups: aqueous coating stabilizers and non-aqueous coating stabilizers. These agents prevent precipitation of refractory powder particles and have an important influence on the coating quality. Small amount of carriers causes a quick precipitation of the filler particles and other solid components contained in the coating composition. If there is an excess of carriers, these will cause an increment of coating density causing difficulties for the coating application on the mould or core surfaces; there will also be a crack risk for the thicker coating layers being dried. It is very important that solid matters used for suspension maintenance have the same mass as the liquid coating stage. In that case, suspension is maintained longer, the coating becomes more efficient and metal penetration is prevented. Aqueous coating stabilizers are the following: bentonite, carboxymethylcellulose, alghynate, and polyacrylamide. Alcohol coating stabilizers are the following: polyvinyl-butyral, polysobutylene, organic types of bentonite. Table 4 shows some suspension maintaining agents (Clegg1978, Svarika 1977, Tomović 1990).

Binding agent	Coating type	Application
Refractory clay	aqueous	Montmorillonite clays, i.e. bentonites, are most significant. As a binding agent, bentonite is mostly used for a water-based quartz coating manufacture, due to electric charge activity on the surfaces of both the binding agent particles and refractory filler particles.
Bentonites	aqueous	Bentonite clays are widely applied, especially Na-bentonite for its swelling and suspension maintaining properties and it is suitable for a gel formation.
Dextrine	aqueous	Good binding agent. It has advantages over clay as there are no cracks on the surface of a dried coating. Starch, molasses and resin have similar properties.
Sodium silicate	aqueous	Binding agents depend on chemical composition defined by a module being determined by the ratio between the number of quartz moles (SiO_2) and sodium oxide (Na_2O). Binding process is based on silica acid colloid-like separation, where condensation is created due to lack of water. Special strength is obtained due to a reaction with the CO_2 gas from the atmosphere.
Oils	aqueous	They might be natural (linen) and process oils (mineral, synthetic oils and alkine resins). They present relatively good binding properties.
Shellac Colophony Resins	alcohol	It doesn't get destroyed or burnt when ignited, therefore the binding agent maintains its properties.

Table 3. Type and properties of binding agents most frequently used

Agent	Influence on coating quality
Sodium alghynate	Sodium alghynates are derivatives of certain seaweed enabling good suspension and they are relatively cheap. They reduce the possibility of heat cracks on the coating after it is dried. They are used in aqueous coatings.
Cellulose derivatives	Carboxymethyl cellulose may replace Na-alghynate, as it is not that sensitive to normal pH variations. It is used in aqueous coatings.
Bentonites	Bentonites, such as organic bentonites where Na ion is either completely or partially substituted by an organic one, are preferable since they create a thyxotropic gel in organic solvents. They are used in alcohol-based coatings.
Metal stearates	Aluminium stearate is used as a suspension forming agent in the coatings where cyclohexane is used as a liquid carrier. In preparation of alcohol coatings, they are used as suspension maintaining agents for resins and oil either.
Sodium bentonite	The agent maintaining a suspension for the longest time, therefore mostly used in aqueous coatings.

Table 4. Suspension maintaining agents

6.4 Liquid carrier or solvent

The role of a liquid carrier or solvent is to dissolve and transport the refractory powder to the sand surface in the form of film. Generally, three types are used: water, alcohol (flammable liquids) and chloric hydrocarbons.

The solvent is chosen depending on several factors, some of which are the following: the type of sand blend, application method, ecological factor, type of production cycle dictating the time allowed for drying and application. Mostly applied solvent is water, for it is the cheapest, but also non-toxic and safe for use. Increment of application costs for this type of coating is due to energy consumption for the coating drying process and for water elimination from the layers applied.

Solvents like alcohol mixtures (methyl, ethyl, isopropyl) with pure water are often flammable liquids. The advantage of alcohol over water is faster drying, which is important for the sand moulds; but, since these coatings are flammable, they require more attention to be paid during production, use and stock-keeping.

Isopropyl-alcohol represents technically most acceptable solvent for the flammable foundry coatings, primarily for its combustion properties being close to perfect. This type of combustion has a reduced erosion risk for the surface observed. Other alcohol solvents are less frequently applied as pure, but mostly together, even when mixed with water (up to 12,0 vol %).

Choloric hydrocarbon is air-dried, eliminating a fire risk when flammable liquids are used. Air-dried coatings are convenient, like flammable liquids: they are self-extinguishable, they have a high steam pressure, they evaporate easily without a usage of heat, but they are also expensive and toxic matters.

7. Recipes

Table 5 shows some recipes of the ceramic coatings used for different casting methods and metal materials for casting production.

8. Application rheology

As far as the effect of ceramic coatings on the quality of the castings obtained is concerned, it is necessary to discuss preparation of the components contained in the coating, coating production and its application rheology which is tightly bound to a proper choice and use of a suitable stabilizer. Generally, it may be defined as follows:

- immersion,
- spraying,
- pouring,
- brush application.

Generally, an ideal rheology suitable for coating application by immersion and pouring is an achievement of the properties of pseudoplastic solution, i.e. an achievement of such rheology enabling an instant viscosity drop during application and an immediate viscosity recovery as soon as the coating application stops. On the other hand, the rheology required for coating application by either spraying or brush is an achievement of thixotropic solution, i.e. an achievement of a light viscosity drop during application and a bit faster recovery to the initial value after application. It practically means that the coating flows well when being applied since recovery to the initial value is not instant, as with pseudo-plastic solutions.

O. No.	Content of filler	Binding agent,	Suspension maintenance agent	Solvent, density 1,8-2 kg/m³
1.	**zircon,** with grain size of 40·10⁻⁶ m, 86-89%	bentonite 3%; bindal H 7,5%	dextrin 0,3-0,5%, carboxymethyl cellulose- (CMC), 0,3-0,5%	water
2.	**mulite,** with grain size of 40·10⁻⁶ m, 91%	bentonite 3%; bindal H 5%	dextrin 0,3-0,5%, CMC, 0,3-0,5%	water
3.	**mica,** with grain size 35-40·10⁻⁶ m, 88-90%	bentonite 4%; bindal H 7%	dextrin 0,5%, CMC, 0,5%	water
4.	**mica,** with grain size <30·10⁻⁶ m, 90 - 94%	bentonite 1,5-2,5%; bindal H 1,3-2%	$Na_3P_3O_3$ 1-3%, carboxymethyl cellulose 0,5%	water
5.	**alumina,** with grain size 45·10⁻⁶ m, 90%	bentonite 3%; Bindal H 6%	dextrin 0,5%, carboxymethyl cellulose 0,5%	water
6.	**cordierite,** with grain size <40·10⁻⁶ m, 93-95%	colophonium ($C_{20}H_{30}O_2$), 1.2-1.5%; dextrine 0,5-1%	BentoneSD-3 1.2-1.5%; phenolphormaldehide resins, 1-1.7 %	alcohol
7.	**chromite,** with grain size 40·10⁻⁶ m, 95-96%	phenolphormaldehide resins, 3-3,5%;	Carboxymethyl cellulose 0,3-0,5%	water
8.	**talc,** with grain size 40·10⁻⁶ m, 85%	bentonite 3,5-5%; bindal H 8%	dextrin 0,3-0,5% carboxymethyl cellulose, 0,3-0,5%	water
9.	**zircon,** with grain size 40·10⁻⁶ m, 98%	bentonite 2%; melasses (1,4 kg/m³), content 3%	Sulphide alkali (1,30 kg/m³), 1,5%	water
10.	**chrome-magnesite sand,** with grain size 40-45·10⁻⁶ m, 99%	bentonite 1%; melasses (1,4 kg/m³), content 5-6%	Sulphide alkali (1,30 kg/m³), 1%	water
11.	**olivine,** with grain size 35-40·10⁻⁶ m, 85,5-90 %	bentonite 1-1,5%;	Dextrin: 1,5-2 %	water

Table 5. Composition of various ceramic coatings

When using a coating, it is necessary to mix it constantly (at the speed of 1 rpm) to get a homogenous suspension which, after being applied, provides for an even and uniform coating layer on the surface of mould, core or pattern. Figure 7 shows the influence of suspension mixing process on the applied deposit weight, while Figure 8 shows the influence of suspension mixing on the coating viscosity.

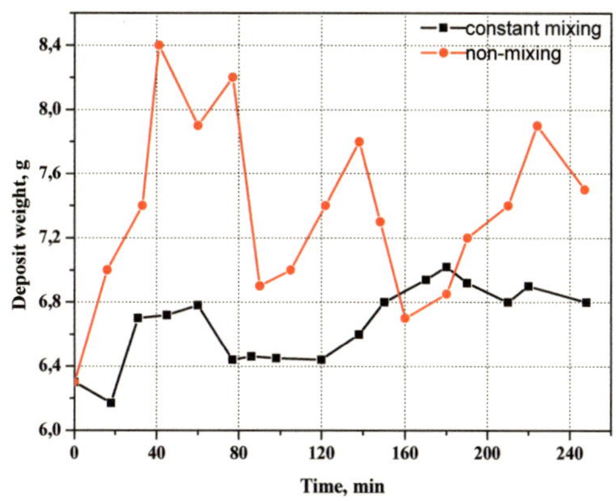

Fig. 7. Influence of mixing on the deposit weight (Brome 1988)

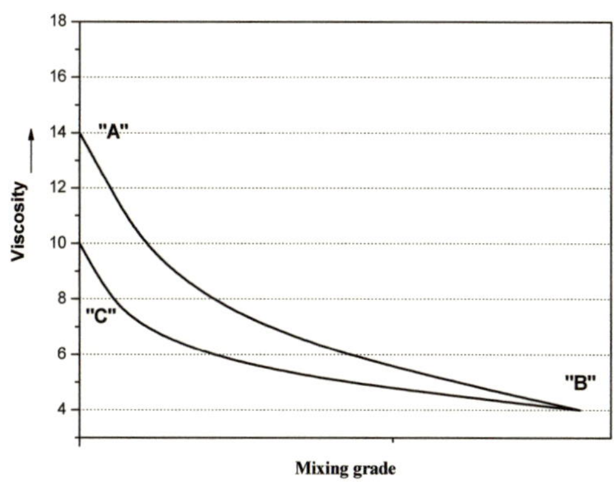

Fig. 8. Influence of mixing extent on the coating viscosity (Brome 1988)

Best results are achieved if certain suspension temperature (usually room temperature) is reached and maintained, apart from suspension stirring when being applied on sand moulds and cores or expendable patterns. Coating suspension temperature influences the quantity of residue and it needs to be constantly controlled (Figure 9). When the temperature rises, coating layer mass drops until certain temperature is reached; afterwards, the mass increases with the temperature growth.

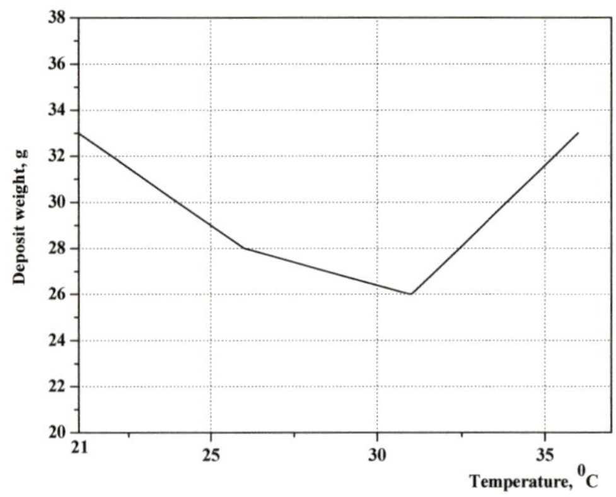

Fig. 9. Temperature influence on deposit weight (Brome 1988)

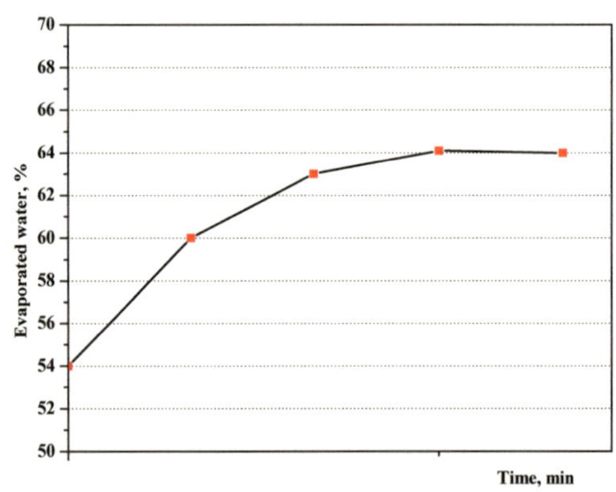

Fig. 10. Influence of drying time on humidity elimination from the coating (Brome 1988)

Coating layers applied must be completely dried either in an oven or in air, Figure10. If the coating is not dry enough, remaining humidity reacts with liquid metal causing porosity defects on castings.

To produce high quality coatings with the properties set in advance, it is necessary to attain a homogenous distribution of refractory filler in the coating suspension. It is also necessary to define the composition of the coating with rheology properties controlled and specially adjusted for a concrete application methods, Figure 11. To attain suspension sedimentation stability, filler particles should be up to $40 \cdot 10^{-6}$ m in size, as it is expected that more tiny filler particles are precipitated slower and that a suspension may homogenize easier and faster. Furthermore, these particles more evenly and fully cover mould and pattern surfaces where the coating is applied. Homogenous coating suspension, Figure11.a, is obtained when the filler particles are rounded and even in size. When larger filler particles are present to a lower extent, with medium grain size of $45 \cdot 10^{-6}$ m, Figure 11.b, it is estimated that the rounded particles with different grain size will also contribute to form a continuous coating layer, due to a better interaction of particles. In case the coating suspension is unevenly stirred during application, coating stratification happens, Figure 11.c (Clegg 1978, Trumbulović 2004).

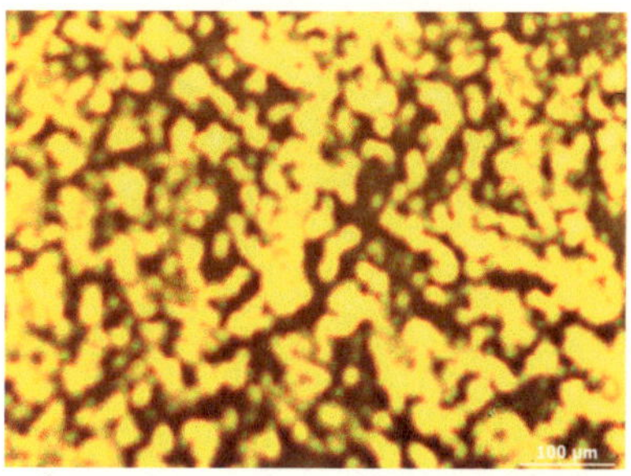

a. homogenous coating suspension

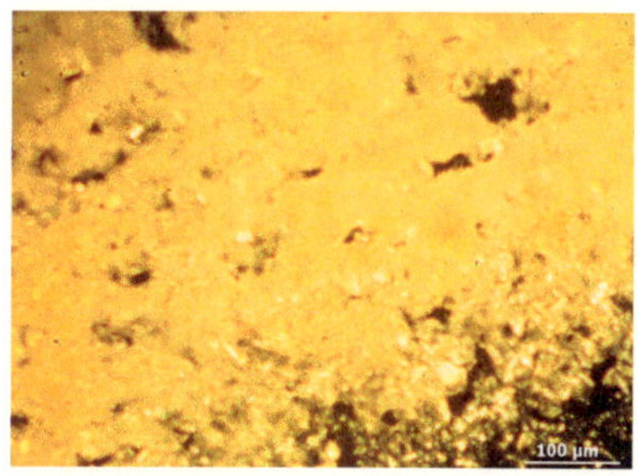

b. coating suspension with larger particles

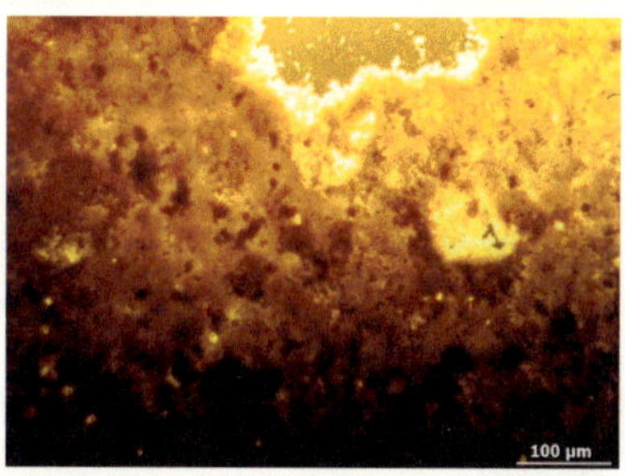

c. non-homogenous coating suspension (drop-like)

Fig. 11. Microphotos of ceramic coating suspensions

One should bear in mind that homogenous distribution of refractory filler in a suspension depends on suspension preparation during coating application. Especially, an important process parameter is the coating suspension density, mostly within the limits from 1,8-2 kg/m^3. In order to attain even coating layers on mould or pattern surfaces during immersion (into a coating suspension tank), the way of eliminating the excess of coating suspension is important. After immersion, „the cluster" (representing a number of patterns stacked at the common inflow system, thus prepared for EPC casting process) is taken out of suspension tank, it is kept in a vertical position for 5-10 s in order to decant the excess of suspension; then, it is inclined for 5s in order for the coating to be evenly distributed along the cluster surface, Figure 12.a-c. After being decanted, "the cluster" is ready for a drying process. When applying the coating on sand moulds by either a brush or rag, by either spraying or pouring, coating density has an important role and it should not exceed the value of 2-2.2 kg/m^3, Figure 12.d. In this way, dried coating layers do not crack, peel or wipe off, and after casting they are easily taken off the castings surfaces, not requiring the castings to be subsequently cleaned, as shown in Figure 13. It has been established that adhesive forces among the filler particles increase with the filler concentration increment. Under the influence of rheology additives and binding agents, continuous and uniform coating layers might be formed on the treated surfaces. Such coating is easily adhered to the treated mould and pattern surfaces. Dried coating layer thicknesses influence gas permeability and they should be as low as possible; their range is from $0{,}5\text{-}1{,}5 \cdot 10^{-3}$ m (Clegg 1978, Tomović 1990).

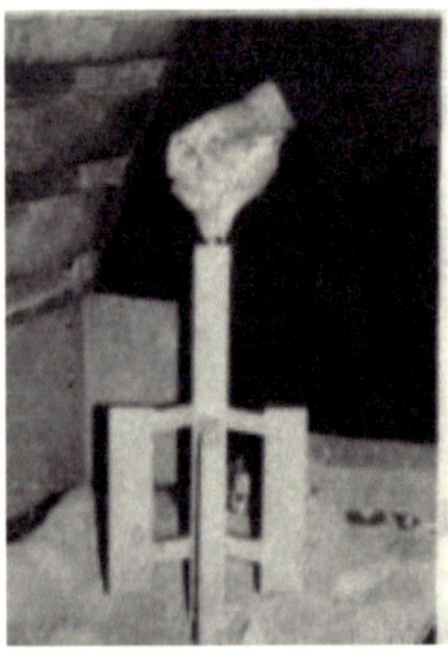

a. coating application by immersion

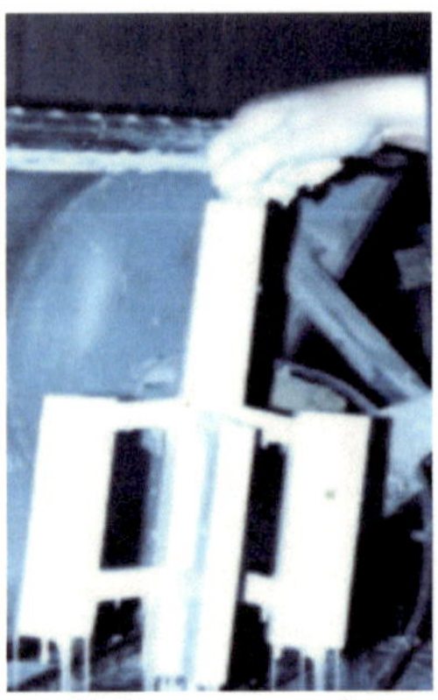

b. taking off the excess of suspension

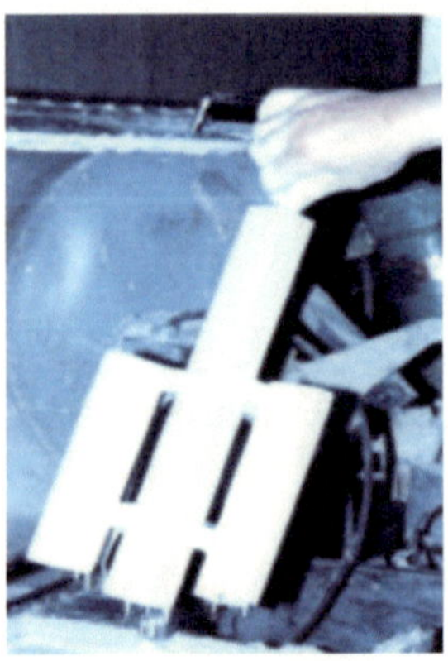

c. balancing the coating layer on the pattern

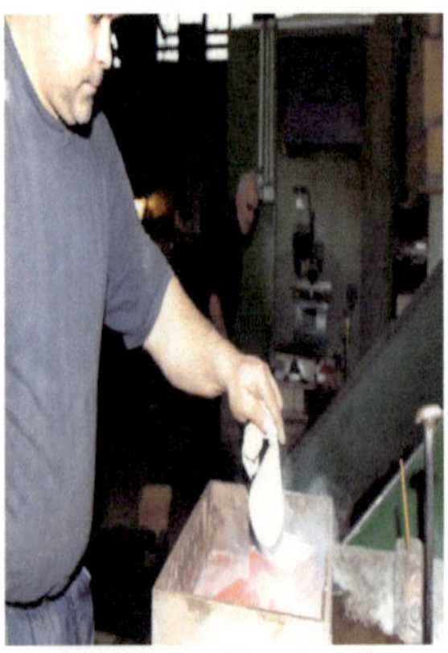

d. coating application by rag

Fig. 12. Coating application stages

a. sand casting b. EPC casting process

Fig. 13. Taking coating layers easily off the casting surface

9. Control and faults

Quality control for foundry coatings and coating application process control as far as sand moulds and cores and expendable or meltable patterns are concerned is important for production of high quality castings. Establishing the quality of foundry coatings is defined by different standards for this type of refractory products. Standards are used to establish both, foundry classification and quality requirements, as well as technical conditions of application, coating sampling methods, coating test methods and marking of coatings and the way of delivery. Most often, the following coating properties are tested:

- application ability,
- drying behaviour,
- wipe-off resistance,
- dry matter amount,
- precipitation,
- penetration (Svarika 1977, Tomović 1990).

Assessment of a coating is done according to the following criteria either:

- the coating must evenly flow down when being applied,
- coating layer must not crack or bubble when being dried,
- coating layer must be easily taken off the castings surface when these are shaken or cleaned,
- coating layer must be wipe-off resistant
- dry matter amount must not vary more than ±2% from the value declared,
- precipitated matters being precipitated for 24 h may amount to max 5%,
- coating penetration depth in the mould wall may be in the range from 0,5 to 2 $\cdot 10^{-3}$ m, depending on use, solvent and binding agent,
- coating layer must be permeable to gases,
- the coating must be compatible with the type and density of polymer pattern, with the type and size of moulding sand grain and with other EPC process parameters (Svarika 1977, Tomović 1990, Brome 1988).

Table 6 shows typical faults detected on the surface of the coated polystyrene patterns. (Aćimović-Pavlović 2010, Clegg 1978)

Uneven surface of coating layers, with quickly dried thicker coating layers on polymer pattern	
Bubbles on the coated surface of polymer pattern, caused by rapidly stirring the coating suspension during application	
Dried ceramic coating layers cracking on the polymer pattern, when the drying is rapid and with thicker coating layers	

Table 6. Typical faults on the polystyrene pattern-coating contact surface

The role of ceramic coatings is to attain a high quality moulding surface. Furthermore, moulding quality is significantly influenced by critical process parameters (casting technology); at first, the choice of material for production of castings, moulds, cores, patterns, ceramic coatings, design faults, i.e. faults concerning pattern construction, inflow system choice and calculation, as well as faults due to disturbance of technological process (human factor). It requires both control and optimization of all process parameters aimed at attainment of the desired structure and utilization properties of castings. Detection, examination and estimate of castings faults should be carried out systematically, at the stage of casting process development and concrete castings production mastering, with an aim to avoid these through preventive measures and to minimize the production costs. Presence of castings faults does not always mean the loss of their utilization value, for it depends on the type, size and position of the fault, as well as on the type of the construction where the castings are installed and on the load character during their exploitation. Therefore, it is necessary to classify the faults with regard to their nature and origin, as well as to the fault outer appearance. This enables the faults to be easily visually detected, estimated and eliminated. Table 7 shows typical moulding surface faults caused by an inadequate ceramic coating application (Acimovic Pavlovic et al., 2007, Cibrik 1977, Svarika 1977, Tsai & Chem, 1988).

a. **Uneven and rough casting surfaces** caused by both grain-like pattern surface and thicker layers of ceramic coating	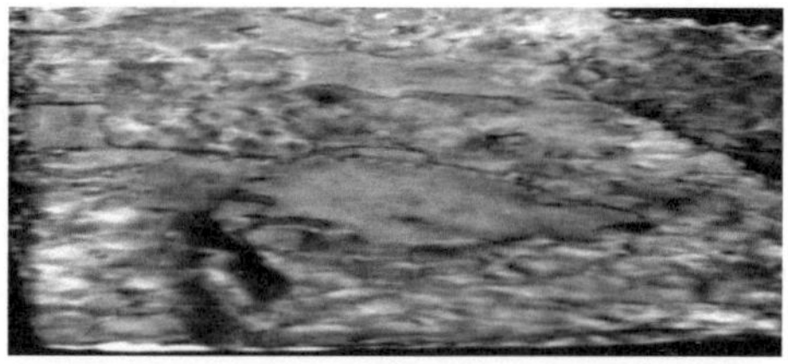
b. **Surface porosity, uneven and wrinkled casting surface** caused by application of thicker coating layers	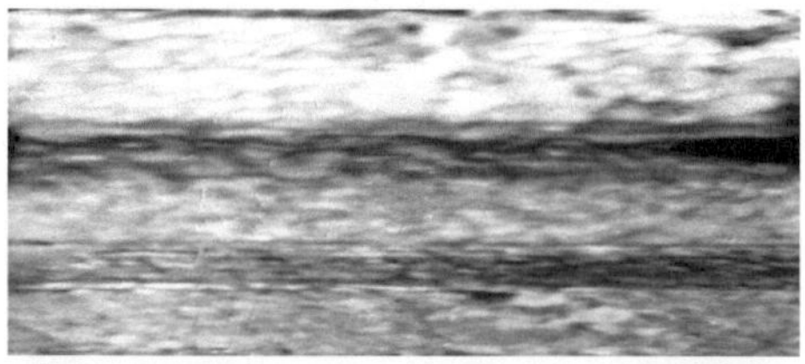
c. **Metal penetrating the mould,** thinner coating layers, with thicker casting walls	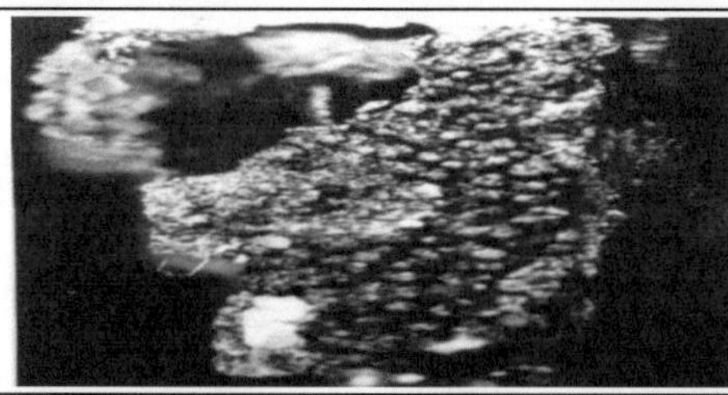
d. **Insufficient replenish fault,** penetration of thinner coating layers, inadequate inflow system design, thick-walled casting shrinkage insufficiently compensated	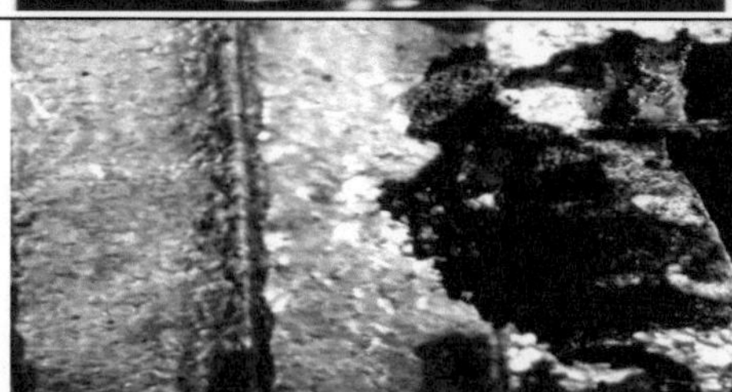
e. **Sintered sand on casting surface** caused by application of thinner layers of diluted coatings	
f. **Pattern surface fault,** fully reproduced on casting surface	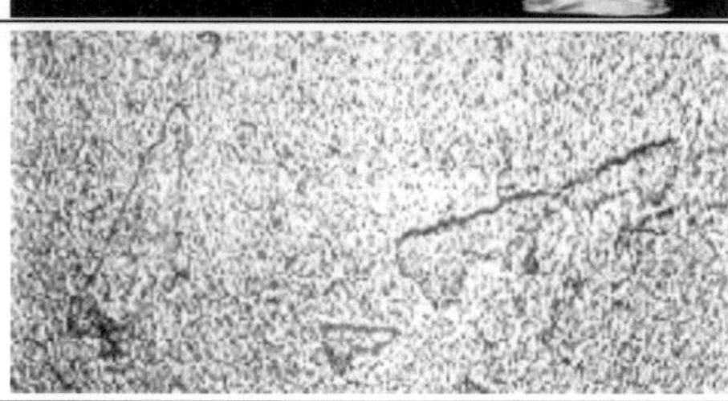

Table 7. Visual control: typical casting surface faults caused by inadequate refractory coating application.

For a detailed analysis of structural and mechanical casting characteristics after a visual examination, it is necessary to apply some other test methods either. Casting test without destruction may reveal the presence of volume faults in the casting. Microstructural tests of castings point to the influence of individual process parameters on the casting cooling and solidification in a mould. Figure 14 shows some faults in the casting structure caused by application of an inadequate ceramic coating in the casting process.

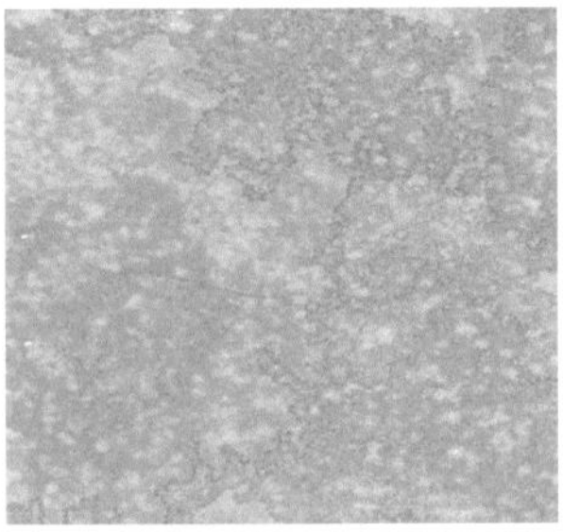

a. casting with no porosity

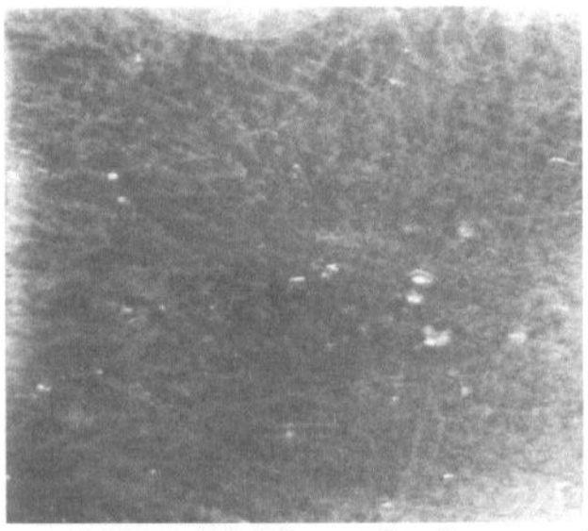

b. casting with classified porosity

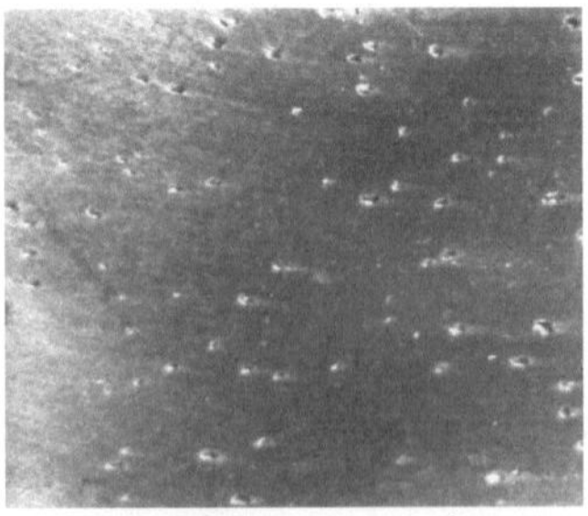

c. intensive porosity over the volume, higher pattern density and thicker coating layers, EPC process

Fig. 14. Casting faults detected by radiographic method

10. Conclusion

Utilization of ceramic coatings with different types of fillers (talc-based, cordierite-based, zircon-based, mulite-based, mica-based, chromite - based and alike) in casting decisively depends on rheology coating properties, i.e. on sedimentation suspension stability. It is necessary to carry out researches referring to optimized coating properties and to their production procedures aimed to obtain the coating properties required. Coating application in practice shows positive effects regarding high quality castings, satisfactory structure and properties, shiny and smooth castings surfaces, with no surface or volume faults whatsoever.

11. References

Aćimović Z., Tomović M., Đuričić M. & Tomović S. (1994) *Litejnoje proizvodstvo*, Vol. 12, ISSN 0024-449x , p.19

Aćimović Z., Pavlović Lj., Trumbulović Lj., Andrić Lj. & Stamatović M. (2003) Synthesis and Caracterization of the Cordierite Ceramics from Non-Standard Raw Materials for Application in Foundry. *Materials Letters*. Vol. 57, No 18, ISSN 0167-577X, pp. (2651-2656).

Aćimović-Pavlović Z., Prstić A. & Andrić Lj. (2007) The characterization of talc-based coating for application for Al-Si alloy casting. *CI&CEQ*. Vol.13, No 1, ISSN 1451-9372,pp. (40-48).

Aćimović-Pavlović Z., Đuričić M., Drmanić S., Đuričić R, (2010) The influence of the parameters of Lost foam process on the quality of the aluminium alloys castings, *Chemical Industry*. Vol.64, No.2, ISSN 0367-598X, pp. (121-127).

Aćimović-Pavlović Z., Andrić Lj., Milošević V., & Milićević S. (2011) Refractory coating based on cordierite for application in new evaporate pattern casting process. *Ceramic International*. Vol.37, No.1, ISSN 0272-8842,pp. (99-104).

Ballman R. (1988) Assembly and coating of polystyrene foam patterns for the Evaporate Pattern Casting Process, *Proceedings of 92nd Casting Congress*, Hartford, Connecticut, USA, 1988,pp. 250.

Brome A.J. (1988) Mould and core coatings and their application. *British Foundrymen*. Vol. 80 No.4, pp.(342-350), ISSN 0007-0718.

Burditt M. (1988) EPC's Promise Belies Complex Process, *Modern Casting*. Vol 8 , August 1988, ISSN 0026-7562,pp.(20-24).

Cho N. D. (1989) Effect of coating materials on fluidity and temperature loss of molten metals in full mould, *Proceedings of 56th World Foundry Congress*, Düsseldorf, Germany, No 7.1.7.10.

Cibrik A.N. (1977) *Fizičko-hemičeskije procesi v kontaktnoj zone metal-forma*, Nauka Dumka Kiev.

Clegg A. (1978) The Full-Mould Process-A Review, Part II: Production of Castings, *Foundry Trade Journal*, Vol. 3, No. 8, ISSN 0015-9042,pp.(383-398).

Davies R.W. (1996) The replacement of solvent based coatings in modern foundries, *Foundrymen*, Vol. 89, No. 9, ISSN 0953-6035, pp.(287-290).

Gorny Z. & Marcinkowski J.(1977) New ideas in investigating and evaluating the sand and core materials, *Transaction*. pp.(893-910).

Josipović Ž., Marković S.,Aćimović Z. & Tomović S. (1994) Foundry linings-present state. *Journal of foundry XLII*, Vol. 94, pp.(17-20).

Monroe R. (1994) *Expandable Pattern Casting*, AFS, USA.

Shivukumar S., Wang L. & Steenhoft B.: Phisico-Chemical aspect of the Full mould casting of aluminium alloys, part I: The Degradation of Polystyrene. *Transaction*. Vol. 95, pp.(791-800).

Svarika A.A.(1977) *Pokritia litejnih form*, Mašinostroenije,Moskva.

Tomović M.N.(1990) *Casting of non-ferrous aloys*, Faculty of Technology and Metallurgy University of Belgrade, Belgrade, Serbia

Trumbulović Lj., Aćimović Z., Gulišija Z. & Andrić Lj. (2004) Correlation of Technological Parameters and Quality of Castings Obtained by the EPC Method. Materials Letters. Vol. 58, No. 11, ISSN 0167-577X , pp.(1726-1731).

Tsai H. & Chem T.S. (1988) Modelling of Evaporative Pattern process, Part I, Metal Flow and Heat Transfer During the Fillings Stage. *Proceedings of 92nd Casting Congress*, Hartford, Connecticut, USA, pp. 300,